CLIMATE CHANGE
ADAPTATION MANUAL

Success in the international negotiations on mitigating climate change seems further away than ever. The importance and urgency of adaptation is becoming more and more apparent. It is now one of the main imperatives of international research and action. However, past and present research on adaptation is mostly not directly applicable to adaptation policy or practice, leaving a disconcerting gap between scientific results and practical advice for decision makers and planners. This book seeks to address this problem. It bridges the gap and provides readers with practical and readily applicable information on climate change adaptation.

Following a context-setting introduction, the book is organised into four main sections. Each reflects an essential component of the adaptation process. Whereas most books in the field focus on adaptation in developing countries, this volume provides an examination of policy and practice in industrialised countries, predominantly in Europe. It offers novel inter-disciplinary insight into cutting-edge knowledge and lessons learned.

Andrea Prutsch is an adaptation policy advisor in the Department of Environmental Impact Assessment and Climate Change at the Environment Agency Austria.

Torsten Grothmann is senior scientist in the Department of Ecological Economics at the University of Oldenburg, Germany.

Sabine McCallum is Head of Department of Environmental Impact Assessment and Climate Change at the Environment Agency Austria.

Inke Schauser is an adaptation policy advisor in the Department of Climate Change Impacts and Adaptation (KomPass) at the Federal Environment Agency, Germany.

Rob Swart is coordinator of international climate change adaptation research at Alterra, Wageningen University and Research Centre, Netherlands.

CLIMATE CHANGE ADAPTATION MANUAL

Lessons learned from European and other industrialised countries

Edited by Andrea Prutsch,
Torsten Grothmann, Sabine McCallum,
Inke Schauser and Rob Swart

Routledge
Taylor & Francis Group

LONDON AND NEW YORK

earthscan
from Routledge

First published 2014
by Routledge
2 Park Square, Milton Park, Abingdon, Oxon, OX14 4RN

and by Routledge
711 Third Avenue, New York, NY 10017

Routledge is an imprint of the Taylor & Francis Group, an informa business

British Library Cataloguing in Publication Data
A catalogue record for this book is available from the British Library

Library of Congress Cataloging-in-Publication Data
Climate change adaptation manual : lessons learned from European
and other industrialised countries / edited by Andrea Prutsch,
Torsten Grothmann, Sabine McCallum, Inke Schauser and
Rob Swart.
     pages cm
   Includes bibliographical references and index.
   1. Climatic changes – Risk management. 2. Climate change
   mitigation. 3. Emergency management. 4. Human beings –
   Effect of climate on. I. Prutsch, Andrea.
   QC903.C55 2014
   363.738′746–dc23   2013032548

ISBN13: 978-0-415-63040-5 (hbk)
ISBN13: 978-0-415-66034-1 (pbk)
ISBN13: 978-0-203-38126-7 (ebk)

Typeset in Baskerville
by HWA Text and Data Management, London

CONTENTS

FIGURES

TABLES

CONTRIBUTORS

Hugo Aschwanden Deputy Head of Water Division, Swiss Federal Office for the Environment, Switzerland

Nathalie Asselman Specialist consultant Flood Risk Management, Deltares, the Netherlands

Anja Bauer Senior Researcher, University of Natural Resources and Life Sciences, Vienna, Austria

Magnus Benzie Research Fellow, Stockholm Environment Institute (SEI), Sweden

Karianne de Bruin Senior Research Fellow, CICERO, Center for International Climate and Environmental Research-Oslo, Norway

Marco Cardinaletti European Project Manager for Sustainable Development, Eurocube srl, Italy

Geoff Darch Principal Scientist, Atkins, UK

Rob B. Dellink Assistant Professor, Environmental Economics and Natural Resources Group, Wageningen University, the Netherlands

Suraje Dessai Professor, Chair in Climate Change Adaptation, Sustainability Research Institute, School of Earth and Environment, University of Leeds, UK

Thomas Dworak Director, Fresh-Thoughts Consulting GmbH, Austria

Megan Gawith Research Scientist, Environment Agency, UK

Louise Grøndahl Policy Advisor, Task Force on Climate Change Adaptation, Danish Ministry of the Environment / Danish Nature Agency, Denmark

Torsten Grothmann Senior Scientist, Department Ecological Economics, University of Oldenburg, Germany

Marjolijn Haasnoot Senior Researcher, Deltares, the Netherlands

Mike Harley Director, Climate Resilience Ltd, UK

Tristan Hatton-Ellis Senior Freshwater Ecologist, Natural Resources Wales, UK

Mattias Hjerpe Assistant Professor, Centre for Climate Science and Policy Research and Water and Environmental Studies, Linköping University, Sweden

Roland Hohmann Senior Scientific Officer, Swiss Federal Office for the Environment, Switzerland

Britta Horstmann Researcher, German Development Institute, Germany

Mark Howden Chief Research Scientist, CSIRO Climate Adaptation Flagship and CSIRO Ecosystem Sciences, Australia

Alistair Hunt Research Fellow, University of Bath, UK

Ekko C. van Ierland Professor, Environmental Economics and Natural Resources Group, Wageningen University, the Netherlands

Stéphane Isoard Project Manager 'Climate change adaptation and economics`, Vulnerability and Adaptation Group, European Environment Agency, Denmark

Kay Johnstone Advisor, Environment Agency, UK

André Jol Head of Group Vulnerability and Adaptation, European Environment Agency, Denmark

Carmen de Jong Professor, Mountain Centre, University of Savoy, France

Susanna Kankaanpää Climate Specialist, Helsinki Region Environmental Services Authority (HSY), Finland

Aleksandra Kazmierczak Research Fellow, University of Manchester, UK

Andrea Knierim Professor, University of Hohenheim, Germany

Pamela Köllner-Heck Senior Scientific Officer, Swiss Federal Office for the Environment, Switzerland

Leena Kopperoinen Senior Researcher, Finnish Environment Institute (SYKE), Finland

Claudia Körner Communication Officer, Sustainability Center Bremen, c/o econtur gGmbH, Germany

Sylvia Kruse Research Fellow, Swiss Federal Institute for Forest, Snow and Landscape Research (WSL), Switzerland

Jaap Kwadijk Director of Science, Deltares, the Netherlands

Kelly Levin Senior Associate, World Resources Institute, US

Andreas Lieberum Head of the Sustainability Center Bremen, c/o econtur gGmbH, Germany

Willem Ligtvoet Senior Policy Advisor, PBL Netherlands Environmental Assessment Agency, the Netherlands

Sabine McCallum Head of Department of Environmental Impact Assessment and Climate Change, Environment Agency Austria

Jelle van Minnen Senior Policy Advisor, PBL Netherlands Environmental Assessment Agency, the Netherlands

Richard H. Moss Senior Scientist, Joint Global Change Research Institute, Pacific Northwest National Laboratory, US

Rohan Nelson Associate Professor, Tasmanian Institute of Agriculture, University of Tasmania, Australia

Christian Pagé Research Engineer (Highly Qualified), Sciences de l'Univers au CERFACS, URA CERFACS/CNRS No1875, Toulouse, France

Lydia Pedoth Researcher, EURAC, Italy

Lasse Peltonen Research Coordinator, Finnish Environment Institute (SYKE), Finland

Marjolein Pijnappels Knowledge communication and visualisation, Studio Lakmoes, the Netherlands

Patrick Pringle Deputy Director, UKCIP, Environmental Change Institute, University of Oxford, UK

Thomas Probst Senior Scientific Officer, Swiss Federal Office for the Environment, Switzerland

Andrea Prutsch Adaptation Policy Advisor, Department of Environmental Impact Assessment and Climate Change, Environment Agency Austria

Marco Pütz Head of Research Group, Swiss Federal Institute for Forest, Snow and Landscape Research (WSL), Switzerland

Bertrand Reysset Technical advisor, Ministry of Ecology, Sustainable Development and Energy, France

Kaj van de Sandt Research Fellow, Wageningen University, the Netherlands

Bruno Schädler Senior Researcher, Institute of Geography, University of Berne, Switzerland

Inke Schauser Adaptation Policy Advisor, Department of Climate Change Impacts and Adaptation (KomPass), Federal Environment Agency, Germany

Stefan Schneiderbauer Senior Researcher, EURAC, Italy

Nobuo Shirai Project Professor, Center for Regional Research Hosei University, Japan

Reinhard Steurer Associate Professor, University of Natural Resources and Life Sciences, Vienna, Austria

Therese Stickler Policy Advisor, Environment Agency Austria

Roger Street Technical Director, UKCIP, Environmental Change Institute, University of Oxford, UK

Rob Swart Coordinator of international climate change adaptation research, Alterra, Wageningen University and Research Centre, the Netherlands

Makoto Tamura Associate Professor, Institute for Global Change Adaptation Science, Ibaraki University, Japan

Mitsuru Tanaka Professor, Faculty of Social Sciences, Hosei University, Japan

Andreas Vetter Policy Advisor, Federal Environment Agency, Germany

Oskar Wallgren Senior Project Manager, Stockholm Environment Institute (SEI), Sweden

Clive Walmsley Senior Climate Change Advisor, Natural Resources Wales, UK

Paul Watkiss Director, Paul Watkiss Associates, Oxford, UK

Thomas J. Wilbanks Corporate Research Fellow, Climate Change Science Institute and Environmental Sciences Division, Oak Ridge National Laboratory, US

Julie Wilk Director Head of Research Area, Centre for Climate Science and Policy Research and Water and Environmental Studies, Linköping University, Sweden

Julian Wright Senior Advisor, Environment Agency, England

Sherry B. Wright Coordinator of Energy and Infrastructure Impacts Programs, Climate Change Science Institute and Environmental Sciences Division, Oak Ridge National Laboratory, US

Kazuya Yasuhara Professor Emeritus, Institute for Global Change Adaptation Science, Ibaraki University, Japan

Marc Zebisch Scientific Head of the Institute for Applied Remote Sensing, EURAC, Italy

FOREWORD

European Commission – DG Climate Action

Climate change is a reality, and its impact is increasingly felt in Europe and around the world. Our first priority remains reducing greenhouse gas emissions to keep global warming below 2°C and avert dangerous climate change. Far-reaching EU legislation on mitigation measures is already in place for this.

However, irrespective of the success of mitigation efforts, climate impacts will increase in the coming decades because of the delayed effect of past and current greenhouse gas emissions. This is why, alongside our mitigation measures, we need policies focused on adaptation. Action taken now will avoid increased costs and losses later.

Adaptation efforts are already underway across the EU at various governance levels and in different sectors. Several EU Member States have adopted a national adaptation strategy, and more are being prepared. Some of these strategies have been followed up by action plans, and there has been some progress in integrating adaptation measures into sectoral policies. Beyond the national level, many transnational cooperation projects on adaptation have been initiated over the last years which receive financing by EU-funds such as the LIFE+ and INTERREG programmes. At the local level, a significant share of adaptation activities takes place at city level. There are already many examples of cities in Europe that have adopted adaptation strategies or action plans or are in the process of developing them.

With the EU Adaptation Strategy adopted in the spring of 2013 we aim to raise the profile of adaptation to climate change and step up action on adaptation across the EU, complementing and supporting the Member States' activities.

The strategy particularly also supports action by promoting greater coordination and information-sharing between Member States, and by ensuring that adaptation considerations are addressed in all relevant EU policies.

Targeted information is crucial for decisions on adaptation and – this being still a fairly new policy field – not always easy to come by. Sharing knowledge and experience in adaptation policy and practice thus plays an important role in facilitating mutual learning from practical examples and supporting effective adaptation action.

This *Climate Change Adaptation Manual* presents a variety of adaptation approaches and practical experiences across Europe and provides a comprehensive overview of the state of the art in other continents. It structures the cases along generic guiding principles for good practice in adaptation, addresses challenges in adaptation processes and how to overcome them, and highlights lessons learned.

I am confident that this book will be useful to adaptation practitioners and of interest to all concerned with adaptation.

Humberto Delgado Rosa
Director of Directorate C
(Mainstreaming Adaptation and Low Carbon Technology),
DG Climate Action, European Commission

Organisation for Economic Co-operation and Development (OECD)

Adaptation to climate change is inevitable. However, successful adaptation will depend on the effectiveness of the actions taken by governments, businesses, civil society and households in response to a changing climate. Supporting this process in a fair, effective and efficient way is a central policy question for OECD members and partner countries.

During the past decade, there has been a progression from research on the potential impacts of climate impacts to the development of strategies to prepare for these changes. Since 2005, 18 OECD countries have produced national strategies that help to plan, co-ordinate and communicate national action on adaptation. The key challenge now is to move from planning to implementation. Our research has identified three priorities for successfully making this transition.

The first priority is to embed adaptation across the public sector as a whole, securing strong engagement beyond environment ministries. The cross-cutting nature of adaptation requires working across traditional sectoral and policy boundaries, often to address existing inefficiencies. For example, improved management of flood risk may require reforms to land-use planning, design standards, insurance markets and innovation policy.

It is also essential to support the private sector's response to climate change. There is currently high awareness of climate change impacts, but sparse action to manage the resulting risks. There is a pressing need to make the business case for targeted action in this area. An integral part of this will be to strengthen links between researchers and end-users to ensure that relevant tools and data are being produced. Regulatory frameworks governing private infrastructure also need to be fit for purpose, removing barriers to cost-effective action on adaptation.

The third priority is to integrate robust monitoring and evaluation from the outset of programme and policy design. By doing so, this will help to ensure that interventions are delivering their expected results and to learn lessons to inform the design of future interventions. It can also help to ensure accountability for mainstreamed approaches to adaptation.

This manual is very timely in offering practical guidance on these key issues. It makes an important contribution in supporting the implementation of high-quality adaptation strategies and measures.

Michael Mullan
Team Leader – Climate Change Adaptation and Development,
Environment Directorate,
Organisation for Economic Co-operation and Development (OECD)

European Environment Agency

Climate change is now a major part of planning for the future. Around the world, the extent and speed of change is becoming ever more evident according to the IPCC Fifth Assessment Report *'Climate Change 2013: The Physical Science Basis'* and, as reported in the EEA's 2012 *'Climate change, impacts and vulnerability in Europe'*, climate change is already causing a wide range of impacts on society and the environment in Europe. While reducing greenhouse gas emissions, there is also a need for society to adapt. Otherwise, damage costs will continue to rise.

EEA's 2013 report on *Adaptation in Europe* shows that 16 European member countries have already developed national adaptation strategies (nine more than in 2008), and that 12 more are in the process of doing so. As the first comprehensive overview on adaptation in Europe, the report also highlights a wealth of regional, transnational and local responses, tailor-made to address specific conditions and needs as well as social and economic contexts. There is no 'one-size-fits-all' approach to adaptation.

National adaptation strategies to date address primarily the water, agriculture and forestry, biodiversity, and human health sectors. Examples of implemented actions show that adaptation of both natural and human systems is already taking place across Europe. At EU level, instruments for implementing adaptation policy include mechanisms such as cohesion, agriculture and infrastructure funds, as well as support under the LIFE+ programme. These are critical in helping to integrate adaptation into EU policy – a process known as 'mainstreaming' of adaptation.

Research has a key role to play in strengthening the knowledge base on climate change adaptation in Europe. National and EU-funded research has improved the understanding of past and current changes in the climate system, scenarios for future climate change and the impacts and vulnerabilities/risks. There are, nevertheless, some areas that require more information and assessments. These include the costs and benefits of adaptation actions and the monitoring and evaluation of adaptation, for example through the development of indicators. In addition, research can help improve the process of planning, implementing and reviewing adaptation policies by further examining the factors for successful adaptation and good practices as well as highlighting remaining knowledge gaps.

In that context, national adaptation portals as well as the European Climate Adaptation Platform (Climate-ADAPT, http://climate-adapt.eea.europa.eu) are important tools for sharing with stakeholders at all levels of governance practical experiences in developing and implementing actions and the results of research (i.e. EU research, INTERREG and ESPON projects) on climate change risks, adaptation practices, national initiatives or decision-support tools.

Policy makers face the challenge of designing and implementing adaptation approaches that are: *coherent* across sectoral domains and levels of governance; *flexible* so that strategies and plans can be progressively adjusted to new conditions as they unfold and are updated with new information from monitoring, evaluation and learning; and *participatory* as involvement of stakeholders (policymakers, NGOs, businesses, citizens) is important in creating a sense of 'ownership' in adaptation policy. These challenges bring opportunities for research to further support the implementation phase of the European Strategy on Adaptation to Climate Change.

The strategy includes a guidance document for development of national adaptation strategies and this *Climate Change Adaptation Manual* provides a wealth of information that complements the approaches presented in the guidance document.

The origin of the *Climate Change Adaptation Manual* was, at least in part, earlier EEA work on guiding principles for good practice in adaptation. EEA is convinced that this publication strengthens the knowledge on adaptation to climate change and that it will support, for a wide range of stakeholders, related policy developments and their implementation.

Dr. Hans Bruyninckx
Executive Director of the European Environment Agency (EEA)

Science

The IPCC in its 5th Assessment report reconfirms most findings of earlier assessments of the risks of climate change. It specifically notes a broadening evidence basis that brings the already observed impacts of climate change to the fore and deepens our understanding of what the future impacts may be.

The stagnant pace of the international negotiations on curbing the increase of greenhouse gas emissions suggests that the options to avoid dangerous impacts of climate change are dwindling. So adaptation is not only unavoidable; indeed, it is also urgent. Since the 2009 Copenhagen UNFCCC Conference of Parties, there is global agreement that climate change is real and that, if major damage is to be avoided, global mean temperature rise should be limited to at most 2°C as compared to pre-industrial levels. However, both the current observed warming, and the future warming to which we are increasingly becoming committed due to past emissions, are at odds with this long-term goal. At the same time, in most regions there is still much uncertainty as to how climate change will manifest itself locally and what the effects will be, in particular with respect to rainfall and crucial extreme events.

Yet despite the remaining uncertainty, it is clear that we must act now to develop policies and allocate resources. This does not only apply to national governments, but also to municipalities, water managers, farmers and firms. But how? And when? What are the options? How should we prioritise them? And which instruments do we have at our disposal to implement them? And when implemented, how do we know that they will be effective?

Because the emphasis at national and international levels has been on mitigation for a long time, actual experiences in adaptation that give answers to these questions are as yet very scarce. Therefore, the publication of this manual is very timely.

Although Europe has, like other regions, started to adapt late to the prospect of climate change, more than half of the EU Member States have now adopted national adaptation strategies, and in 2013 a European adaptation strategy was released. At the same time, cities, river basin management institutions, and energy and agricultural organisations have started to consider boosting their climate resilience. Europe is, therefore, leading the way in experimenting with climate change adaptation, and it is thus appropriate that the examples in this manual originate from that continent.

The manual organises its guidance around ten principles for adaptation that are broadly applicable in any region. Thus, although adaptation can be highly localised in its character, it is possible to transfer knowledge about adaptation from one context to another; and the manual makes abundantly clear that meaningful adaptation is possible, now, regardless of current uncertainty.

In the coming years, more experiences will be developed with actual adaptation to climate change in Europe and elsewhere. Research and practice will further increase our understanding of what works and what does not, and in which circumstances. Through the work of **UNEP PROVIA** (Programme of Research on Vulnerability, Impacts and Adaptation) and many other initiatives, research results and practical experiences can be shared around the world to help reduce the number of people and assets at risk. This manual is critical reading for all who want to engage in reducing vulnerability to climate change, from practitioners to policy makers and researchers.

Professor Martin Parry
Grantham Institute and Centre for Environmental Policy,
Imperial College London

ACKNOWLEDGEMENTS

The idea of developing this book was triggered by the publication of 'Guiding principles for adaptation to climate change in Europe' in 2010. We are grateful to the European Environment Agency (EEA), in particular to André Jol and Stéphane Isoard, for giving us the opportunity to develop those guiding principles under the work programme of the European Topic Centre on Air and Climate Change (ETC/ACC). We acknowledge the feedback from 252 experts on earlier versions of the guiding principles responding to an internet survey, overall agreeing that they integrate the most important aspects of good practice in adaptation.

We are thankful to the publishers at Earthscan, now Routledge, for enabling us to compile this book based on the earlier ETC/ACC publication and we also recognise the dedicated team at Routledge, in particular our main contact Helen Bell, for support throughout the production of the manual. Extended thanks also go to all contributing authors, who agreed to enrich this book with a wide variety of case studies reflecting their experiences with adaptation policy and practice around the world. We very much appreciated the high quality and timely input which indeed substantially helped to finalise the manuscript.

Claire Bacher, John James O'Doherty, Ulrike Hofer and Ingrid Klaffl deserve gratitude for their English proofreading and giving us detailed comments on the wording and harmonising the referencing.

Finally, the authors are also grateful to Humberto Delgado Rosa (European Commission), Hans Bruyninckx (European Environment Agency), Michael Mullan (OECD) and Martin Parry (Grantham Institute and Centre for Environmental Policy, Imperial College) for kindly providing forewords to this book.

PART I

Introduction and overview

1

GUIDING PRINCIPLES FOR GOOD ADAPTATION AND STRUCTURE OF THIS BOOK

Inke Schauser, Sabine McCallum, Andrea Prutsch, Torsten Grothmann and Rob Swart

1.1 Why this book?

Climate change is already affecting our environment, our economy, and our way of living: evidence from all continents and most oceans shows that many natural and social systems are being affected by regional climate change (Parry et al. 2007). A series of extreme weather events in the last decade, such as the heat wave in Europe in 2003, flooding of the rivers Elbe and Danube in 2002 and 2013, and the drought and associated forest fires in 2010 in Russia, have also shaped public awareness of climate change. These events have all highlighted the need for Europe to adapt itself to the realities of climate change, especially considering that the frequency and magnitude of extreme events such as these are projected to increase in the coming years (IPCC 2012).

Adaptation is necessary to minimise negative impacts resulting from current and expected climate change, and in order to maximise our ability to benefit from any opportunities that climate change may bring. The question is, how to adapt? Simple advice, such as 'save energy' or 'use more renewable energy', is suitable and effective when it comes to preventing climate change. However, when it comes to adapting to climate change, there is no equivalent simple advice. Nevertheless, some common recommendations and guiding principles do exist, and they are equally valid in different countries, for different economic sectors, and for different climatic threats.

Adaptation to climate change is already taking place in Europe and across the world. In fact, adaptation to changing conditions is not a new phenomenon. Both traditional and industrialised societies have adapted their environments to alleviate risks associated with climate variability throughout human history, a process known as autonomous adaptation. It is to be expected that some autonomous adaptation by economies and societies will continue to take place even without any policy intervention. Autonomous responses to a changing environment by individuals, groups or organisations will take place whenever actors perceive that people or assets are at risk, or where they perceive that action will provide economic and societal benefits.

However, autonomous adaptation on its own will not be fully adequate for coping with climate change (EEA 2013). The complexity of climate change, and the range and magnitude of risks we might face, may overwhelm a process of autonomous adaption. Instead, it will be necessary to implement planned adaptation, which proactively addresses potential risks and opportunities. Planned adaptation will help policy makers and societies to take decisions that will remain both robust (to cover all possible climate change scenarios) and flexible (so the measures can be changed if conditions change) to cope with an uncertain future.

Adapting to climate change is a critical challenge for the future. However, it must be remembered that climate change is only one of many stresses that influence decision making, which is also affected by short-term political or economic challenges, among other things. Climate change must therefore be considered alongside – and not separate from – environmental, social and economic issues affecting human and natural systems. Thus, adaptation requires a balanced approach that addresses both climate and non-climate risks (e.g. globalisation, demographic change).

In many cases, adaptation options do not solely target adaptation to climate change, but also aim at other societal or policy objectives. Planning for adaptation provides the opportunity for actors from different sectors to cooperatively address future risks and opportunities, while acknowledging the different backgrounds and values of all involved. Engaging stakeholders, and in particular stakeholders affected by climate change, offers an opportunity to identify innovative solutions to climate change that also have legitimacy and broad support.

The past decade has seen a great accumulation of knowledge on climate change adaptation, both in terms of the policy responses, and in terms of practical, on-the-ground measures. This manual brings together in one volume a variety of adaptation approaches and a broad spectrum of lessons learned in climate adaptation policy and practice up to now. It provides adaptation planners and decision makers with scientifically based information that is also practical and user-friendly. It elaborates on key aspects of successful adaptation by giving examples of adaptation policy and practice that have already been implemented in Europe. Experience from previous adaptation actions can also be very relevant in other regions outside Europe.

The title 'Adaptation Manual' is not intended to imply that this volume provides readers with detailed step-by-step instructions on how to plan for adaptation. Instead, 'manual' should be understood in a broader sense. This work shows different approaches for each adaptation phase, building on generic guiding principles, and highlighting lessons learned to support sharing of existing knowledge and experiences.

TABLE 1.1   Phases of adaptation processes and guiding principles for good adaptation

Prepare the ground for adaptation

Explore potential climate change impacts and vulnerabilities and identify priority concerns (cf. Chapter 5)
Initiate adaptation, ensure commitment and management (cf. Chapter 6)
Build knowledge and awareness (cf. Chapter 7)
Identify and cooperate with relevant stakeholders (cf. Chapter 8)

Plan for adaptation

Explore a wide spectrum of adaptation options (cf. Chapter 9)
Prioritise adaptation options (cf. Chapter 10)
Work with uncertainties (cf. Chapter 11)

Implement adaptation and review results

Avoid maladaptation (cf. Chapter 12)
Modify existing and develop new policies, structures and processes (cf. Chapter 13)
Monitor and evaluate systematically (cf. Chapter 14)

1.2   Structure of the book

This manual is divided into four parts. Part I provides an overview of the general challenges of adaptation, and the state of current adaptation research and action.

Part II is the core component of the book. It presents the ten guiding principles for good adaptation that were developed for the EEA by the editors of this book (Prutsch et al. 2010) (cf. Table 1.1). These ten principles cover the different interlinked phases of adaptation, from preparing the ground, to the selection and planning of measures, to the implementation of these measures and the monitoring and assessment of their progress. In every chapter, the guiding principle is first explained and updated based on new scientific literature. The principle is then illustrated by three to four cases representing experiences from adaptation policy and practice in several European countries. These cases reflect adaptation experiences from various regions, sectors and levels of decision making. They have been chosen to illustrate how specific guiding principles can be realised in practice and not every case can illustrate all ten guiding principles. Finally, the lessons learned from the cases are identified for each chapter and guiding principle.

It must be remembered that the guiding principles are not organised in order of importance. The guiding principles are strongly interlinked and mutually influenced. Together, they give a common basis for cooperative adaptation activities across sectors and for all decision-making levels. While developed with a focus on Europe, the guiding principles are also relevant elsewhere in the world.

In Part III, the focus of the manual is broadened, turning to the state of adaptation policy and practice in the US, Australia, Japan and developing countries. These contributions reflect on the relevance of the guiding principles for countries outside Europe. Part IV, the final chapter of the manual, summarises the main lessons learned from the various practical experiences presented in this book.

With this manual, we aim to provide readers with inter-disciplinary insights into cutting-edge knowledge and on climate change adaptation, including knowledge from adaptation research, policy and practice. We believe that learning from past experience and basing future decisions on what works will increase long-term adaptive capacity and resilience. Thus, we hope that the lessons learned in this volume will add to the discussion on how to reach good practice in adaptation and encourage taking further proactive steps towards a climate-resilient society.

References

EEA (2013) *Adaptation in Europe. Addressing risks and opportunities from climate change in the context of socio-economic developments*, EEA Report 3/2013, Copenhagen.

IPCC (2012) *Managing the Risks of Extreme Events and Disasters to Advance Climate Change Adaptation.* A Special Report of Working Groups I and II of the Intergovernmental Panel on Climate Change [Field, C.B., V. Barros, T.F. Stocker, D. Qin, D.J. Dokken, K.L. Ebi, M.D. Mastrandrea, K.J. Mach, G.-K. Plattner, S.K. Allen, M. Tignor and P.M. Midgley (eds.)], Cambridge University Press, Cambridge, UK, and New York, NY, USA, 582 pp.

Parry, M.L., Canziani, O.F., Palutikof, J.P., van der Linden, P.J. and Hanson C.E. (ed.) 2007 *Climate Change 2007: Impacts, adaptation and vulnerability.* Contribution of working group II to the fourth assessment report of the Intergovernmental Panel on Climate Change, Cambridge, UK: Cambridge University Press.

Prutsch, A., Grothmann, T., Schauser, I., Otto, S. and McCallum, S. (2010) *Guiding principles for adaptation to climate change in Europe.* ETC/ACC Technical Paper 2010/6, European Topic Centre on Air and Climate Change. http://air-climate.eionet.europa.eu/reports/ETCACC_TP_2010_6_guid_princ_cc_adapt

2

FACING THE SPECIFIC CHALLENGES OF ADAPTATION

Andrea Prutsch, Sabine McCallum,
Torsten Grothmann, Inke Schauser
and Rob Swart

2.1 Introduction

Despite similarities to other planning processes, adaptation to climate change is characterised by several challenges that set it apart from other planning processes and must be acknowledged and proactively addressed. The analysis of these challenges and consideration of how they could be overcome led to the ten guiding principles for good adaptation presented in Chapter 1 and elaborated in Chapters 5 to 14.

Adaptation is complex due to the fact that climate change affects all regions, most sectors, all levels of decision making and many actors from various backgrounds differently (Grothmann 2011). This diversity and the long-term and large-scale nature of the risks involved imply that current practice in planning will be insufficient for comprehensive adaptation. Planning for adaptation also involves dealing with imperfect knowledge and uncertainties; this can conflict with the expectation that decision making should be based on agreed-upon "hard" evidence and robust information. In addition, non-climatic factors such as social and economic issues must be taken into account in comprehensive adaptation decision making.

In this chapter, we present a concise summary of the main challenges that can occur in adaptation processes.

2.1.1 Adaptation faces regional disparities

Climate impacts emerge at the regional and local levels in manifold ways. Due to the diversity of bio-physical and socio-economic situations in different regions, the impacts of climate change will differ from region to region. For example, the European Alps are characterised by small-scale climatic conditions, which in turn result in various climate change impacts reflecting the scattered natural landscapes. The impacts of climate change also vary according to the degree of socio-economic development and the adaptive capacity of the area.

2.1.2 Adaptation affects sectors differently, and there are cross-sectoral effects

Climate change affects most sectors; thus, adaptation is a multi-sectoral issue (Burton et al. 2006, Agrawala & Frankhauser 2008). The sectors involved might follow different objectives, and a certain adaptation measure in one sector could create negative side-effects for another sector. In addition, actors from a variety of sectors will be involved in adaptation, representing diverse values and interests; some of these might be controversial and generate resistance (de Bruin et al. 2009). The improved horizontal integration of adaptation policies across sectors within and beyond the environmental domain will be needed. Additionally, horizontal integration will also require mechanisms that facilitate the interactions between state, business and civil society actors.

2.1.3 Adaptation concerns various levels of decision making

The "sphere of competence of authorities in charge of environmental protection [...] does not always match with the boundaries of the affected environment" (Liberatore 1997: 116). This applies to the authorities responsible for adaptation, as adaptation pressures and responses cut vertically across different levels of decision and policy making, from the EU level to national, regional and local levels (Klein et al. 2007). These different levels and actors interact with each other within hierarchical structures (Adger et al. 2005). The need for appropriate adaptation extends beyond the local and regional scales (Adger et al. 2005, Paavola & Adger 2006, Swart et al. 2009) and cannot be the sole responsibility of any single institution.

2.1.4 Adaptation affects many stakeholders with possibly conflicting interests

Climate change will (and already does) affect a range of actors and stakeholders (e.g. public authorities, businesses, NGOs, scientific organisations) in different ways; this makes adaptation imperative for most actors. Multi-actor collaboration can be challenged by questions regarding, e.g. roles, power, authority and responsibility, as well as by the multiple and possibly conflicting interests of actors collaborating on adaptation measures (Lebel et al. 2010). Thus, new mechanisms are often needed to facilitate cooperation between stakeholders from different fields and with different competencies and levels of experience.

2.1.5 Adaptation despite imperfect information and uncertainties

The uncertainty of future climate change and its impacts is often perceived to be the most significant challenge facing adaptation, since changes in the future climate cannot be accurately projected. This is due to uncertainties with regard to changes in greenhouse gas emissions and related economic and social developments (e.g. human behaviour, demographic and technological developments), incomplete scientific knowledge of the climate system (e.g. tipping points) and insufficient long-term data for many of the impact indicators on appropriate spatial and temporal scales. Uncertainties in projected global climate change impacts can often increase when downscaled to the regional level, and it is not clear as yet whether more reliable high-resolution projections can ever be developed.

Nevertheless, climate change projections provide clear trends and sufficient concerns to justify adaptation actions (Adger et al. 2007). In addition, it must be acknowledged that climate change impacts can already be seen, e.g. glaciers melting and permafrost thawing (cf. the Europe-wide summary in EEA 2012). Further changes can be expected due to greenhouse gas emissions in the past and to lagging progress in climate mitigation policy. Thus, the uncertainties associated with climate change projections should not be an excuse for inaction or a reason to delay adaptation. Relying entirely on reactive adaptation is undesirable, since the costs of impact damages and thus reactive adaptation are expected to be much higher than the costs of proactive adaptation and residual damages (Frankhauser et al. 1999, Stern 2006). Furthermore, a wait-and-see attitude would have serious drawbacks in the case of irreversible climate change consequences (Smith 1997). Finally, and perhaps most importantly, as several cases in this manual illustrate, there are many opportunities for actions that will decrease vulnerability that would make sense even if climate change does not materialise as projected.

2.1.6 Long-term global projections versus information needs for short-term decision making

In the long term and at the global level, the results from climate change models provide a reasonably clear picture, as the patterns of change are very similar for all models (Hallegatte 2009), in particular for temperature projections but also for precipitation in various regions (including the Mediterranean). However, this information does not necessarily meet the requirements of most decision-makers and planners for shorter planning horizons, nor is it always provided at the appropriate scale to fine-tune adaptation measures for site-specific conditions in a proactive manner (Frankhauser 2009).

Reacting to changes in the short term will result in poor adaptation because climate change by its nature is a long-term problem, and the worst effects may only become visible after decades have passed. In addition, certain adaptation options have long lead times and might only be fully effective in the long term (e.g. the introduction of new trees in forest management, the development of new crop types). Thus, timing and the sequence of response actions are important (Frankhauser 2009, Horstmann 2008); the most appropriate solutions may transcend traditional ways of thinking and current practices in decision making.

2.1.7 Adaptation embedded within a broader context

Addressing climate change is only one of many stresses that influence decision making (Scheraga & Grambsch 1998). In many situations, short-term political or economic challenges will have a greater influence on decisions than climate change. Thus, in the case of adaptation, non-climatic developments (e.g. globalisation, demographic developments) must also be considered. For example, the level of heat-related fatalities also depends on age distribution and the quality of public health care. Thus, other processes must be taken into account as well when planning for adaptation.

In addition to the consideration of economic and demographic changes, the achievement of ecological and social justice presents a further challenge for adaptation. This issue relates to disparities in climate change impacts between different regions, sectors, actors, population groups and species. In Europe, mountain regions, coastal areas, flood-prone river systems, urban areas, the Mediterranean and the Arctic are likely to be more severely affected than other regions (EEA 2010). In addition, climate change is expected to have negative consequences for biodiversity, whereas agriculture may benefit, at least in the short term and in regions that are disadvantaged by cold temperatures at present. The elderly, ill and very young are more vulnerable to heat events than average, healthy, middle-aged people. The challenge is to reach ecological and social (including intergenerational) balance – also in terms of the costs of climate change and adaptation – in order to prevent social conflicts or ecological damage. In this context, Paavola and Adger (2006: 607) propose that "adopting the principles of avoiding dangerous climate change, forward-looking responsibility, putting the most vulnerable first and equal participation of all would be a step towards fairer adaptation".

2.1.8 Identifying and overcoming potential barriers

Multiple types of barriers may hinder successful adaptation action. These include financial (e.g. lack of financial resources for adaptation), cognitive (e.g. lack of motivation for adaptation), behavioural (e.g. lack of leadership qualities in critical positions), social (e.g. lack of political action) and cultural (e.g. cultural traditions that restrict cooperation between different stakeholders) barriers (Adger et al. 2007, Lorenzoni et al. 2007, Hulme et al. 2007, EEA 2009, Burch 2010, Grothmann 2011). Further barriers can arise from hindering regulations, opposing political interests, impeding organisational or managerial structures or the lack of technology (Grothmann 2011, Grothmann et al. 2009, Clar et al. 2013). Missing information regarding climate change and its impacts might also pose a major barrier to adaptation. In addition, the lack of information exchange and communication between sciences and policy as well as limited experience and competence in dealing with climate-related impacts could hinder the adaptation process (Amundsen et al. 2010, Clar et al. 2013, Grothmann & Siebenhüner 2012).

However, it is important to note that even sufficient information and awareness of the need for adaptation do not necessarily lead to action (Baron 2006, Weber 2006). In fact, decisions are influenced by personal experiences, individual perceptions and values, emotions and hidden agendas (Grothmann & Patt 2005, van de Kerkhof 2006). Hence, every decision-making process includes a certain level of "surprise" that is difficult to estimate in advance and thus to influence. Nevertheless, to the extent possible, barriers should be identified and openly confronted so that they can be removed or transcended (Grothmann 2011, Grothmann et al. 2009, Lim et al. 2004, Smith et al. 2009).

2.2 Addressing challenges in the guiding principles for good adaptation

Addressing these various challenges is not an easy task, and because every adaptation case is unique, it is impossible to provide generalisable step-by-step instructions for how to tackle them. Nevertheless, the ten guiding principles for good adaptation presented in Chapter 1 and elaborated in Chapters 5 to 14 attempt to address these challenges by articulating important elements that should be taken into consideration when preparing the ground for adaptation, planning and implementing adaptation and reviewing its success.

Certain guiding principles focus on a particular challenge. For example, the guiding principle "Work with uncertainties" specifically addresses the challenge of realising adaptation despite imperfect information and uncertainties. Other guiding principles tackle a wide range of challenges. For example, the principle "Identify and cooperate with relevant stakeholders" is useful for addressing the challenge entailed by adaptation's effects on a variety of stakeholders with possibly conflicting interests. However, it also addresses the challenges of dealing with potential cross-sectoral effects, including various levels of decision making, embedding adaptation within a broader context and identifying and overcoming potential barriers.

Future research and practice will determine whether additional challenges of adaptation should be added and new guiding principles developed to address them.

References

Adger, W.N., Arnell, N.W. and Tompkins, E.L. (2005) "Successful adaptation to climate change across scales", *Global Environmental Change* 15 (2), pp. 77–86.
Adger, W.N., Agrawala, S., Mirza, M.M.Q., Conde, C., O'Brien, K., Pulhin, J., Pulwarty, R., Smit, B. and Takahashi, K. (2007) "Assesment of adaptation practices, options,

constraints and capacity", in Parry, M.L., Palutikof, J.P., van der Linden, P.J. and Handon, C.E. (ed.) *Climate Change 2007: Impacts, Adaptation and Vulnerabilty,* Working Group II to the Fourth Assessment Report of the Intergovernmental Panel on Climate Change, Cambridge University Press, Cambridge, UK, pp. 717–743.

Agrawala, S. and Frankhauser, S. (ed.) (2008) *Economic Aspects of Adaptation to Climate Change. Costs, Benefits and Policy Instruments,* OECD, Paris.

Amundsen, H., Berglund, F. and Westskog, H. (2010) "Overcoming barriers to climate change adaptation – a question of multilevel governance?", *Environment and Planning C: Government and Policy* 28 (2), pp. 276–289.

Baron, J. (2006) "Thinking about global warming", *Climatic Change* 77 (1–2), pp. 137–150.

Burch, S. (2010) "Transforming barriers into enablers of action on climate change: Insights from three municipal case studies in British Columbia, Canada", *Global Environmental Change* 20 (2), pp. 287–297.

Burton, I., Diringer, E. and Smith, J. (2006) *Adaptation to Climate Change: International Policy Options,* prepared for the Pew Center on Global Climate Change.

Clar, C., Prutsch, A. and Steurer, R. (2013) Barriers and guidelines for public policies on climate change adaptation: A missed opportunity of scientific knowledge-brokerage, *Natural Resources Forum* 37 (1): pp. 1–18.

De Bruin, K., Dellink, R.B., Ruijs, A., Bolwidt, L., van Buuren, A., Graveland, J., de Groot, R.S., Kuikman, P.J., Reinhard, S., Roetter, R.P., Tassone, V.C., Verhagen, A. and van Ierland, E.C. (2009) "Adapting to climate change in The Netherlands: An inventory of climate adaptation options and ranking of alternatives", *Climate Change* 95 (1–2), pp. 23–45.

EEA – European Environment Agency (2009) *Regional Climate Change and Adaptation. The Alps facing the challenge of changing water resources,* Technical Report, Copenhagen.

EEA – European Environment Agency (2010) *The European Environment – State and Outlook 2010: Adapting to Climate Change.* European Environmental Agency, Copenhagen.

EEA – European Environment Agency (2012): *Climate Change, Impacts and Vulnerability in Europe 2012.* European Environmental Agency, Copenhagen.

Frankhauser, S. (2009) *A Perspective Paper on Adaptation as a Response to Climate Change,* Copenhagen Consensus Centre, Denmark.

Frankhauser, S., Smith, J.B. and Tol, R.S.J. (1999) "Weathering climate change: Some simple rules to guide adaptation decisions", *Ecological Economics* 30 (1), pp. 67–78.

Grothmann, T. (2011) "Governance recommendations for adaptation in European urban regions: Results from five case studies and a European expert survey", in Otto-Zimmermann, K. (ed.) *Resilient Cities – Cities and Adaptation to Climate Change – Proceedings of the Global Forum 2010,* Hamburg: Springer, pp. 167–175.

Grothmann, T. and Patt, A. (2005) "Adaptive capacity and human cognition: the process of individual adaptation to climate change", *Global Environmental Change* 15 (3), pp. 199–213.

Grothmann, T. and Siebenhüner, B. (2012) "Reflexive governance and the importance of individual competencies: The case of adaptation to climate change in Germany", in Brousseau, E., Dedeurwaerdere, T. and Siebenhüner, B. (ed.) *Reflexive Governance and Global Public Goods,* Cambridge (MA): MIT Press, pp. 299–314.

Grothmann, T., Nenz, D. and Pütz, M. (2009) "Adaptation in vulnerable alpine regions – lessons learnt from regional case studies", in European Environment Agency (ed.) *Regional Climate Change and Adaptation – The Alps Facing the Challenge of Changing Water Resources,* EEA Report No 8/2009, pp. 96–108.

Hallegatte, S. (2009) "Strategies to adapt to an uncertain climate change", *Global Environmental Change* 19 (2), pp. 240–247.

Horstmann, B. (2008) *Framing Adaptation to Climate Change – A Challenge for Building Institutions*, Discussion Paper 23, Deutsches Institut für Entwicklungspolitik, Bonn.

Hulme, M., Adger, W.N., Dessai, S., Goulden, M., Lorenzoni, I., Nelson, D., Naess, L.O., Wolf, J. and Wreford, A. (2007) *Limits and Barriers to Adaptation: Four Propositions*, Tyndall Briefing Note No. 20, Tyndall Centre for Climate Change Research, University of East Anglia, Norwich, UK.

Klein, R., Huq, S., Denton, F., Downing, T.E., Richels, R.G., Robinson, J.B. and Toth, F.L. (2007) *Interrelationships Between Adaptation and Mitigation. Climate Change 2007: Impacts, Adaptation and Vulnerability*, Contribution of Working Group II to the Fourth Assessment.

Lebel, L., Grothmann, T. and Siebenhüner, B. (2010) "The role of social learning in adaptiveness: Insights from water management", *International Environmental Agreements: Politics, Law and Economics* 10 (4), pp. 333–353.

Liberatore, A. (1997) "The integration of sustainable development objectives into EU policymaking: Barriers and prospects", in Baker, S., Kousis, M., Richardson, D. and Young, S. (ed.) *The Politics of Sustainable Development: Theory, Policy and Practice within the European Union*, London: Routledge, pp. 108–126.

Lim, B. and Spanger-Siegfried, E. (ed.) (2004) *Adaptation Policy Frameworks for Climate Change. Developing Strategies, Policies and Measures*, UNDP, Cambridge: Cambridge University Press.

Lorenzoni, I., Nicolson-Cole, S. and Whitmarsh, L. (2007) "Barriers perceived to engaging with CC among the UK public and their policy implications", *Global Environmental Change* 17, pp. 445–459.

Paavola, J. and Adger, W.N. (2006) "Fair adaptation to climate change", *Ecological Economics* 56 (4), pp. 594–609.

Scheraga, J.D. and Grambsch, A.E. (1998) "Risks, opportunities, and adaptation to climate change", *Climate Research* 10, pp. 85–95.

Smith, J.B. (1997) "Setting priorities for adapting to climate change", *Global Environment* 7 (3), pp. 251–264.

Smith, J.B. and Lenhart, S. (1996) "Climate change adaptation policy options", *Climate Research* 6 (2), pp. 193–201.

Smith, J.B., Vogel, J.M. and Cromwell II, J.E. (2009) "An architecture for government action on adaptation to climate change. An editorial comment", *Climatic Change* 95, pp. 53–61.

Stern, N. (2006) *The Economics of Climate Change*, Cambridge: Cambridge University Press.

Swart, R., Biesbroek, R., Binnerup, S., Carter, T.R., Cowan, C., Henrichs, T., Loquen, S., Mela, H., Morecroft, M., Reese, M. and Rey, D. (2009) *Europe Adapts to Climate Change: Comparing National Adaptation Strategies*, PEER Report No. 1, Helsinki.

Van de Kerkhof, M. (2006) "Making a difference: On the constraints of consensus building and the relevance of deliberation in stakeholder dialogues", *Policy Sciences* 39 (3), pp. 279–299.

Weber, E.U. (2006) "Experienced-based and description-based perceptions of long-term risk: Why global warming does not scare us (yet)", *Climatic Change* 77 (1–2), pp. 103–120.

3

ADAPTATION RESEARCH

Where do we stand and where should we go?[1]

Torsten Grothmann

3.1 Introduction

Since the publication of the Stern Review (Stern 2007) and the 4th assessment report of the IPCC (Parry et al. 2007, Solomon et al. 2007), there has been a broad consensus – not only among climate change researchers but also among many decision-makers – about the reality of climate change and the necessity to reduce greenhouse gas emissions and realise measures to adapt to the unavoidable change. There is less consensus regarding probable future climate change impacts and vulnerabilities, and even less on which adaptation measures are necessary and suitable. Research on climate change impacts, vulnerabilities and adaptation (in short: IVA research), regularly summarised by Working Group II of the IPCC, represents a relatively young and fragmented field. Nevertheless, the precautionary principle (UNFCCC 1992) states that a lack of complete scientific certainty and consensus should not be used to justify postponing adaptation measures where there is the threat of serious (e.g. deaths due to heatwaves) or irreversible damage (e.g. extinction of species).

Because of the initial framing of climate change as an environmental problem and the emphasis on questions related to the reality of climate change and the nature and magnitude of climatic risks, climate change science and policy has long given more weight to the natural sciences, which have strongly contributed to the recognition of climate change as a serious problem by rigorous quantitative analyses. The search for solutions that would facilitate mitigation in industrialised countries led to the involvement of the engineering and economic sciences. During recent years, the social sciences have become increasingly involved because their expertise is necessary for an understanding of the social impacts and vulnerabilities as well as the socio-economic aspects of adaptation (e.g. legal, economic, cultural and behavioural barriers to adaptation). Interdisciplinary research involving both natural and social scientists is therefore necessary.

Furthermore, transdisciplinary research (research involving stakeholders from society and practical experts) is required. Researchers alone cannot answer questions related to societal values and norms such as: What does "dangerous anthropogenic interference with the climate system" (UNFCCC 1992: 4) mean? Which potential impacts of climate change should be regarded as too risky to be acceptable? Determining the measures that can and should be taken to adapt to or prevent unacceptable climate change risks also requires the involvement of members of the affected social systems.

Fortunately, a variety of studies on climate change impacts, vulnerabilities and adaptation already fulfil these interdisciplinary and transdisciplinary requirements. Unfortunately, IVA research to date has primarily focused on *problems* (potential climate change impacts and vulnerabilities) and barriers to solutions (e.g. barriers to the realisation of adaptation measures). Rigorous empirical research on adaptation *solutions* (e.g. how to plan, implement and evaluate adaptation measures) is still relatively rare. However, various individuals and organisations have started to develop and implement adaptation actions. This book builds on these experiences and makes them accessible to a wider audience so that others can learn from them.

The overall aim of this chapter is to reflect on IVA research, primarily from a practice-oriented and a social-science perspective, and to develop recommendations for its further improvement. First, some central concepts of IVA research are described. Second, general methodologies of IVA research are considered. The chapter concludes with a summary of recommendations for future IVA research.

3.2   Theoretical concepts in IVA research

3.2.1   *Vulnerability*

Due to the traditional natural-science focus in climate change research, most IVA assessments were initially framed from a natural-science perspective, concentrating on the potential impacts of climate change on natural systems. Nowadays, IVA studies are expanding to include assessments of vulnerabilities and adaptation potential in social-ecological systems, which implies a shift in the balance towards the social sciences. The IPCC defines *vulnerability* to climate change as "a function of the character, magnitude, and rate of climate change and variation to which a system is exposed, its sensitivity, and its adaptive capacity" (Parry et al. 2007: 883). Thus, sensitivity (i.e. will climate change probably result in impacts?) and adaptive capacity (i.e. will the affected system (e.g. a city) be able to adapt to the impacts or avoid them?) are important aspects of climate change research. Both aspects are strongly related to socio-economic factors (e.g. institutional settings and capacities, access to resources, perceptions, skills and knowledge). Future IVA research should try to address these socio-economic factors more comprehensively (including their uncertainties) in order to enable more accurate projections of potential loss and damage due to climate change.

Notwithstanding the different interpretations of the concept, the IPCC's definition of vulnerability can help decision-makers to understand the kinds of factors they must consider in assessing the vulnerability of the system for which they are responsible, be it a city, a business organisation or an entire nation. The concept shows that the loss and damage resulting from climate changes very much depend upon the sensitivity of a system and how well it is prepared for these changes. The assessment of a system's adaptive capacity is particularly useful, as the identification of weaknesses in adaptive capacities (e.g. lack of knowledge regarding heatwave risks) can be used to plan adaptation measures (e.g. communication of heatwave risks) (cf. Grothmann et al. 2013). In general, because of the vagueness of the vulnerability concept, it is helpful to focus on more precisely articulated adaptation questions related to its components (Hinkel 2011).

3.2.2 Resilience, robustness, adaptive management and iterative risk managment

Future climate change and its impacts can only be assessed in terms of ranges (e.g. ranges of temperature or precipitation change). These uncertainties in the projections of future climate change and its impacts are due to the limited scientific understanding of the climate system and social-ecological systems, but also – sometimes primarily – to the uncertainties in future greenhouse gas emissions (the major drivers for climate change and main input variables in assessments of future climate change and its impacts).

Due to these uncertainties, scientific concepts that describe processes or qualities that allow social-ecological systems to function under a wide range of possible future conditions have become popular in IVA research and are especially useful for decision-makers seeking to adapt to climate change. In IVA research, the most prominent concept that deals with the challenge of uncertainty is *resilience*: "The resilience approach emphasizes non-linear dynamics, thresholds, uncertainty and surprise, how periods of gradual change interplay with periods of rapid change and how such dynamics interact across temporal and spatial scales" (Folke 2006: 253). Resilient systems are flexible and can quickly adapt to changed conditions (e.g. communities with diversified livelihoods). In particular, the consideration of potential surprises (e.g. climate changes or impacts not included in climate (impact) projections, such as the heatwave in Europe in 2003) differentiates the resilience approach from the robustness approach. The *robustness* approach uses a wide range of projections of climate change and its impacts to identify robust adaptation actions (e.g. Dessai & Hulme 2007, Hallegatte et al. 2012). Other concepts used in IVA research for dealing with uncertainties in future changes are *adaptive management* (e.g. Pahl-Wostl 2007) and *iterative risk management* (e.g. Carter et al. 2007).

Future IVA research should seek to (i) better integrate the various concepts used in IVA studies (e.g. vulnerability, resilience, adaptive capacity), (ii) refrain from inventing new concepts for aspects that have been sufficiently described in existing concepts and (iii) better relate concepts to theoretical frameworks (such as social-ecological systems or sustainable development). This will hopefully result in a more consistent and comprehensible use of concepts, avoiding some of the common misunderstandings among scientists and between scientists and practitioners that often hinder the generation of new insights and the development of adaptation solutions.

3.2.3   *Adaptation and transformation*

Many of the concepts mentioned above (e.g. robustness, adaptive management) are generally used to describe, explain and support the planning and implemention of adaptation rather than to assess climate change impacts or vulnerabilities. This brings us to the next central concept in IVA research: *adaptation*. Depending on the conceptual perspective (vulnerability, resilience, robustness, etc.), adaptation to climate change is conceptualised as reductions in vulnerability, the realisation of adaptive/iterative risk management or the reinforcement of robustness and/or resilience. More generally, the IPCC defines *adaptation* as "adjustment in natural or human systems in response to actual or expected climatic stimuli or their effects, which moderates harm or exploits beneficial opportunities" (Parry et al. 2007: 869).

A recent theoretical development in IVA research is the conceptualisation of adaptation as a transformative process. *Transformations* are fundamental cultural, political, economic, infrastructural or technological changes (WBGU 2011). Increasingly, publications have argued the case for transformative adaptation (e.g. Kates et al. 2012, O'Brien 2012, Park et al. 2012), particularly in a potential future world in which the global temperature target of 2°C is exceeded. A recent study on the impacts of a possible global temperature increase of 4°C projects that the impacts on water availability, ecosystems, agriculture and human health will lead to the large-scale displacement of populations and adverse consequences for human security and economic and trade systems (PIK 2012). In light of these severe projected impacts of climate change, many more transformative adaptation measures would appear to be required.

3.3   Some general methodological considerations

IVA research not only focuses on generating or testing theoretical concepts; it also seeks to generate and test methods. These methods are often useful for later research purposes, as well as for practitioners and decision-makers who want to analyse or prioritise climate change impacts, vulnerabilities or adaptation options in their respective fields of responsibility.

A comprehensive overview of all of the specific methods and tools that are or could be used in IVA research would go beyond the scope of this chapter. Many useful methods and tools are presented in the topic-specific chapters of this book. Thus, this section will focus on some general methodological procedures used in IVA research.

3.3.1 Interdisciplinary research and assessments of climate change impacts and vulnerabilities

Most impact studies on climate change focus on analysing potential climate changes and their direct biophysical impacts on natural systems (e.g. water availability) using a scenario- or model-driven approach. They try to represent system dynamics and their potential consequences in the future, mostly at large spatial scales. These kinds of studies are often called *"top-down" assessments* (Dessai & Hulme 2004).

"Bottom-up" assessments, that evolved out of social vulnerability research (often used in natural hazards research; cf. Wisner 2009), focus on analysing socio-economic factors at the local scale that allow or hinder people to react to changes in their environment (Dessai & Hulme 2004). Social vulnerability research, based heavily on stakeholder engagement methods, begins by analysing present-day restrictions in social systems (e.g. poverty, social inequality) that might increase vulnerabilities to climate change or become barriers to adaptation. This approach is useful for the identification of important socio-economic factors and current problems and developments that will need to be integrated into climate change adaptation in order to make use of synergies and avoid conflicts.

Top-down and bottom-up assessments have different strengths and weaknesses and their integration can avoid some of the weaknesses. Current IVA research on climate change is already integrating both approaches, primarily at the local and regional levels. Nevertheless, improved methods of integration will be required to overcome the problems arising from the different temporal and spatial scales of analyses (cf. example Ghimire et al. 2010) and to better tailor assessments to the scales used in decision-making (Brown et al. 2012).

In order to develop an understanding of social vulnerabilities, social-science expertise is necessary. This knowledge can contribute concepts and methods that permit the *social differentiation* of the affected population along dimensions such as income, education, social or income inequality, gender, racial, ethnic or religious background and access to information. Furthermore, the *social context* must be considered, including culture, history, political and economic systems and especially the existence and effectiveness of institutions that could plan and implement adaptation measures (Gupta et al. 2010). Perceptual aspects are also important in an understanding of social vulnerabilities. If other problems (e.g. economic problems) are perceived as more important or more urgent than adaptation to climate change, the chances of implementing adaptation measures will decrease, which might in turn increase vulnerability (Grothmann et al. 2013). Furthermore, in order to assess a social system's future vulnerability to climate change, probable *societal changes* in the future must be considered in addition to probable climate change. Here, not only changes in demographics, land use and level of education should be taken into account, but also future social changes that can be more difficult to predict, such as lifestyle changes, individualisation, deregulation and potential changes in political stability. In addition, *second-order climate change risks* (risks that result from the realisation of mitigation or adaptation measures) should be considered in assessments of future climate change-related impacts and vulnerabilities. Inclusion of these factors will contribute to the development of more realistic scenarios of the future that cover the broad range of possible future changes (in natural and social systems).

Hence, IVA research must be interdisciplinary, involving both natural scientists and social scientists. Without the expertise of the social sciences, important social factors, differentiations, contexts and societal changes are likely to be overlooked in IVA studies. Without natural scientists, climate change and its impacts on natural systems cannot be understood in sufficient detail. The specific natural and social sciences most important to a study will depend on the case-specific impacts, vulnerabilities and adaptation options as well as the social-ecological system under investigation.

It is often very difficult to integrate results from the natural sciences with results from the social sciences. In addition, interdisciplinary coordination is often complex because these disciplines apply very different paradigms in their research efforts (e.g. quantitative versus qualitative). Even interdisciplinary coordination between different social-science disciplines can be troublesome. One positive example of an obviously successful interdisciplinary coordination is the CLIMAS project (Climate Assessment for the South-West, USA). Future IVA research should seek to improve interdisciplinary coordination so that IVA research results will truly be interdisciplinary, rather than simply multi-disciplinary.

3.3.2 *Transdisciplinary research and assessments of adaptation*

For the purpose of realising bottom-up assessments of climate change impacts and vulnerabilities, but even more for the assessment of adaptation options and implemented adaptation measures, stakeholder engagement is essential. Many IVA studies – for example, ATEAM (Advanced Terrestrial Ecosystem Analysis and Modelling) and CLIMAS, which has also been cited as a model for interdisciplinary coordination – have already applied transdisciplinary research methods. Transdisciplinary research involves cooperation between scientists and non-scientists (e.g. stakeholders affected by climate change impacts or adaptation policies) in order to generate new knowledge. Lebel et al. (2010) identify six advantages of "social learning" for adaptation to climate change that can also be seen as the potential advantages of transdisciplinary methods in IVA research:

1   Transdisciplinary research can reduce informational uncertainty. *Informational uncertainty* refers to deficits in knowledge with regard to future developments. By involving stakeholders and local experts who have knowledge about a studied system, uncertainties regarding potential climate change impacts, vulnerabilities and feasible adaptation options can be reduced.[2]

2   Transdisciplinary research reduces normative uncertainty. *Normative uncertainty* is defined as uncertainty about values, norms and goals and also relates to perceptions of acceptable risk. For example, strong stakeholder participation in a water-sensitive region can clarify priorities (for instance, tourism) and acceptable risks (for instance, agricultural losses).

3   Transdisciplinary research facilitates consensus-building regarding the criteria used for monitoring and evaluation; these are essential elements of adaptive management and the adaptive governance schemes often used in adaptation to climate change.

4   Transdisciplinary research can empower stakeholders to influence adaptation and to take appropriate actions themselves through the sharing of knowledge and responsibility in participatory processes.

5   Transdisciplinary research can reduce conflicts and identify synergies between the adaptation activities of various stakeholders, thus improving the overall chances of success. Coordination is crucial to overcome fragmentation across sectors, regions and decision levels.

6   Transdisciplinary research can improve the probable fairness, social justice and legitimacy of adaptation decisions and actions by addressing the concerns of all relevant stakeholders. Deliberative processes bring together alternative perspectives and forms of knowledge, reducing the likelihood that adaptation responses address only the interests of influential and powerful actors.

Furthermore, by involving stakeholders and practitioners, transdisciplinary research leads to more practice-oriented research results and can actually contribute to solving problems, rather than simply analysing them. One particular methodological framework within transdisciplinary research goes beyond mere problem analysis (the focus of much previous IVA research) and seems to hold promise for the development of adaptation solutions. This framework, *Action Research*, refers to research initiated to solve a problem or a reflective process of progressive problem-solving (cf. example, Burns 2007). By focusing on developing solutions, the application of this framework could help shift the focus of IVA research from adaptation problems to adaptation solutions.

With regard to specific methods for transdisciplinary procedures in IVA research, experiences from the fields of environmental, natural hazards and developmental research can also be of great use (Allen 2003, Smit & Wandel 2006). Workshop formats and also stakeholder interview procedures are important social-science methods in this respect. These methods have been adapted and improved for the purposes of IVA research – for example, in participatory integrated assessments (e.g. Salter et al. 2010), participatory vulnerability research (e.g. Smit & Wandel 2006) and community-based adaptation (e.g. van Aalst et al. 2008).

To date, transdisciplinary IVA studies in industrialised countries have primarily included governmental and business stakeholders, often neglecting stakeholders and experts from civil society. This is due in part to the fact that many civil society organisations in industrialised countries do not perceive that they have any "stakes" in the adaptation issue (diverging from their perceptions of strong stakes with regard to climate change mitigation). However, this lack of involvement of stakeholders and experts from civil society seems also to be due to the lack of personal and financial resources for such involvement. Future IVA research might consider reimbursing these stakeholders in order to increase their involvement and thereby also the legitimacy of transdisciplinary adaptation processes.

3.4 Final remarks

Climate change impacts, vulnerabilities and adaptation cannot be analysed by the natural and engineering sciences alone; social-science expertise and methods are also necessary for an understanding of social impacts and vulnerabilities, as well as the social aspects of adaptation. Future IVA research should improve interdisciplinary coordination so that IVA research results will be not only multi-disciplinary, but truly interdisciplinary as well. Within this interdisciplinary coordination, an important aim is the further clarification and theoretical integration of concepts such as vulnerability, resilience, robustness, adaptive capacity, adaptive management, adaptation and transformation.

Future IVA studies should also improve the methods and procedures for transdisciplinary research. Transdisciplinary research designs can reduce informational and normative uncertainties, build consensus regarding the criteria used to monitor and evaluate adaptation, empower stakeholders to realise adaptation, reduce conflicts and identify synergies between adaptation activities, improve the fairness, social justice and legitimacy of adaptation decisions and lead to more practice-oriented research that will contribute to solving problems, not merely analysing them. One particularly promising methodological framework in this respect is Action Research. By focusing on developing solutions, the application of Action Research could help the focus of IVA research to shift from adaptation problems to adaptation solutions.

It would be desirable for more practical adaptation processes – such as those presented in this manual – to be included in IVA research activities, preferably in comparative case studies that apply similar conceptual and methodological frameworks. This would allow a more systematic identification of the lessons learned from such adaptation processes and facilitate the wider distribution of results via scientific publications.

Notes

1 This chapter is an updated and extended version of a previous publication (Grothmann et al. 2011) that was written with the intention to advance research on climate change impacts, vulnerabilities and adaptation, particularly in Germany.
2 Here, the knowledge of practitioners and stakeholders with regard to probable barriers to climate change adaptation that could limit adaptive capacities is also important, as these barriers can vary widely from region to region (Grothmann et al. 2009).

References

Allen, K. (2003) 'Vulnerability reduction and the community-based approach', in Pelling, M. (ed.) *Natural Disasters and Development in a Globalising World*, London: Routledge, pp. 170–184.

Brown, C., Ghile, Y., Laverty, M.A. and Li, K. (2012) 'Decision scaling: Linking bottom-up vulnerability analysis with climate projections in the water sector', *Water Resources Research* 48. DOI:10.1029/2011WR011212.

Burns, D. (2007) *Systemic Action Research: A strategy for Whole System Change*, Bristol: Policy Press.

Carter, T.R., Jones, R.N., Lu, X., Bhadwal, S., Conde, C., Mearns, L.O., O'Neill, B.C., Rounsevell, M.D.A. and Zurek, M.B. (2007) *New Assessment Methods and the Characterisation of Future Conditions. Climate Change 2007: Impacts, Adaptation and Vulnerability*. Contribution of Working Group II to the Fourth Assessment Report of the Intergovernmental Panel on Climate Change, Parry, M.L., Canziani, O.F., Palutikof, J.P., van der Linden, P.J. and Hanson, C.E. (ed.), Cambridge: Cambridge University Press, pp. 133–171.

Dessai, S. and Hulme, M. (2004) 'Does climate adaptation policy need probabilities?' *Climate Policy* 4 (2), pp. 107–128.

Dessai, S. and Hulme, M. (2007) 'Assessing the robustness of adaptation decisions to climate change uncertainties: A case study on water resources management in the East of England', *Global Environmental Change* 17 (1), pp. 59–72.

Folke, C. (2006) 'Resilience: The emergence of a perspective for social-ecological systems analyses', *Global Environmental Change* 16 (3), pp. 253–267.

Ghimire, Y.N., Shivakoti, G.P. and Perret, S. (2010) 'Household-level vulnerability to drought in hill agriculture of Nepal: Implications for adaptation planning' *International Journal of Sustainable Development and World Ecology*, 17 (3), pp. 225–230.

Grothmann, T., Nenz, D. and Pütz, M. (2009) 'Adaptation in vulnerable alpine regions – lessons learnt from regional case studies', in *Regional Climate Change and Adaptation – the Alps Facing the Challenge of Changing Water Resources*, Technical Report nr. 9/2009, Copenhagen: European Environment Agency (EEA), pp. 96–108.

Grothmann, T., Grecksch, K., Winges, M. and Siebenhüner, B. (2013) 'Assessing institutional capacities to adapt to climate change – integrating psychological dimensions in the Adaptive Capacity Wheel', *Natural Hazards Earth System Science* 1, pp. 793–828.

Grothmann, T., Daschkeit, A., Felgentreff, C., Görg, C., Horstmann, B., Scholz, I. and Tekken, V. (2011) 'Anpassung an den Klimawandel – Potenziale sozialwissenschaftlicher Forschung in Deutschland', GAiIA 20 (2), pp. 84–90. http://edoc.gfz-potsdam.de/pik/rawdisplay.epl?mode=doc&id=4869

Gupta, J., Termeer, K., Klostermann, J., Meijerink, S., van den Brink, M., Jong, P., Nooteboom, S., and Bergsmaa, E. (2010) 'The Adaptive Capacity Wheel: a method to assess the inherent characteristics of institutions to enable the adaptive capacity of society', *Environmental Science Policy* 13 (6), pp. 459–471.

Hallegatte, S., Shah, A., Lempert, R., Brown, C., and Gill, S. (2012) *Investment Decision Making Under Deep Uncertainty: Application to Climate Change*, Policy Research Working Paper 6193, Washington, DC: The World Bank.

Hinkel, J. (2011) 'Indicators of vulnerability and adaptive capacity: Towards a clarification of the science-policy interface', *Global Environmental Change* 21 (1), pp. 198–208.

Kates, R.W., Travis, W.R. and Wilbanks, T.J. (2012) 'Transformational adaptation when incremental adaptations to climate change are insufficient', Proceedings of the National Academy of Sciences of the USA 109 (19), pp. 7156–7161.

Lebel, L., Grothmann, T. and Siebenhüner, B. (2010) 'The role of social learning in adaptiveness: Insights from water management. International Environmental Agreements', *Politics, Law and Economics* 10 (4), pp. 333–353.

O'Brien, K. (2012) 'Global environmental change II: From adaptation to deliberate transformation', *Progress in Human Geography* 36 (5), pp. 667–676.

Pahl-Wostl, C. (2007) 'Transitions towards adaptive management of water facing climate and global change', *Water Resources Management* 21 (1), pp. 49–62.

Park, S.E., Marshall, N.A., Jakku, E., Dowd, A.M., Howden, S.M., Mendham, E. and Fleming, A. (2012) 'Informing adaptation responses to climate change through theories of transformation', *Global Environmental Change* 22 (1), pp. 115–126.

Parry, M.L., Canziani, O.F., Palutikof, J.P., van der Linden, P.J. and Hanson, C.E. (ed.) (2007) *Climate Change 2007: Impacts, Adaptation and Vulnerability. Contribution of Working Group II to the Fourth Assessment Report of the Intergovernmental Panel on Climate Change*, Cambridge, UK: Cambridge University Press.

PIK – Potsdam Institute for Climate Impact Research (2012) *4°C Turn Down the Heat – Why a 4°C Warmer World Must be Avoided. A Report for the World Bank by the Potsdam Institute for Climate Impact Research and Climate Analytics*, Washington, DC: The World Bank.

Salter, J., Robinson, J. and Wiek, A. (2010) 'Participatory methods of integrated assessment – a review', *Climate Change* 1 (5), pp. 697–717.

Smit, B. and Wandel, J. (2006) 'Adaptation, adaptive capacity and vulnerability', *Global Environmental Change* 16 (3), pp. 282–292.

Solomon, S., Qin, D., Manning, M., Chen, Z., Marquis, M., Averyt, K.B., Tignor, M. and Miller, H.L. (ed.) (2007) *Climate Change 2007: The Physical Science Basis. Contribution of Working Group I to the Fourth Assessment Report of the Intergovernmental Panel on Climate Change.* Cambridge, UK: Cambridge University Press.

Stern, N. (2007) *The Economics of Climate Change – The Stern review*, Cambridge, UK: Cambridge University Press.

UNFCCC – United Nations Framework Convention on Climate Change (1992). United Nations.

van Aalst, M.K., Cannon, T. and Burton, I. (2008) 'Community level adaptation to climate change: The potential role of participatory community risk assessment', *Global Environmental Change* 18 (1), pp. 165–179.

WBGU – German Advisory Council on Global Change (2011) *World in Transition. A Social Contract for Sustainability*, Berlin, GE: WBGU.

Wisner, B. (2009) 'Vulnerability', in Kitchin, R. and Thrift, N. (ed.) *International Encyclopedia of Human Geography*, Oxford, UK: Elsevier Publishing, pp. 176–182.

4

ADAPTATION POLICY INITIATIVES IN EUROPE

Sabine McCallum and Stéphane Isoard

4.1 Introduction

Adaptation can be implemented both in preparation for future risks or in response to already occurring impacts generated by a changing climate. Both dimensions of adaptation aim at building adaptive capacity and thereby increasing the ability of individuals, groups or organisations to adapt to changes. Thus adaptation includes a continuous variety of research, actions and decisions about all aspects of life that can be affected by climate change (Adger et al 2005: 78).

Adaptation can be triggered by various factors (such as extreme weather events, water scarcity, and drought) and is often embedded in a broader context of a changing environment. Thus many classifications of adaptation options exist (e.g. summarised in Smit et al. 2000) based on their purpose, mode of implementation, or on the institutional form they take. One of the most common ways of categorising types of adaptation is to look at the different decision-making processes that lead to their implementation (EEA 2013). According to this approach, there are two basic types of adaptation, planned adaptation and autonomous adaptation, explained below:

- *Planned adaptation* aims at taking measures to counteract current or expected impacts of climate change within the context of ongoing and expected societal change. It is the result of a deliberate decision, based on an awareness that conditions have changed (reactive adaptation) or are about to change (anticipatory adaptation) and that action is required to return to, maintain, or achieve a desired state.
- *Autonomous adaptation* is a spontaneous response in natural or human systems to a variety of factors, including climatic stimuli, socio-economic developments, and market forces.

These definitions are helpful as a broad outline, but in practice, the difference between planned and autonomous adaptation is not clear-cut. Many actions are interrelated and depend on each other. This is to underline that we need to take account of plenty of forms of adaptation to sketch the whole variety of initiatives and actions that can be captured by the term adaptation.

Information is still scarce on autonomous adaptation – both for the private sector and the public sector. Thus this chapter focuses mostly on planned adaptation and related policy responses, which are framed to complement individual action by proactively taking decisions in response to current or future climatic risks.

In the private sector, businesses are increasingly aware of the need to respond to climate change, both in terms of the way they run their operations today, and in terms of their plans to develop their business in the future. Climate change will have a range of impacts on businesses. Among other things, it will disrupt business operations, increase costs of maintenance and materials, and raise insurance prices. In some instances, companies are already starting to realise new commercial opportunities for products and services to help people and companies adapt to the impacts of climate change. These new products and services are being introduced to sectors as diverse as agriculture, water management, healthcare, logistics and transport, industry and manufacturing, finance, and business consulting services. Other commercial opportunities arising from climate include new shipping possibilities due to reduced sea ice, increased agricultural production in areas that were previously not economically viable (such as wine production in more northern countries), and the opening of new regions for tourism (OECD 2012). Further, we notice an increasing demand for climate adaptation consultancy services from countries with expertise in adaptation technologies such as the Netherlands.

In most countries, the private sector does not yet seem to be fully integrated into adaptation policy processes. This is because national frameworks and research activities often do not explicitly prioritise topics related to the economy and business. There is therefore limited information about adaptation measures being taken by the private sector. An exception to this is the insurance sector, where the level of awareness and action on adaptation is relatively high. Therefore there is a need to review adaptation actions planned and implemented by both individual businesses (e.g. utilities) and by broader sections of the private sector (e.g. the insurance and banking sectors). Utilities are particularly well placed to plan for adaptation. They are often large companies with substantial financial resources that can be used to effectively mainstream adaptation. Private actions in response to climate change usually aim at preventing negative impacts or using new opportunities to make profit for a small enterprise or group. Nevertheless, the way they plan and provide their services (e.g. energy, water, transport, and ICT) can substantially impact the resilience of European society and reduce the need for the public sector to invest in adaptation measures (EEA 2013, McCallum et al. 2013a).

However, the Stern Review (Stern 2007) argues that market forces alone are unlikely to lead to efficient adaptation because of the uncertainty in the climate projections and lack of financial resources. Moreover, marginal adjustments in markets and individual behaviour will mainly be driven by a sectoral/individual and short-term perspective. Public intervention will therefore be needed to provide a comprehensive approach aimed at improving the adaptive capacity of natural and human systems over a medium- and long-term perspective. Hence the major objectives of public policy in this area are:

1   to protect those least able to cope, by addressing the causes of vulnerability;
2   to provide information for planning and initiating adaptation by non-government actors;
3   to protect important public goods such as ecosystem services, coastal defence, and early warning systems for extreme events; and
4   to consider measures to take advantage of possible benefits arising from climate change.

Adaptation is a challenge that cuts across all economic sectors, and affects all geographical scales, from the local level to the regional and national level, and all the way up to the international level. In Europe, it therefore requires a response across all levels of governance from the municipal level all the way up to the European level (EEA 2013).

Over the last decade, there have been significant achievements in the way Europe has responded to climate change. These have affected both policy development and adaptation action (planned and autonomous) across various governance levels and in most affected sectors. The following section (4.2) briefly summarises the achievements so far at EU, national, transnational, regional, and local level.

4.2   EU-level efforts

In view of the specific and wide-ranging nature of climate change impacts across the EU's territory, the European Union has recognised its important role in developing an EU-wide framework for adaptation. Thus, the European Commission has adopted a Green Paper on Adaptation (EC 2007) followed by a White Paper "Adapting to climate change: Towards a European framework for action" (EC 2009). These efforts have recently been complemented by the adoption of the EU strategy on adaptation to climate change on 16 April 2013 (EC 2013a).

In the following sub-sections (4.2.1–4.2.3) this step-by-step development of adaptation strategy in Europe is briefly described to provide an overview on what has been considered and achieved so far.

4.2.1   2007 Green Paper on adapting to climate change in Europe – options for EU action

In 2007, the European Commission adopted a Green Paper on adapting to climate change in Europe (EC 2007), recognising that all parts of Europe will increasingly feel the adverse effects of climate change. The Green Paper was followed by a broad stakeholder consultation. There was general agreement among the stakeholders on the need to exploit the synergies between climate change mitigation efforts and adaptation efforts and to do more research on the vulnerabilities and risks of climate change. In particular, the following issues were highlighted in the stakeholder consultations:

- the need to exchange best practices,
- the importance of the subsidiarity principle,
- the crucial role of local and regional authorities,
- the necessity to integrate adaptation into all relevant EU policies,
- the need for greater awareness raising,
- the importance of active participation by civil society and all governance levels in tackling this issue,
- the importance of protecting ecosystems and biodiversity,
- the need to analyse current and future funding mechanisms in order to make them more suited to addressing adaptation needs,
- the need for early action and the benefits that early action can bring.

In the stakeholder consultations, there was also a common agreement that although a 'one size fits all' approach would not be an adequate response, EU action coupled with action at all lower administrative levels would yield significant added value compared to initiatives exclusively taken by the Member States.

Through consulting widely with stakeholders and the public after publication of the Green Paper, it also became clear that some essential issues were not addressed. According to the feedback received, the Green Paper gave too little attention to the following areas: communication and awareness raising; the importance of additional research and monitoring systems; an overarching strategic vision demonstrating the need for EU leadership; the importance of migration and social cohesion aspects; the need to coordinate mitigation and adaptation strategies; and the mechanisms for financing adaptation efforts. Most stakeholders therefore considered that the EU had a role to play in providing climate change information, appropriate guidance, financial support, and technical expertise. They suggested that the EU should develop a clear and strong strategic framework for adaptation across all sectors, addressing cross-border issues and coordinating action across Europe, strengthening or extending existing EU frameworks and directives to include adaptation while updating and harmonising existing structures.

4.2.2   2009 White Paper on adapting to climate change: Towards a European framework for action

Responding to the feedback gathered from the broad stakeholder involvement for the Green Paper, the EU adopted an Adaptation White Paper in 2009 (EC 2009). This White Paper set out the steps to be taken in preparing the 2013 EU strategy on adaptation to climate change.

The White Paper highlighted five main reasons for the EU to take action on climate change adaptation:

- Many climate change impacts and adaptation measures have cross-border dimensions;
- Climate change and adaptation affect EU policies;
- Solidarity mechanisms between European countries and regions might need to be strengthened because of climate change vulnerabilities and adaptation needs;
- EU programmes could complement Member State resources for adaptation;
- Economies of scale can be significant for research, information and data gathering, knowledge sharing, and capacity building (EEA 2013).

The 2009 White Paper was framed to complement and ensure synergies with actions by Member States. It focuses on four 'pillars' to reduce the EU's vulnerability and improve its resilience:

- Pillar 1: develop and improve the knowledge base at regional level on climate change impacts, vulnerabilities mapping, costs and benefits of adaptation measures to inform policies at all levels of decision-making;
- Pillar 2: integrate adaptation into EU policies (mainstreaming);
- Pillar 3: use a combination of policy instruments – market-based instruments, guidelines, and public–private partnerships – to ensure effective delivery of adaptation;
- Pillar 4: work in partnership with the Member States and strengthen international cooperation on adaptation by mainstreaming adaptation into the EU's external policies.

With regards to Pillar 1 (developing and improving the knowledge base), the White Paper noted that information (including research results) already exists, but is not shared across Europe. Furthermore, the White Paper and various assessment reports underlined that information on climate change impacts and vulnerability – and information on the costs and benefits of adaptation measures in Europe – remained both scarce and fragmented. They stressed that more spatially detailed information was needed to develop adequate adaptation policies. Substantial efforts have therefore been undertaken since then to establish a comprehensive information platform. This platform was launched on 23 March 2012 as the 'European Climate Adaptation Platform' (Climate-ADAPT[1]) (McCallum et al. 2013a). Hosted by the European Environment Agency (EEA), Climate-ADAPT contains information on adaptation policy across Europe, and also includes adaptation case studies from across Europe as well as a number of software tools to facilitate accessing this information. Climate-ADAPT is the EU entry point to information on adaptation, and it complements other knowledge generation and dissemination efforts being implemented or planned at national and sub-national levels (EEA 2013).

The task of Pillar 2, mainstreaming climate change adaptation into EU policies, has been particularly important, and much work has already been done on mainstreaming. The Climate-ADAPT platform provides an up-to-date overview of the main initiatives for integrating adaptation into EU sectoral policies.[2] The key policy initiatives subject to mainstreaming concentrate on the following nine sectors: water management, marine and fisheries, coastal areas, agriculture and forestry, biodiversity, infrastructure, finance and insurance, disaster risk reduction, and health. Mainstreaming initiatives in these nine sectors concentrate on the most vulnerable areas in Europe: the Arctic; northern, north-western and central-eastern Europe; the Mediterranean basin; cities and urban areas; mountain areas; coastal areas; areas prone to river floods; islands; and outermost regions (EEA 2013). The EU's Seventh Framework Programme for Research and Technological Development (FP7), as well as several European Commission service contracts, have played an important role in informing mainstreaming activities and potential policy intervention.

In addition to these mainstreaming initiatives in the EU's sectoral policies, work on mainstreaming has expanded to include strategic financial planning. The 2011 Commission proposal for the next Multiannual Financial Framework[3] (MFF) 2014–2020 recognises mainstreaming as the EU's favoured approach to both building the resilience of sectors and policy domains and facilitating the necessary transition to a low-carbon economy. The MFF also includes a minimum contribution of 20% for climate related expenditure and stipulates that all EU funds will need to take climate change into account in their funding allocation decisions. For example, 35% of funds allocated under the European future research policy have been earmarked for climate change research (EC 2013b).

With regard to Pillar 3, the Commission has launched several studies to both identify policy instruments suited for adaptation purposes and to develop specific guidelines. Furthermore, the Commission has consulted stakeholders in the private sector on specific issues, such as standards and insurance (McCallum et al. 2013 a, b).

Pillar 4 concerns working in partnership with the Member States and strengthening international cooperation. To develop this Pillar, the Commission created an Adaptation Steering Group (ASG) in September 2010. The ASG brought together Member States and a diverse range of stakeholders, including business organisations and NGOs, to support the Commission in implementing the White Paper's actions.

For strengthening the international cooperation on adaptation both the proposed review of the EU Environment Integration Strategy and the Mid-Term Review of European Commission cooperation strategies present a good opportunity to emphasise the need for integrating adaptation needs into international cooperation activities. In addition, the Commission's activities put emphasis on early warning systems and integrating climate change into existing tools such as conflict prevention mechanisms and security sector reform. Adaptation is also being brought into the EU's dialogue with European Neighbourhood

Policy (ENP) partner countries. The regular 'Energy, Transport, Environment' sub-committees that bring together ENP countries and the EU offer a forum for structured dialogue on adaptation. Finally, under the UNFCCC, the EU is taking an active role in the negotiations to ensure adaptation issues are adequately dealt with in a post-2012 agreement (McCallum et al. 2013a).

4.2.3 2013 EU strategy on adaptation to climate change

Building on the actions announced in the 2009 White Paper (which have been mostly implemented or were in the process of being so by the end of 2012), the Commission adopted an EU strategy on adaptation to climate change on 16 April 2013 (EC 2013a). Overall, the strategy aims at contributing to a more climate-resilient Europe by enhancing the preparedness and capacity to respond to the impacts of climate change at local, regional, national, and EU levels. Furthermore, it supports the development of a coherent approach for adaptation and improve coordination. The EU Adaptation Strategy consists of a package of various documents.[4] The main political document is a Communication 'An EU Strategy on adaptation to climate change' (EC 2013a). This communication sets out eight actions to be taken in the strategy's three priority areas: (1) Promoting action by Member States, (2) Better-informed decision-making and (3) Climate-proofing EU action: promoting adaptation in key vulnerable sectors. The Communication is complemented by accompanying documents concerning: the Impact Assessment;[5] a Green Paper on the insurance of natural and man-made disasters; and Commission staff working documents on adaptation in specific sectors and policy areas (coastal and marine issues, impacts on human, animal and plant health, infrastructure, environmental degradation and migration). Furthermore, the EU Adaptation Strategy package provides guidelines and recommendations for the integration of climate change adaptation into different EU programmes and investments, such as the Cohesion Policy and the 2014–2020 rural development programmes under the Common Agricultural Policy (CAP). Guidelines on developing adaptation strategies are also included.

The Impact Assessment (EC 2013b) that accompanies the strategy recognises that further action is needed to reach the main objectives identified in the 2009 White Paper. More specifically, the following is proposed:

- On *knowledge gaps*, additional EU-funded and national research is needed to fill gaps on methods, models, data sets, and forecasting tools. This research will improve the understanding of current and expected climate impacts, vulnerabilities, and adaptation options.
- On assessing the *cost and benefit of adaptation options*, some progress has been made at microeconomic level, but important gaps remain on the macroeconomic approach best suited for modelling adaptation and assessing its implications.
- No detailed assessment is available yet on the impacts of climate change and adaptation policies on *employment* and on the *well-being of vulnerable social groups,*

though some progress has been made in the context of the recently adopted Employment package, especially through work on green jobs. *Mainstreaming adaptation* needs to be reinforced in some of the areas already highlighted as being of key importance in the 2009 White Paper. These areas include the EU's energy policy, the climate-proofing of EU-funded infrastructure projects, and the potential for insurance and other financial products to complement adaptation measures.

Contrary to the scope of the White Paper and Pillar 4, the EU Adaptation Strategy does not consider international issues concerning climate change adaptation in the rest of the world. This is now covered under the development and cooperation policy and through the UNFCCC negotiations.

The EU Adaptation Strategy sets a timeframe for implementing its actions that lasts until 2018. It states that in 2017, the Commission will report to the European Parliament and the Council on the state of implementation of the Strategy and propose a review if needed. The Strategy also encourages all Member States to adopt a comprehensive national adaptation strategy. To monitor progress in this regard the Commission will develop an adaptation preparedness scoreboard by 2014, identifying key indicators for measuring Member States' level of readiness. In 2017, the Commission will assess whether adaptation action being taken in the Member States is sufficient, by referring to the coverage and quality of the national strategies. If it deems progress to be insufficient, the Commission will consider introducing a legally binding instrument (EC 2013a).

4.3 National efforts

As of June 2013, 16 European countries have adopted a national adaptation policy (strategy and/or plan), namely Austria, Belgium, Denmark, Finland, France, Germany, Hungary, Ireland, Lithuania, Malta, the Netherlands, Portugal, Spain, Sweden, Switzerland, and the UK. The remaining European countries are in the process of developing one. Up-to-date information on how countries in Europe prepare, develop, and implement national adaptation strategies can be found on the country pages of Climate-ADAPT.[6] The National Adaptation Strategies (NASs) mostly mark the first attempt to coordinate the issue of adaptation (Bauer et al. 2012). Each of the NASs in place has been directed by government committing themselves to take action on climate change adaptation (McCallum et al. 2013a). Nevertheless, most NASs are non-binding frameworks, and only a few countries have a binding requirement. One example is the UK's Climate Change Act from 2008. The Act requires the creation of an Adaptation Sub-Committee (ASC) within the Committee on Climate Change (a scientific advisory body), and it demands that Climate Change Risk Assessments and the National Adaptation Programme be renewed every 5 years, starting with 2012.

Most of the existing strategies include very little information on implementation (e.g. responsibilities, concrete adaptation measures, financing of adaptation action). Some countries have therefore set out concrete national action plans (NAPs). Countries with NAPs include Austria, Belgium, Bulgaria, Denmark, Finland, France, Germany, Hungary, Lithuania, the Netherlands, Poland, Spain, Sweden, and the UK (EEA 2013). The Finnish NAS as the first adaptation strategy in Europe assigns timeframes and owners to the adaptation actions, which are categorised by sector. The Portuguese, Irish, and Lithuanian NASs do not include action plans but provide a framework for developing such a plan.

All NASs appear to be intended as evolving documents which will be reviewed and updated to take account of advancing climate change science, research, and technology (McCallum et al. 2013a).

Also, in all cases, European countries have based their adaptation policies (whether NASs and/or NAPs) on numerous assessments and studies. These studies were concerned with (expected) climate change impacts; climate change vulnerabilities; and adaptation needs and options. Most countries have established nationwide research programmes that aim to inform climate change policies. For example, the Spanish Meteorological Service provides regional climate change scenarios based on downscaled global climate model projections. These regional climate change scenarios form the basis for all adaptation policies and actions in Spain (Bauer et al. 2012).

National web portals related to adaptation are also becoming more common across Europe. These portals reveal large differences in the availability of information between countries. Some portals provide broad sectoral information as well as guidance and links to various sources, organisations, and local examples of adaptation. Other web portals are less broad, focusing more on climate change mitigation than adaptation, or consisting mainly of one organisation's website to which some rather general information on adaptation has been added with little coverage of individual sectors. Of the countries that stated they had an online portal, Austria, Denmark, Finland, Germany, Norway, Sweden, Switzerland, and the UK have portals with a broad coverage of adaptation, including sectoral information and examples of adaptation action at various levels of governance (EEA 2013).

In addition to the studies and web portals mentioned above, scientific advisory bodies and (climate) services have also been established in a number of countries to provide specialised adaptation-related advice. The bodies provide information, guidance, tools, and individual consulting services to policy-makers (and frequently also to non-state actors). Examples of these bodies include UKCIP in the UK, KomPass in Germany, and Klimatilpasning in Denmark.

Some countries have started to work on monitoring and evaluating their adaptation actions. Some of this work is focused on the development of progress indicators that could express in relatively simple terms some of the complexity of implementing adaptation. However, this process is only in its infancy, and official indicators of adaptation have not yet been established. Germany, France and the UK are the countries that are most advanced in their work on developing formal indicators to monitor progress in adaptation. Finland has chosen a slightly different route. It has reviewed its adaptation strategy, and requires government departments and agencies to monitor progress in adaptation without using formally adopted indicators. Other approaches for monitoring and evaluation could also be valuable as alternatives to setting up an indicator system, for example using existing monitoring systems in various sectors that already have a methodology in place.

4.4   Transnational, regional, and local efforts

Many transnational cooperation projects on adaptation have been initiated over the last years. For example, in the Danube region a macro-regional adaptation strategy for the water sector was adopted in December 2012 by the member countries of the International Commission for the Protection of the Danube River (ICPDR).

Many other initiatives are partially financed by EU funds such as the LIFE programme[7] and the Cohesion funds to support European Territorial Cooperation.[8] The European web platform Climate-ADAPT features a dedicated section on strategies and actions that have been developed or are currently under development for the EU's transnational regions and for other regions and countries.[9]

Transnational initiatives differ in scope and according to the specific adaptation challenge at hand, but they all share a pan-regional perspective, allowing them to develop adaptation responses that do not stop at national borders. They all focus on involving stakeholders at regional and local level in order to learn about the specific needs of these communities and jointly develop feasible adaptation responses. Many of these transnational projects create case study regions within the greater transnational cooperation area, where project results can be tested and discussed with regional and local stakeholders towards their practical applicability (McCallum et al. 2013a).

Results of EU-funded projects often also inform other strategic initiatives and programmes. UN Conventions such as the Alpine and Carpathian Conventions are highly engaged with INTERREG projects and make use of their outcomes in a political context (e.g. fostering the implementation of the Climate Action Plan under the Alpine Convention) (McCallum et al. 2013a). Also, in the Baltic region, several applied transnational adaptation projects, including prominently the INTERREG project BaltAdapt, led to the development of a proposal for a 'Baltic Sea Region Adaptation Strategy', which is now subject to political endorsement.

In the Mediterranean, CYPADAPT[10] is an EU LIFE-funded project, in which partner organisations from Greece and Cyprus aim to produce a national adaptation strategy for Cyprus. The Italian-Greek-Spanish ACT project (funded by the EU's LIFE programme)[11] is a joint project between local authorities in these three countries, which aims at preparing local adaptation strategies in three cities each in Italy, Greece, and Spain. Another EU-funded project is 'Green and Blue Space Adaptation for Urban Areas and Eco Towns' (GRaBS), which involves 14 partner organisations from eight European countries.[12] The aim of the project is to facilitate the exchange of knowledge and good practice on climate change adaptation strategies between local and regional authorities. In addition to these individual transnational joint projects, there are also important regional groupings of transnational projects such as SIC-adapt![13] and C3-Alps.[14] SIC-adapt! is a 'strategic initiative cluster' comprising eight transnational projects in northwest Europe, while C3-Alps is a cluster of Alps-based adaptation projects that promote knowledge sharing in the Alps region (EEA 2013).

Several regional adaptation initiatives have also been set up, steering research activities and developing sectoral actions. For example, the Brittany region in France carried out a study[15] about the different climate change impacts that could develop between now and 2030 in order to analyse different adaptation possibilities. The French region of Bourgogne has a regional adaptation project that is studying possible climate change impacts as far ahead as 2040, and has developed an adaptation strategy for each sector.[16] Similar work is being carried out in five other regions in France. Regional-level adaptation and mitigation strategies exist in 10 regions in Spain, and each of the federal states in Germany have regional-level adaptation strategies that are already in place or are in the process of development. The Pyrenees region also launched a process to develop a regional adaptation strategy, involving both sides of the Pyrenees (France and Spain) (McCallum et al. 2013a).

With regard to local efforts, some information has been provided by EEA member countries to the country webpages on Climate-ADAPT upon request from the European Commission at the end of 2011. Those pages are regularly updated in close interaction with country representatives. However, information on adaptation action at the local level is still scarce. Moreover, the available information often does not specify if these are spontaneous 'bottom-up' actions, or if they are initiated/encouraged by 'top-down' long-term planning. More detailed information on the implementation of policy measures is needed before we can ascertain the precise relationship between local/regional adaptation actions and national-level planning (EEA 2013). Many European countries have established databases on their national climate change websites that collect and present examples of good practice from regional and local adaptation. Denmark is especially focused on local-level adaptation; all municipalities in the country are obliged to prepare climate change adaptation plans within the next two years (McCallum et al. 2013a).

A significant share of local adaptation activities takes place at city level. City-level adaptation has been addressed in detail in the EEA report *Urban adaptation to climate change in Europe* (EEA 2012), which provides a wide range of examples of local adaptation action in various European countries. An overview of country initiatives is also available on Climate-ADAPT.[17]

Many cities in Europe have adopted adaptation strategies or action plans, or are in the process of developing them. A significant number of cities have developed strategies that cover both mitigation and adaptation (EEA 2013). A number of cities or city regions have also initiated specific adaptation measures, and a significant number of these specific measures are part of existing climate strategies. Some of these cities have been motivated to work on climate adaptation because they were encouraged to do so by the national government or because they participated in EU-funded projects. Other cities have started working on adaptation to address local risks independent of such higher-level incentives. Often cities develop adaptation strategies by adding them to already established climate change mitigation strategies.

In several countries (for example, France, Germany, Hungary, Norway, Romania, Spain, and Switzerland), cities form collaborative networks for climate change mitigation and adaptation activities with other cities in the same country (e.g. Norway's 'The Cities of the Future'[18] and Spain's RECC[19]). The purpose of these networks is to share experiences between cities and provide them with practical guidance on how to reduce their greenhouse gas emissions, and how to better adapt to the impacts of climate change. Some of these networks have been created as a result of international projects, while some were initiated by national government bodies or grow from existing networks. These networks are not solely composed of local governments; they can also involve researchers or NGOs as partners (EEA 2013). There are also international networks of cities that address climate change adaptation and mitigation issues. These networks include ICLEI, the Covenant of Mayors, and C40.[20]

4.5 Conclusion

Planned adaptation – as a response to current and future climate change risks – is on the rise in Europe. This is happening at a number of governance levels: local, regional, national, EU, and transnational. Equally, there has been a growing recognition among stakeholders and policy-makers that adaptation needs to be addressed and integrated at various levels, but must differ in scope and detail. In the coming years, it will be important to ensure coherence in adaptation policy responses in a multilevel policy environment, make best use of synergies, and learn from practical experiences. The EU Adaptation Strategy will provide an important framework for fostering adaptation action, collecting and disseminating relevant information, and engaging with European countries to facilitate the exchange of knowledge and experiences.

Public policy intervention must be based on sound scientific knowledge and targeted at identified risks at each specific governance level. It is critically important that we recognise the inter-connections between science, policy development, and implementation at all levels. Emerging policy fields such as climate change adaptation are particularly dependent on research results as the knowledge base for better-informed decisions. Scientific knowledge on adaptation-relevant issues is growing constantly and rapidly, but it is often being developed in parallel to the ongoing policy process without directly informing it. Linking this knowledge with public policy formulation is in many cases still a challenge.

Although climate change impacts mostly unfold on the local level, there are still very few examples of implemented adaptation actions (with the exception of cities). From the information available, it appears that the regional and local levels are especially limited in their implementation of on-the-ground adaptation measures. At these levels, adaptation activities are mostly limited to research. With a growing number of European countries establishing a national framework for adaptation, we expect an increased implementation of concrete adaptation responses at regional and local levels in the coming years.

However, adaptation will not only occur in a hierarchical and top-down manner by the implementation of policies from the European level down to the regional and local levels. Several other factors (such as the occurrence of extreme weather events, socio-economic circumstances, and demographic development) will continue to trigger adaptation actions. Efficient adaptation therefore requires improved coordination and information sharing about ongoing and planned activities.

Notes

1 http://climate-adapt.eea.europa.eu
2 http://climate-adapt.eea.europa.eu/web/guest/eu-sector-policy/general
3 The Multiannual Financial Framework (MFF) shall ensure that European Union expenditure develops in an orderly manner and within the limits of its own resources. It shall be established for a period of at least five years. The annual budget of the Union shall comply with the multiannual financial framework (EC 2008). The MFF de facto sets political priorities for future years and constitutes therefore a political as well as budgetary framework ('in which areas should the EU invest more or less in the future?') (EC 2008).
4 http://ec.europa.eu/clima/policies/adaptation/what/documentation_en.htm
5 Before the European Commission proposes new initiatives, it assesses the potential economic, social, and environmental consequences that they may have. Impact assessment is a set of logical steps which helps the Commission to do this. It is a process that prepares evidence for political decision-makers on the advantages and disadvantages of possible policy options by assessing their potential impact.
6 http://climate-adapt.eea.europa.eu/web/guest/countries
7 The LIFE programme is the EU's funding instrument for the environment. The general objective of LIFE is to contribute to the implementation, updating, and development of EU environmental policy and legislation by co-financing pilot or demonstration projects with European added value. LIFE began in 1992, and to date there have been three complete phases of the programme (LIFE I: 1992–1995, LIFE II: 1996–1999, and LIFE III: 2000–2006). During this period, LIFE has co-financed some 3,104 projects across the EU, contributing approximately €2.2 billion to the protection of the environment.
8 European Territorial Cooperation is central to the construction of a common European space, and a cornerstone of European integration. It has clear European added value: helping to ensure that borders are not barriers, bringing Europeans closer together, helping to solve common problems, facilitating the sharing of ideas and assets, and encouraging strategic work towards common goals.
   In the period 2007–13, the European Territorial Co-operation objective (formerly the INTERREG Community Initiative) covers three types of programmes:
   • 53 cross-border co-operation programmes along internal EU borders. ERDF contribution: €5.6 billion.

- 13 transnational co-operation programmes cover larger areas of co-operation such as the Baltic Sea, Alpine, and Mediterranean regions. ERDF contribution: €1.8 billion.
- The interregional co-operation programme (INTERREG IVC) and three networking programmes (Urbact II, Interact II, and ESPON) cover all 27 Member States of the EU. They provide a framework for exchanging experience between regional and local bodies in different countries. ERDF contribution: €445 million.

9 http://climate-adapt.eea.europa.eu/web/guest/transnational-regions
10 http://uest.ntua.gr/cypadapt
11 http://www.actlife.eu/EN/index.xhtml
12 http://www.grabs-eu.org
13 http://www.sic-adapt.eu
14 http://www.c3alps.eu
15 http://www.cseb-bretagne.fr/index.php?option=com_remository&Itemid=92&func=fileinfo&id=117
16 http://www.bourgogne.ademe.fr/adaptation-au-changement-climatique-en-bourgogne-boite-outils
17 The national profiles in Climate-ADAPT include many examples of local action: http://climate-adapt.eea.europa.eu/web/guest/countries
18 http://www.regjeringen.no/en/sub/framtidensbyer/cities-of-the-future-2.html?id=551422
19 http://www.redciudadesclima.es/index.php/
20 http://www.iclei.org/; http://www.covenantofmayors.eu/index_en.html; http://www.c40cities.org/

References

Adger, W.N., Arnell, N.W. and Tompkins, E.L. (2005) 'Successful adaptation to climate change across scales', *Global Environmental Change* 15 (2), pp. 77–86.

Bauer, A., Feichtinger, J. and Steurer, R. (2012) The governance of climate change adaptation in 10 OECD countries: Challenges and approaches, *Journal of Environmental Policy & Planning* 14 (3), pp. 279–304.

EC – European Commission (2007) *Green Paper from the Commission to the Council, the European Parliament, the European Economic and Social Committee and the Committee of the Regions: Adapting to Climate Change in Europe – Options for EU Action*, 354 final, Commission of European Community, Brussels. http://eur-lex.europa.eu/LexUriServ/site/en/com/2007/com2007_0354en01.pdf.

EC – European Commission (2008) *Consolidated Version of the Treaty on the Functioning of the European Union*, Chapter 2, Article 312, Brussels. http://eur-lex.europa.eu/LexUriServ/LexUriServ.do?uri=OJ:C:2008:115:0047:0199:en:PDF.

EC – European Commission (2009) *Adapting to Climate Change: Towards a European Framework for Action*, White Paper, 147, Brussels. http://eur-lex.europa.eu/LexUriServ/LexUriServ.do?uri=COM:2009:0147:FIN:EN:PDF.

EC – European Commission (2013a) *An EU Strategy on Adaptation to Climate Change*, COM (2013) 216 final, *Communication from the Commission*, Brussels. http://ec.europa.eu/clima/policies/adaptation/what/docs/com_2013_216_en.pdf.

EC – European Commission (2013b) *Commission Staff Working Document – Accompanying Document to the EU Strategy on Adaptation to Climate Change. Impact Assessment – Part 1.* COM (2013) 132 final http://ec.europa.eu/clima/policies/adaptation/what/docs/swd_2013_132_en.pdf.

EEA – European Environment Agency (2012) *Urban Adaptation to Climate Change in Europe: Challenges and Opportunities for Cities Together with Supportive National and European policies,*

EEA Report 2/2012, Copenhagen. http://www.eea.europa.eu/publications/urban-adaptation-to-climate-change.

EEA – European Environment Agency (2013) *Adaptation in Europe. Addressing Risks and Opportunities from Climate Change in the Context of Socio-economic Developments*, EEA Report 3/2013, Copenhagen. http://www.eea.europa.eu/publications/adaptation-in-europe

McCallum, S., Dworak, T., Prutsch, A., Kent, N., Mysiak, J., Bosello, F., Klostermann, J., Dlugolecki, A., Williams, E., König, M., Leitner, M., Miller, K., Harley, M., Smithers, R., Berglund, M., Glas, N., Romanovska, L., van de Sandt, K., Bachschmidt, R., Völler, S. and Horrocks, L. (2013a) *Support to the Development of the EU Strategy for Adaptation to Climate Change: Background Report to the Impact Assessment, Part I – Problem Definition, Policy Context and Assessment of Policy Options*, Environment Agency Austria, Vienna, Austria. http://ec.europa.eu/clima/policies/adaptation/what/docs/background_report_part1_en.pdf.

McCallum, S., Prutsch, A., Berglund, M., Dworak, T., Kent, N., Leitner, M., Miller, K. and Matauschek, M. (2013b) *Support to the development of the EU Strategy for Adaptation to Climate Change: Background Report to the Impact Assessment, Part II – Stakeholder Involvement*, Environment Agency Austria, Vienna, Austria. http://ec.europa.eu/clima/policies/adaptation/what/docs/background_report_part2_en.pdf.

OECD (2012) *Policy Forum on Adaptation to Climate Change in OECD Countries*, Summary Note, Paris, France. http://www.oecd.org/env/cc/OECD%20Adaptation%20Policy%20Forum%2010-11%20May%202012%20-%20Summary%20Note.pdf.

Smit, B., Burton, I., Klein, R.J.T. and Wandel, J. (2000) 'An anatomy of adaptation to climate change and variability', *Climatic Change* 45 (1), pp. 223–251.

Stern, N. (2007) *The Economics of Climate Change – the Stern Review*, Cambridge, UK: Cambridge University Press.

PART II

Phases of adaptation processes and guiding principles for good adaptation

5

EXPLORE POTENTIAL CLIMATE CHANGE IMPACTS AND VULNERABILITIES AND IDENTIFY PRIORITY CONCERNS

Inke Schauser, Andrea Prutsch, Torsten Grothmann, Sabine McCallum and Rob Swart

Explanation of the guiding principle

Due to the diversity of biophysical and socio-economic situations, the impacts of climate change will vary from region to region, thus affecting sectors, actors and decision-making levels differently (Scheraga & Grambsch 1998). Furthermore, impacts in one region or sector may have consequences for other regions or sectors, and impacts on one actor or decision-making level may have consequences for other actors or levels. Identifying and ranking the potential direct and indirect impacts of climate change helps to structure the process of developing specific adaptation actions. In the 1990s, to support early climate policy discussions, the Intergovernmental Panel on Climate Change (IPCC) produced technical guidelines for scientists (Carter et al. 1994, Parry et al. 1998); the United Nations Environment Programme (UNEP) similarly issued its *Handbook on Methods for Climate Change Impact Assessment and Adaptation Strategies* to support national studies (Feenstra et al. 1998). The main focus of this material was on increasing the scientific understanding of potential climate change impacts and vulnerability and on raising awareness. Due to recent advancements UNEP has taken the initiative in the context of PROVIA (Programme of Research on Climate Change Vulnerability, Impacts and Adaptation) to develop new comprehensive guidance released in 2013 and expected to be further developed over the next few years. This guidance will also provide detailed and comprehensive information on appropriate methods and tools for the assessment of impacts, vulnerability and adaptation efforts.

A concise and general approach (modified from Ribeiro et al. 2009, HM Government 2010, Willows & Connell 2003, Wilby & Dessai 2010, Dessai et al. 2009, AEA et al. 2005) involves the following steps:

1   Identify key systems

In this context, key systems may include spatial areas within a certain region (e.g. inner cities), ecosystems (e.g. mountain ecosystems), socio-economic groups (e.g. the poor, ill or elderly), economic sectors (e.g. tourism) or critical assets and infrastructure (e.g. roads). The identification of the key systems affected by climate change can be supported by asking questions (Ribeiro et al. 2009) such as:

What systems and sectors
- are highly influenced by weather and the climate?
- are of ecological, social and/or economic importance for the planning area?
- involve long-term planning or infrastructure with long lifespans?
- involve critical infrastructure?
- have little experience in managing climate impacts?

2   Learn from the impacts of past weather events and recent climate trends

The analysis of how a key system is currently affected by weather events and climatic conditions and how it is responding can facilitate a better understanding of the potential impacts of climate change. A simple matrix can be developed to indicate the magnitude of impacts, their distribution and variation, their likelihood and persistence, etc. In addition, cooperation with stakeholders (including practitioners) may provide relevant insights into the current vulnerability of key systems (cf. Chapter 8).

3   Identify sets of possible climatic and socio-economic conditions

Based on past experiences and various projections of future climate and socio-economic conditions, a range of plausible representations of future climate change impacts can be developed (Schröter et al. 2005). At this point, the adaptive capacity of key systems should also come into play, as this provides a picture of the space within which adaptation actions are feasible (Adger & Vincent 2005).

4   Rank potential impacts by the order in which they should be addressed

Based on the outcome of the above assessment, the impacts should be ranked to indicate which problems need to be addressed first by adaptation. Agreement among stakeholders (cf. Chapter 8) regarding the criteria used to rank the importance of various impacts can facilitate the identification of priority adaptation needs.

The exploration of potential impacts should also consider non-climatic factors and developments (e.g. demographic and socio-economic changes, habitat destruction by land use), as these will also influence climate change impacts (e.g. heat-related fatalities, biodiversity loss). Therefore, an inter- and transdisciplinary approach should be favoured (cf. Chapter 3).

Chapter overview

The three cases in this chapter show different approaches to the assessment of potential impacts and the prioritisation of adaptation needs. The regional and local vulnerability assessment approaches presented in the first two cases (Sections 5.1 and 5.2) focus on potential climatic impacts and also consider adaptive capacities. *Adaptive capacity* is understood as the ability of a system to adjust to climate change (Parry et al. 2007). The third case study (5.3) focuses on the identification of key systems, introducing an approach used to rank potential climate change impacts at the national level in Switzerland.

5.1 Regional vulnerability assessment in the Alps

Marc Zebisch, Stefan Schneiderbauer and Lydia Pedoth

5.1.1 Climate change in the European Alps

The Alps are one of the most vulnerable regions to climate change in Europe (Parry et al. 2007), with recent climate change trends above the European average (Auer et al. 2007), highly climate-sensitive natural systems (e.g. glaciers, Alpine ecosystems) and equally sensitive human systems (e.g. agriculture, winter tourism, energy production by hydropower). In preparing for the adverse impacts of climate change on such systems, the timely planning of appropriate adaptation measures is crucial. As one of the first steps of this planning process, the potential impacts of climate change and possible adaptive reactions must be investigated by means of a vulnerability assessment. In this section, we present an interdisciplinary regional vulnerability assessment framework that helps actors to identify regional hotspots of vulnerability and provides a decision basis and support for the planning of adaptation measures.

The results presented here were elaborated within the framework of the CLISP project – Climate Change Adaptation by Spatial Planning in the Alpine Space (EC Alpine Space Programme, Territorial Cooperation IV). The overall objective of CLISP was to develop climate change adaptation strategies in collaboration with regional authorities in the field of spatial planning.

5.1.2 General methodology of a regional vulnerability assessment

According to the IPCC (Parry et al. 2007), *vulnerability* is the degree to which a system is susceptible to or unable to cope with the adverse effects of climate change. In this definition, vulnerability is a function of the exposure of a system and is dependent on the character, magnitude and rate of climate variation, the system's sensitivity, the potential impact on the system (due to exposure and sensitivity) and its adaptive capacity, which includes potential additional adaptation activities (cf. Figure 5.1.1).

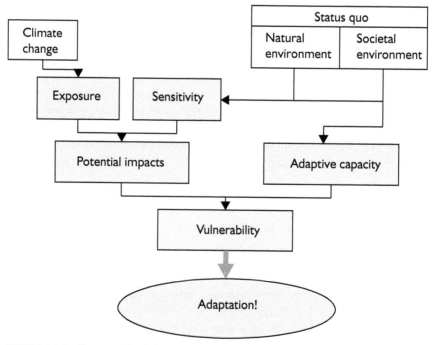

FIGURE 5.1.1 Framework of the vulnerability assessment based on Parry et al. (2007)

Within the CLISP project, we developed a consistent framework for a vulnerability assessment in five steps that integrates both quantitative and qualitative information (cf. Table 5.1.1).

In the implementation of this regional vulnerability assessment in test regions, quantitative information was limited to the analysis of climate scenarios and the output of selected impact models. Qualitative information was acquired from stakeholders by means of questionnaires, structured interviews and workshops. Typical stakeholders were representatives from regional authorities and partners of the CLISP project. Qualitative and quantitative information was integrated into a semi-quantitative scheme of five classes. Individual aspects of vulnerability (sensitivity, impact and adaptive capacity) were aggregated following a rule-based assessment.

The final results were communicated in a report supplemented by maps of climate scenarios and selected impacts. To illustrate the assessment framework, we present here the analysis for one of the six sectors (agriculture) in one of the nine case-study regions (the Autonomous Province of Bolzano).

Step 1: Definition of the system: Sectors of concern and impact chains

The first task of any vulnerability assessment is the definition of the system of concern. The guiding questions in this respect are: What or who might be vulnerable? What is the spatial dimension? Which impacts are relevant?

TABLE 5.1.1 Five-step regional vulnerability assessment

Steps	Description	Quantitative	Semi-quantitative	Qualitative / Narrative
Step 1: Definition of the system	Definition of spatial reference units Selection of systems and sectors of concern Definition of impact chains for relevant impacts			×
Step 2: Exposure	Analysis of climate change scenarios for the spatial reference units and calculation of exposure indicators	×		
Step 3: Sensitivity and potential impacts	Stakeholder-based sensitivity assessment (questionnaire, reviewed by experts)			×
	Stakeholder-based assessment of potential impacts based on information on exposure and their estimates of sensitivity (questionnaire, reviewed by experts)			×
	Model- and indicator-based analysis of potential impacts based on exposure and sensitivity for selected chains	×		
	Integrated expert- and rule-based assessment of impacts (semi-quantitative, reviewed by stakeholders) Narrative description of potential impacts		×	×
Step 4: Adaptive capacity	Stakeholder- and rule-based assessment of adaptive capacity (questionnaire, semi-quantitative and narrative elements, reviewed by experts)		×	×
Step 5: Vulnerability	Rule-based aggregation of impacts and adaptive capacity in relation to vulnerability (semi-quantitative) Narrative assessment of vulnerability		×	×

Source: Authors' research

For the case study of the Autonomous Province of Bolzano, we chose the NUTS3 level as the spatial reference unit and six sectors of concern as subsystems for the assessment (water management, agriculture, forestry, tourism, health, energy). For each sector, we generated impact chains based on a literature review and exchanges with stakeholders; these visualise the relationship between climate signals, direct potential impacts on biophysical systems and impacts on socio-ecologic systems. All further assessment steps relate back to these impact chains.

Step 2: Exposure: Analysis of climate change scenarios

According to the IPCC's definition, *exposure* is "the nature and degree to which a system is exposed to significant climate variations" (IPCC 2001: 987). To investigate exposure in our case study, climate signals from eight projections based on four different regional climate models from the FP6 ENSEMBLES project (van der Linden & Mitchell 2009) and other sources (Lautenschlager et al. 2009) were analysed for the time periods 2011–2030 and 2031–2050, including a pessimistic case (A1B) and an optimistic case (B1) scenario. All parameters were calculated as absolute change from the reference period (1961–1990).

Mean temperatures in all projections show clear warming trends in all seasons, with a more pronounced warming after 2030. Mean annual temperatures increase by between +1°C and +2°C (up to +2.9°C in summer) by the year 2050 for the Bolzano region.

Monthly and seasonal precipitation forecasts are fairly heterogeneous, with no clear trend emerging. However, all scenarios indicate a shift from snow to rain in winter and early spring. These results reveal the wide ranges and deep uncertainties in climate projections; particularly for the Alps and Central Europe, there is no reliable trend for precipitation that can definitively predict whether the future climate will be drier or wetter.

Step 3: Analysis of sensitivity and potential impacts

Sensitivity, by definition, determines how a system reacts to a given exposure; the *potential impact* is the effect resulting from exposure at a given sensitivity. The sensitivity of a system depends on its physical and socio-economic properties, including its entire history of adaptation. For example, the sensitivity of agriculture to droughts with respect to crop yield was operationalised as a function of soil properties but also in terms of the presence or absence of irrigation infrastructure.

In this study, we used both quantitative and qualitative approaches to assess sensitivity and potential impacts for each of the potential impacts identified in the impact chains in Step 1:

1   Impact models were applied where the relationship between weather/climate and impact is relatively well understood and can be represented by models – for example, snow cover and changes in the forest line.
2   Quantitative indicators were used where the processes are too complex to be modelled but simple indicators can act as proxies for potential impacts. Examples include the climatic water balance as a proxy for the impact 'water stress on agriculture'.
3   Qualitative information on sensitivity and potential impact was collected in a series of expert interviews and a stakeholder workshop, a process that can uncover potential impacts that cannot be captured with models or indicators. Often this knowledge is derived from past experiences (e.g. impacts of the summer heat wave/drought in 2003).

Together with the stakeholders, the results of the qualitative and quantitative assessments for each potential impact were clustered into five impact classes (very high positive impact, high positive impact, no likely significant effect, high negative impact, very high negative impact). The results of the quantitative and qualitative investigations were summarised, creating one assessment per impact by averaging the quantitative and qualitative results; experts were then given the chance to modify the assessment by one class up or down. We initially intended to use fixed thresholds based on normalised values for the assessment of the quantitative indicators; however, we realised that for in-depth assessments at the regional level, the experts' evaluations often overrode the more formal, indicator-based assessment. For instance, for evaluating the potential impact of 'drought damage in agriculture', changes in the climatic water balance are a good indicator. However, according to the regional experts, dry regions in the case study can expect less damage from droughts, since irrigation systems have been installed there for hundreds of years; in contrast, in relatively wet regions without irrigation systems, the damages could be severe.

In the Autonomous Province of Bolzano, potential negative impacts for agriculture can be expected in terms of water availability for certain regions, the effects of extreme events and increasing problems with pests and diseases. Possible positive impacts include a longer growing season, the chance for new crop varieties and the expansion of certain agricultural crops (apples, wine grapes) into higher altitudes.

Step 4: Assessment of adaptive capacity

In the context of a vulnerability assessment, adaptive capacity is the element that is most difficult to evaluate operationally. In our conceptualisation, *adaptive capacity* describes the potential space for adaptation activities and measures, as well as constraints and framing conditions. We differentiate between three levels of adaptive capacity:

1   impact-specific adaptive capacity (inside a sector),
2   sector-specific adaptive capacity, and
3   the generic adaptive capacity of a region or test case.

For each level of adaptive capacity, a number of indicators were defined. For the assessment of the test region, data were collected from regional experts using structured questionnaires. The assessment framework utilises a scheme of five classes (very high capacity, high capacity, moderate capacity, low capacity, very low capacity). Indicators were aggregated by a simple averaging procedure.

At the level of impact-specific adaptive capacity, the evaluation is based on the assessment of the following indicators: availability of specific adaptation options for each potential impact, the efficiency of the adaptation options and the degree of their implementation within the region.

At the sector-specific level, adaptive capacity is based on an evaluation of more general adaptation options for the entire sector, in addition to an assessment of the sector's general adaptive capacity that takes into account factors such as its institutional, legislative and economic settings.

The assessment of the generic adaptive capacity is based on factors that support adaptation options across sectors (such as institutional, legislative and economic settings), applying a cross-sectoral perspective. For a more detailed discussion, see Schneiderbauer et al. (2011).

For this case study, impact-specific adaptive capacity in the field of agriculture was rated as moderate to high due to the high flexibility of the agricultural sector with respect to adapting crops and management strategies in the short term. The sector's adaptive capacity was ranked as moderate at the sector-specific and generic levels. The more detailed findings for the agricultural sector included:

- Positive: good implementation of adaptation measures, even though the measures had not been undertaken in order to adapt to climate change
- Negative: gap in adaptation to weather extremes
- Potential for improvement in awareness-raising and legislative and institutional measures
- Relatively good socio-economic situation due to financial subsidies and the cultural and demographic situation, but high dependency on subsidies.

Step 5: Integration of vulnerability

In this final step, the results of the various assessments are aggregated into a vulnerability description for each impact following a decision matrix. This matrix describes vulnerability as a function of the strength of the potential impact and the adaptive capacity to respond to this impact; for example, a very high potential negative impact (– –) cannot be compensated by a high adaptive capacity (++), and thus vulnerability would be assessed as very high (– –). The results are displayed in a final vulnerability table (cf. Table 5.1.1).

In the case study, vulnerability in the sector agriculture was moderate (o) or low (+) for most impacts, with the exception of vulnerability with regard to weather extremes and pests and diseases; these were rated as high (–). The table is complemented by vulnerability maps highlighting local disparities within the overall assessment. Regional disparities primarily indicate a high vulnerability for intensive agriculture and grasslands in dry and hot valleys, and a low vulnerability (including potential positive effects) for grassland agriculture in higher regions, where the water supply is sufficient.

5.1.3   Concluding thoughts

Based on the results of our study, but more significantly on the process of assessing regional vulnerabilities in a continuous dialogue with stakeholders, we conclude:

TABLE 5.1.2 Results of the vulnerability assessment for the agricultural sector per impact

Type of potential impact (PI)	Growing season and phenology	Water demand	New crop varieties and location of production	Crop yields	Meteorological extremes/ natural hazards	Pests and diseases
Potential impact	+	−	+	0	−	−
Adaptive capacity	0	+	+	0	0	0
Vulnerability per PI	+	0	+	0	−	−

Source: Authors' research

- *When assessing vulnerability, use climate change scenarios with care, and don't rely too much on quantitative approaches:* Climate change scenarios are a useful and necessary input for vulnerability assessments, but regional climate scenarios generally show a wide range of results and often have high uncertainties (particularly for precipitation). Information on potential changes in extreme events (other than temperature extremes) is not reliable. The quantitative assessment of impacts (i.e. by means of models or indicators) is only possible for a small number of potential impacts (e.g. hydrology, ecology, agricultural impacts) and is also subject to uncertainty.
- *Closely examine the status of the system at risk and learn from history:* To a large extent, a system's vulnerability to climate change can be understood by analysing its sensitivity to climate and weather extremes in the past (such as the heat wave of 2003). Regions that are already sensitive to the current extreme climate conditions will be the most vulnerable. Most stakeholders are primarily interested in current conditions and future conditions over the next 20 years, but climate change signals are expected to be weak within this time frame.
- *Don't be afraid of qualitative and narrative data, and respect stakeholders as experts:* Involve stakeholders at all levels (farmers, water managers, decision-makers, etc.) in the vulnerability assessment, and respect them as experts. They know their system best and are the people who will ultimately have to adapt to climate change. Qualitative and narrative information can often reveal more details relevant to a vulnerability assessment than a purely quantitative approach.
- *In assessing future vulnerability, consider the development of non-climate aspects of the system at risk and other pressures as well:* In most cases, climate change is just an additional pressure. Other pressures – such as changes in land use, demographic changes, urbanisation, globalisation, an increase in traffic and a decrease in biodiversity – should also be considered. Vulnerability often arises from a combination of the expected socio-economic changes within the system and climate change. For example, the risk of damage to buildings or infrastructure via natural hazards might increase in the future due to the expansion of settlements into hazard zones, combined with an increase in the frequency and intensity of natural hazards.

- *Plan and implement appropriate adaptation measures in a timely fashion, and be ready to act despite uncertainty:* Often, technical measures (torrent protection, irrigation, etc.) have already been well implemented. What is frequently missing are 'softer' adaptation strategies, such as measures intended to adequately 'climate-proof' spatial planning processes, improve the coordination of actions within institutions or improve the communication of risks.

5.2 Practical guidance for vulnerability assessments at the regional and local scale (BalticClimate)

Mattias Hjerpe and Julie Wilk

5.2.1 The BalticClimate stepwise vulnerability assessment approach

The main objective of the BalticClimate project was to enhance capacity to manage vulnerability to climate change at the local level among government officials and in small and medium-sized enterprises. A stepwise, participatory vulnerability assessment approach (henceforth 'vulnerability approach') was tested and improved in seven localities in countries around the Baltic Sea in order to ensure that the approach would be useful for practical adaptation activities and adaptable to different local challenges. The vulnerability approach and each individual exercise are thoroughly described in Wilk et al. (2013). To test the effectiveness of the self-assessment aspect of the vulnerability approach, local groups were instructed to carry out the assessment by themselves, with support from the project's scientific team available on request. Within the project, both the positive and negative consequences of a changing climate were targeted under the term *vulnerability*.

The vulnerability approach is organised into three blocks: 1) getting started, 2) assessment of exposure, sensitivity and adaptive capacity and 3) managing vulnerability (cf. Figure 5.2.1). Each block consists of a number of exercises that reflect the key elements of integrated vulnerability assessments. Exercise instructions include descriptions of the exercise's aim, outcome and relation to other exercises, as well as a step-by-step description of how the assessment should be conducted.

The crucial *getting started* block seeks to (i) ensure that all local groups are familiar with the concepts of climate change, vulnerability and adaptation (by means of an exercise and reference to regionalised scenarios of climate change), (ii) create a local climate inventory (by gathering locally relevant climate change and socio-economic data), (iii) describe the object of analysis and (iv) establish a strong and diverse stakeholder group.

The *assessment* block is organised around the three elements of vulnerability: exposure, sensitivity and adaptive capacity (E, S and A in Figure 5.2.1). The approach emphasises the necessity to evaluate vulnerability from a variety of angles, particularly in terms of different climatic and socio-economic futures. Local groups are encouraged to indicate what they perceive to be the most critical

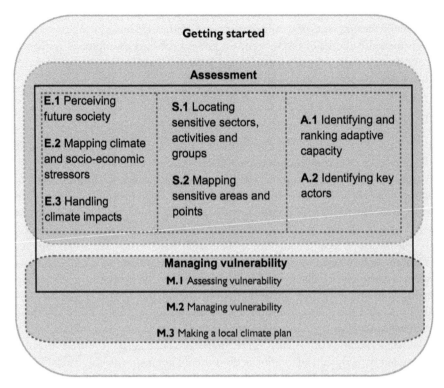

FIGURE 5.2.1 A chart illustrating the three blocks and the specific exercises in the vulnerability approach used in the BalticClimate project (E=exposure, S=sensitivity, A=adaptive capacity, M=managing vulnerability) (source: Wilk et al. 2013)

socio-economic and climatic stresses and impacts, as well as to identify the weakest points in their current activities in relation to these stresses and impacts.

In the *managing vulnerability* block (M in Figure 5.2.1), after having synthesised the outcomes of the assessment block exercises into a vulnerability assessment for their local area and identifying and prioritising adaptation measures, the local groups decide what to include in their local climate plan.

5.2.2 Local assessment in practice: Handling complexity and heterogeneity

After completing the initial exercises in the 'getting started' block, local groups conducted their vulnerability assessments (cf. top dark box in Figure 5.2.1). This assessment consisted of the following exercises:

1 Exposure: Perceiving future society, mapping climate and socio-economic stressors and handling climate impacts
2 Sensitivity: Locating sensitive sectors, activities and groups, mapping sensitive areas and points
3 Adaptive capacity: Identifying factors of adaptive capacity and key actors.

Finally, the outcomes of these exercises were synthesised into a local vulnerability assessment. In the following section, we present selected examples of outcomes from the exposure, adaptive capacity and local vulnerability assessment exercises to illustrate three recommendations.

Assessment of exposure

In the exposure portion of the assessment block, a common vision of possible climatic and socio-economic futures was first developed. Provided with four Special Report on Emissions Scenarios (SRES), participants selected what they viewed as the most plausible socio-economic scenario for their region and determined what their local policy responses to such a world would be. Interestingly, the seven groups studied ended up selecting the two extreme opposite scenarios: B1 (environment before economy, globalisation before regionalisation) and A2 (economy before environment, regionalisation before globalisation). With respect to socio-economic challenges, all groups agreed on the importance of ageing populations and increasing competition from abroad, but disagreed regarding the likelihood that their local governments' policies would be able to overcome these challenges. In some regions, other political challenges were perceived as affecting local vulnerability significantly more than climate change. In these regions, groups felt strongly that adaptation actions should not aggravate the critical socio-economic challenges; as a result, actions that also entailed beneficial social effects were favoured.

In step two of the exposure section, the groups selected what they believed to be the most critical climate and socio-economic stressors. The most critical climatic stressors were selected from a generic list based on current climate literature; the groups had access to regionalised climate projections (based on SRES B1 and A2 scenarios) to inform their selections. This list included changes in climatic parameters and first-order impacts on the biophysical environment (such as coastal erosion). The evaluation criteria were: relevance for the region, direction of change and level of significance. A German region selected seasonal shift in precipitation from summer to winter, poleward shifting of species (animals, plants, pests), the impact of forest fires on monocultures and coniferous forests and sea-level rise as the most significant climatic stressors for agriculture. The most critical socio-economic stressors were selected by means of a self-assessment of current social challenges and their expected changes over the next twenty or fifty years. Questionnaires were used to evaluate each stressor's degree of significance for local vulnerability on a scale ranging from 1 (low) to 5 (high); this served as a basis for group discussions. Groups generally agreed that when climate change stressors aggravated the already critical socio-economic stressors and challenges identified in the 'perceiving future society' exercise, vulnerability was (or would become) high and thus adaptive actions would be necessary. In the German region, the following three socio-economic stressors were deemed to be the most significant:

- *Globalisation.* Due to increasing worldwide competition, globalisation can result in emigration in general and of skilled labour in particular. This affects vulnerability primarily through the associated reduction in adaptive capacity.
- *Ageing of the population in the region.* This is enhanced by emigration and affects vulnerability primarily by increasing sensitivity: the elderly are generally considered to be more vulnerable to extreme temperatures and weather events.
- *Problems with maintaining local and regional financial capacity to provide welfare services.* Financial resources affect adaptive capacity by impacting the ability to fund adaptation measures.

In the third step of the exposure block, groups assessed how they were handling current climate variability and how they would manage the potential impacts resulting from future climate change.

Assessment of sensitivity

Subsequently, the sensitivity block clarified the importance of considering the views of those affected by the changes and of those with expert knowledge regarding local technical infrastructure and economic activities. Sensitivity was approached from both a spatial perspective (identifying sensitive areas) and a socio-economic impact perspective (identifying activities and sectors for which the anticipated climate changes would result in enormous impacts on economic and social activity, as well as groups of people that would be severely impacted by the changing climate).

Assessment of adaptive capacity

In the adaptive capacity section of the assessment block, two exercises were conducted. First, the groups identified the key factors affecting their ability to adapt to climate change, ranking generic factors in terms of significance (low to high, using a deck of cards from 2 to ace) and convertibility (ranging from 'easy to change' to 'impossible'). The most significant factors in decreasing order affecting adaptive capacity, based on the averages of all group rankings, were identified as: political support, skilled staff, knowledge about climate change, financial resources and collaboration. Participants were then encouraged to concentrate on the hard-to-change factors. Interestingly, the Finnish case reveals that the most commonly identified factors were viewed as only moderately difficult to change (cf. Figure 5.2.2). The Finnish group assessed the following factors as hard to change: influencing public opinion on climate change, reducing energy intensity in production, minimising additional environmental impact and changing logistical models. This group stressed that new energy-efficient models of transport for goods, services and people were badly needed.

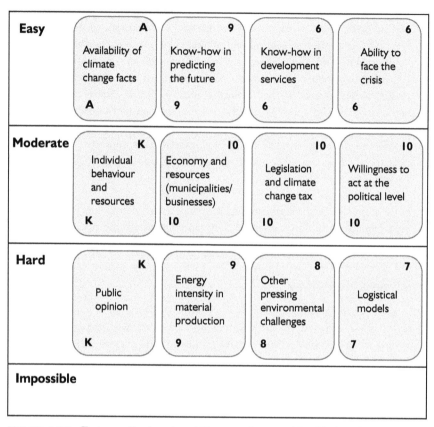

FIGURE 5.2.2 Factors affecting the ability to adapt, as identified and ranked in a Finnish region (using values from a deck of cards ranging from 6 (low) to ace (high), A = ace, K = king) (source: authors' research)

Second, the local groups identified additional key actors who should be invited to participate in the assessment (defined as the actors necessary to carry out the key measures identified by the groups). The adaptive capacity of relevant organisations and their importance in enabling local climate adaptation were ranked. For these key organisations, their current and potential collaborative activities and responsibilities for specific tasks were assessed. For example, in a Swedish case, the key actors whom the group determined to be necessary for the successful integration of climate change concerns in regional planning were ranked based on the objectives of promoting the regional development plan to the general public, establishing closer collaboration between municipal planners, utilities and service providers and increasing the transfer of experiences between developers and local government.

Assessment of vulnerability

The final synthesis of the outcomes of the evaluations of exposure, sensitivity and adaptive capacity into an overall local vulnerability assessment proved to be problematic. The outcomes of all exercises were compiled in a table with columns for exposure, sensitivity and adaptive capacity. Participants then first individually reflected on the following questions: What types of vulnerability factors are most important? What relationships exist between them? Which of these factors will result in harm, negative effects or the inability to cope with the adverse effects of climate change? What scales are relevant for different factors (local, regional, national, etc.)? Are there any factors missing or that are redundant? The groups then selected the most important factors that influenced their vulnerability, currently and in the future. In general, these outcomes were characterised by the groups as separate vulnerabilities rather than an integrated vulnerability. The participants were reluctant to explicitly aggregate the various elements, or they did not see the point in doing so; instead, they began to identify adaptation actions to deal with each vulnerability element – that is, exposure, sensitivity and adaptive capacity – and each stressor separately.

5.2.3   Conclusion

Our experiences with the application of the stepwise vulnerability assessment in the seven localities demonstrate the importance of the local groups clarifying at the start of the process (i) the purpose of the assessment, (ii) the system components and drivers that should be included in the investigation and (iii) the problem the assessment is expected to solve (Preston et al. 2011, Hinkel 2011). In conclusion, the following recommendations for achieving a successful outcome with the vulnerability assessment can be made.

Lesson 1: Encourage assessment from many angles for more robust outcomes

Because the exercises are designed to work like a funnel, our participants first identified a large number of factors; subsequently, they ranked stressors, adaptive capacity and other relevant factors according to different dimensions (e.g. significance, timescale and ability to influence) based on the aim of the exercise. They were also instructed to approach topics from a generic global/regional/aggregate economic perspective and then to consider how that would be applied to their specific local context or economic sector. By stepwise prioritisation, the complex issues were condensed into a few key challenges and opportunities. Initially, the participants were sceptical, but as the assessment went on, they became more accustomed to making prioritisations. Heterogeneity of perspectives did not prevent the local groups from agreeing on a common perspective in their vulnerability assessment. Occasionally, they took note of differences in order to be able to analyse them in more depth at a later date.

Encouraging assessment from multiple perspectives and the continuous prioritisation of a large set of options resulted in the systematic reduction of heterogeneity and complexity to a more manageable level.

Lesson 2: Actor collaboration is essential for local adaptive capacity

Involving actors early on in the process has been proven to effectively foster a sense of ownership in local adaptation efforts (André 2013). In our vulnerability assessments, we have seen that actor collaboration is a key ingredient for local adaptive capacity and thus for the ability to influence local vulnerability to climate change. Inadequate collaboration was found to limit adaptive capacity due to (i) the currently unclear distribution of responsibilities for climate change adaptation, (ii) the lack of collaboration within and across organisations that were seen as essential for enabling local climate adaptation and (iii) the lack of overall coordination for local climate change adaptation. This means that successful adaptation will require not only clear mandates, but also platforms and resources that can facilitate systematic interactions between key organisations. In addition, clearly assigning coordination responsibility for climate adaptation to one organisation or asserting the accountability of all organisations in this regard was deemed to be essential for the success of local climate adaptation. All local groups were able to identify potential collaborations between several actors; these interactions were seen as vital to ensuring a successful societal response to climate change. Enhanced collaboration between actors was also highlighted in the 'managing vulnerability' block, as different adaptation needs and options require different sets of actors for successful governance.

Lesson 3: You will need to create your own data

Data availability and previous experiences with climate change efforts determined our starting points, which differed significantly across the seven local groups. We found that a lack of data is the norm, even after considerable effort has been devoted to obtaining it. This was one of the reasons we decided to add the 'getting started' block, which was not initially planned as part of the vulnerability assessment framework. To fill the gaps in data, local groups had to rely on their own qualitative and quantitative judgements, even though they were generally reluctant to do so at first. As the assessment proceeded, participants became familiarised with assigning weights to the factors being evaluated. Ultimately, the groups were able to assign quantitative weights to the socio-economic and climate stressors, as well as qualitatively assessing the degree of influence of various factors on local adaptive capacity and identifying key actors. This is of particular importance because vulnerability assessments require a great deal of data (Malone & Engle 2011).

5.3 An ad-hoc prioritisation methodology applied in the Swiss national adaptation strategy

Roland Hohmann, Thomas Probst, Pamela Köllner-Heck, Hugo Aschwanden and Bruno Schädler

5.3.1 The Swiss adaptation strategy

On 2 March 2012, the Swiss Federal Council adopted its strategy for climate change adaptation (Swiss Confederation 2012a), which provides the basis for the coordination of adaptation activities on the federal level in Switzerland. The strategy focuses on the nine most important sectors that are affected by climate change and that fall under the federal government's jurisdiction. Within these sectors, the strategy identifies fields of action (i.e. topics for which adaptation actions are required) and assigns priorities to these fields.

The Swiss adaptation strategy defines the application of a risk approach as one of its key principles, implying that adaptive action shall be undertaken first where the climate change-induced risks and opportunities are greatest. In order to ground this principle on a sound quantitative basis, Switzerland is undertaking a nationwide, integral, cross-sectoral analysis of climate change-induced risks and opportunities. However, this assessment is still ongoing, and the results were not available when the adaptation strategy was developed. Therefore, an ad-hoc methodology was developed and applied in the identification and prioritisation of the fields of action within the adaptation strategy. This methodology will be described in the sections below.

5.3.2 An ad-hoc prioritisation methodology with six fields of action in the water management sector

Sectoral strategies at the national level for the most important sectors that are affected by climate change are a key element of the Swiss adaptation strategy. These sectors were selected on the basis of a comprehensive impact assessment published by the Swiss Advisory Body on Climate Change (OcCC) in 2007 (OcCC 2007). In this assessment, the impacts of climate change on the most important environmental and socio-economic systems in 2050 were qualitatively discussed. From these environmental and socio-economic systems, nine sectors were selected for the adaptation strategy and further investigation: water management, natural hazards prevention, agriculture, forestry, energy, tourism, biodiversity management, human and animal health and spatial planning.

The aim of each sectoral strategy is to identify the main topics for which adaptation action is needed (fields of action) and subsequently to define adaptation goals for these fields of action. To identify the fields of action, the ad-hoc methodology was applied in all sectors as follows:

1   In the first step, all of the topics in a sector that could potentially be affected by climate change were identified. The reference point for the assessment was a regional climate scenario for Switzerland for the year 2050 provided by the OcCC (OcCC 2007). Other non-climatic factors were not explicitly considered; these factors are covered in other sectoral or cross-sectoral strategies, e.g. the Swiss Strategy on Natural Hazards (PLANAT 2004) and the Swiss Strategy on Biodiversity (Swiss Confederation 2012b).

2   In a second step, these topics were assessed by dedicated experts from science and the public sector with respect to the following dimensions:
   • Impact of climate change: How strongly is the specific topic projected to be (positively or negatively) affected by climate change?
   • Relative importance of the change: How important are the climate change-induced variations in the specific topic from the overall view of the sector?
   • Need for action: How urgent is it to react to the expected changes, and how much needs to be done? The need for action takes into account the measures that have already been implemented and possible measures yet to be undertaken. Needs for action that are not related to climate change are not considered.

3   The three dimensions are partially correlated. In general, the need for action is greatest for those topics for which the impact of climate change is large and the relative importance of the change is considered to be high. Conversely, it is smallest for those topics for which the impact of climate change is small and the relative importance of the change is considered to be low. Nevertheless, in some cases, the need for action might be small despite large and important impacts because all the necessary adaptation measures have already been implemented and no further measures are available.

4   All topics within a sector were evaluated with respect to the three dimensions above using a simple three-step scale (low – moderate – high). This evaluation was based on sectoral stakeholder and expert judgements, since no comprehensive quantitative basis was available and there is no common numerical scale for the assessed sectors. This implies that the ratings of topics can be compared within a sector but not between different sectors. For validation, the experts' assessments were double-checked by other stakeholders and experts in several independent reviews.

5   Based on the evaluation, topics for which all three dimensions were assessed as moderate or high were selected as fields of action in adaptation. These fields were then discussed in detail within the sectoral strategies. All other topics with a low sensitivity to climate change, with climate change impacts of little importance or with a low need for action were not considered to be fields of action and were not further explored in the sectoral strategies.

In the rest of this section, the application of this methodology in the water management sector is discussed as an example.

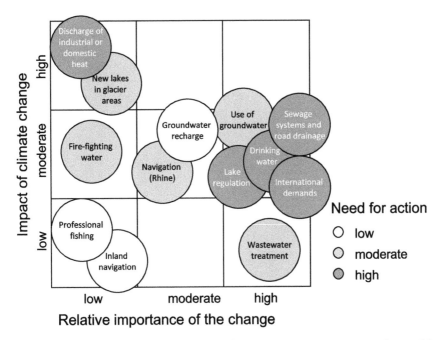

FIGURE 5.3.1 Assessment of the topics in the water management sector that could potentially be affected by climate change (source: authors' research)

The impacts of climate change on the water cycle affect most natural and anthropogenic systems. Thus, most sectors have an interest in adaptation to climate change-induced variations in the water cycle. In fact, the water management sector identified thirty topics that could potentially be affected by climate change. The majority of these topics (e.g. water cooling of thermal power plants or artificial snowmaking) are cross-sectoral in nature, and their assessment falls under the responsibility of other sectors. For thirteen of the identified topics, the water management sector has sole responsibility. The assessment of these topics with respect to the impact of climate change, the relative importance of the change and the need for action is depicted in Figure 5.3.1 and summarised in Table 5.3.1.

Based on the assessment of the water management topics, six fields of action with respect to adaptation were identified:

• International demands
• Drinking water
• Sewage systems and road drainage
• Lake regulation
• Use of groundwater
• Navigation (Rhine).

TABLE 5.3.1 Assessment of topics in the water management sector that could potentially be affected by climate change (in terms of the impact of climate change, the relative importance of the change and the need for action)

Topics in the water management sector	Impact of climate change	Relative importance of the change	Need for action
Drinking water			
Climate change will lead to local and temporary shortages in the availability of ground and surface water, which may affect small drinking-water supply systems.	moderate	high	high
Use of groundwater			
Warmer temperatures and an increase in droughts and floods due to climate change may affect groundwater temperature, oxygen content and contamination, with consequences for the use of groundwater.	moderate	high	moderate
Groundwater recharge			
Groundwater recharge near small rivers will be impaired during longer-lasting and more intense periods of drought.	moderate	moderate	low
Fire-fighting water			
The potential increase in forest fires will lead to an increase in the amount of water used in fire-fighting efforts.	moderate	low	moderate
Wastewater treatment			
Low river discharge during periods of drought places high quality demands on wastewater treatment in order to prevent contamination due to insufficient dilution.	low	high	moderate
Sewage systems and road drainage			
The dimensions of the sewage system and the road drainage system may be insufficient to cope with an increase in extreme precipitation events resulting from climate change.	moderate	high	high

Lake regulation			
An increase in floods and droughts due to climate change may lead to changes in the demand for lake regulation.	moderate	high	high
Navigation (Rhine)			
An increase in periods with small or large river discharge will impair transportation on the Rhine.	moderate	moderate	moderate
Inland navigation			
An increase in low water levels in lakes during periods of drought will affect inland navigation.	low	low	low
New lakes in glacier areas			
New lakes that evolve below melting glaciers may be used for hydropower production or as tourist attractions. Glacier lake outbursts may become a threat to lower-lying regions.	high	low	moderate
Professional fishing			
Fishing resources may be affected by rising water temperatures in lakes and rivers.	low	low	low
Discharge of industrial or domestic heat			
Climate change may lead to an increase in the use of groundwater and surface water for industrial and domestic cooling.	high	low	high
International demands			
Coordination demands with respect to international waters will increase due to climate change.	moderate	high	high

Source: Authors' research

These fields of action were further discussed in the sectoral strategy for the water management sector. The first four fields of action were prioritised because of their high need for action, whereas the need for action was considered to be moderate for the last two topics. The other topics did not qualify as fields of action because their sensitivity to climate change, the importance of the change or the need for action were assessed as low.

5.3.3   Conclusion

Because the results of a comprehensive impact assessment were not yet available, an ad-hoc methodology was applied to all sectors included in the Swiss adaptation strategy, and a total of forty-eight fields of action were identified. The methodology proved to be a simple and effective way to assess and prioritise potential climate change impacts and needs for action, using the judgements of stakeholders and experts. The application of the methodology was restricted to the influence of climate change; as a result, other factors were neglected, even though these factors might actually be more important for certain sectors. Furthermore, the methodology applies a risk approach, one of the key principles defined by the Swiss strategy.

The application of a common methodology had several positive effects: (1) It enabled a learning process among the involved stakeholders and experts in terms of distinguishing between adaptation and other important actions not related to climate change, (2) it allowed all topics in a sector to be assessed in a transparent fashion on the basis of expert knowledge and (3) it forced the stakeholders and experts to assess and prioritise the fields of action within a sector.

However, the methodology also has some shortcomings. The most significant is that the assessment scales applied are different in each sector and have no quantitative basis. Consequently, the fields of action cannot be compared across sectors in terms of the impact of climate change, the importance of the change and the need for action, and thus no cross-sectoral prioritisation of fields of action is possible.

In order to provide a common quantitative base for this type of cross-sectoral comparison of fields of action, Switzerland is undertaking a nationwide, integral, cross-sectoral assessment of climate change-induced risks and opportunities. However, this assessment has proven to be difficult and labor-intensive, and the first results will become available only in a few years. The ad-hoc methodology described in this section has been shown to be an efficient and pragmatic substitute.

5.4   Lessons learned for the assessment of climate change impacts and vulnerabilities

Inke Schauser

Assessments of impacts, risks and vulnerability are often undertaken at the beginning of an adaptation process. These assessments are necessary in order to obtain an overview of possible climate change impacts and to identify the most important impacts or the most vulnerable groups, sectors or regions. They can further be used to raise awareness about climate change, as well as to create a common understanding of potential climate change effects and their interdependencies. Furthermore, they can help researchers to monitor the effectiveness of adaptation policies and advance scientific knowledge (Hinkel 2011). Practical experience with impact and vulnerability assessments has demonstrated the importance of clarifying at the start of the process the purpose of the assessment, the vulnerable object and the problem that the assessment is expected to solve. To maximise the assessment's utility in practice, it should be connected to the tasks or the mandates of the organisations involved in the assessment. At the local level in particular, it should also be used to develop adaptation options.

Because climate change will have different impacts across various regions and sectors, multi- or cross-sectoral and spatially differentiated assessments are needed. In Switzerland (Section 5.3), an expert-based assessment seeking to identify key systems and rank potential impacts was accomplished by asking questions about the importance of the potential impact to the sector and the need for action. Knowledge of current and future climatic conditions was used as background information. According to Hohmann and his colleagues, the process achieved several positive results: (1) It enabled a learning process among the involved stakeholders and experts in terms of distinguishing between adaptation and other important actions not related to climate change, (2) it allowed an assessment of all of the topics in a sector to be conducted in a transparent fashion on the basis of expert knowledge and (3) it forced stakeholders and experts to assess and prioritise the fields of action within a sector.

However, any assessment of future conditions (such as potential climate change impacts) involves various types of uncertainties and must overcome methodological challenges. The approach taken in the Alps by Project CLISP (described by Zebisch et al. in Section 5.1) focuses on the second step proposed in the guiding principle: because the system's status already reveals existing vulnerabilities to current climate variability, historical climate data was considered in addition to projected climate change scenarios. The system's biophysical and socio-economic conditions must also be investigated, as they are likely to worsen as a result of climate change.

Zebisch et al. point out that stakeholders play a crucial role in the development of impact and vulnerability assessments at the local and regional levels as preparation for adaptation action, as they can provide profound knowledge regarding the system at risk and because they are the actors, who will ultimately have to adapt to climate change. Their qualitative and narrative information can often uncover the details relevant for a vulnerability assessment more effectively than a purely quantitative approach.

The approach developed for a local impact and vulnerability assessment in the BalticClimate project (described by Hjerpe and Wilk in Section 5.2) emphasises the third step of the guiding principle: in addition to identifying the point of departure defined by data availability and available knowledge on climate change adaptation, it is also important to consider different development pathways and perspectives. This entails not only including a range of stakeholders but also elaborating on the local implications of various socio-economic, climate and policy circumstances and potential futures. Vulnerability often arises from a combination of socio-economic conditions and climate impacts. Local experiences show that adaptation is contextualised and varies across regions based on social and political circumstances. This implies that assessment methodologies must be flexible in order to be able to address the most critical local challenges. By stepwise prioritisation of the heterogeneous and complex issues, these challenges can be condensed into only a few manageable key problems and opportunities.

All three cases conclude that the incorporation of experts and stakeholders plays an important part in identifying and prioritising climate change impacts and vulnerabilities. The cases emphasise that narrative knowledge and (semi-)qualitative approaches should be considered in addition to quantitative methods. Experiences in the Alps (Section 5.1) reveal the importance of considering historical knowledge of past weather events as well as climate projection data. The practical experience in the BalticClimate project (5.2) illustrates that potential climate impacts and adaptive capacities are deeply connected to social processes and the practices and culture of local organisations. Even in situations in which climate change appears to be a minor issue in comparison to more severe socio-economic stressors, it is necessary to investigate how climate change adaptation activities could mitigate socio-economic problems as well. The methodologies presented in the Swiss case (5.3) and the Alps case (5.1) also provide good examples of how simple aggregation rules can be used to rank different impacts.

Thus, the following lessons learned overall can be synthesised:

- Impact and vulnerability assessments serve many purposes and enable a learning process with regard to adaptation among the experts and stakeholders involved.
- The participation of stakeholders and their qualitative and historical knowledge are important aspects in assessing the current and future vulnerability of a system; their specialised knowledge can fill in data gaps, especially at the regional and local levels.
- Current vulnerabilities can predict future vulnerability to climate change to a great extent: regions that are already sensitive to current climate extremes can often be expected to be among the most vulnerable regions in the future.
- Future vulnerability also depends on the socio-economic development of the system and other pressures. Vulnerability frequently arises from a combination of expected socio-economic changes (inside and outside the system) and climate change.

References

Adger, W.N. and Vincent, K. (2005) 'Uncertainty in adaptive capacity', *External Geophysics, Climate and Environment* 337 (4), pp. 399–410.

AEA, SEI and Metroeconomica (2005) *Objective Setting for Climate Change Adaptation Policy.* Project report commissioned by Defra.

André, K. (2013) *Climate Change Adaptation Processes: Regional and Sectoral Stakeholder Perspectives,* Linköping, Sweden: Linköping Studies in Arts and Science, No. 579, Linköping University Press.

Auer, I., Böhm, R., Jurkovic, A., Lipa, W., Orlik, A., Potzmann, R., Schöner, W., Ungersböck, M., Matulla, C., Briffa, K.R., Jones, P., Efthymiadis, D., Brunetti, M., Nanni, T., Maugeri, M., Mercalli, L., Mestre, O., Moisselin, J.M., Begert, M., Müller-Westermeier, G., Kveton, V., Bochnicek, O., Stastny, P., Lapin, M., Szalai, S., Szentimrey, T., Cegnar, T., Dolinar, M., Gajic-Capka, M., Zaninovic, K., Majstorovic, Z. and Nieplova, E. (2007) 'HISTALP – historical instrumental climatological surface time series of the Greater Alpine Region', *International Journal of Climatology* 27 (1), pp. 17–46.

Carter, T.R., Parry, M.L., Harasawa, H. and Nishioka, S. (1994) *IPCC Technical guidelines for assessing climate change impacts and adaptations,* London, Department of Geography, University College London.

Dessai, S., Hulme, M., Lempert, R. and Pielke, R. (2009) 'Climate prediction: a limit to adaptation?', in Adger, N., Lorenzoni, I. and O'Brien, K. (ed.) *Living with Climate Change: Are There Limits to Adaptation?* Cambridge: Cambridge University Press.

Feenstra, J.F., Burton, I., Smith, J.B. and Tol, R.S.J. (1998) *Handbook on Methods for Climate Change Impact Assessment and Adaptation Strategies,* UNEP/IVM, Nairobi and Amsterdam.

Hinkel, J. (2011) 'Indicators of vulnerability and adaptive capacity: Towards a clarification of the science–policy interface', *Global Environmental Change* 21 (1), pp. 198–208.

HM Government (2010) *Climate Change: Taking Action. Delivering the Low Carbon Transition Plan and Preparing for a Changing Climate,* London: Defra

IPCC (2001) 'Climate change 2001: impacts, adaptation, and vulnerability', in McCarthy, D., Canziani, O.F., Leary, N.A., Dokken, D.J. and White, K.S. (ed.) *Contribution of Working Group II to the Third Assessment Report of the Intergovernmental Panel on Climate Change,* Cambridge: Cambridge University Press.

Lautenschlager, M., Keuler, K., Wunram, C., Keup-Thiel, E., Schubert, M., Will, A., Rockel, B. and Boehm, U. (2009) *Climate Simulation with CLM, Climate of the 20thCentury run no.1, Data Stream 3: European Region MPI-M/MaD,* World Data Centre for Climate. DOI: 10.1594/WDCC/CLM_C20_1_D3.

Malone, E.L. and Engle, N.L. (2011) 'Evaluating regional vulnerability to climate change: purposes and methods', *Wiley Interdisciplinary Reviews-Climate Change* 2 (3), pp. 462–474.

OcCC (2007) *Climate Change and Switzerland 2050, Expected Impacts on Environment, Society and Economy,* Bern, Switzerland: OcCC/ProClim.

Parry, M.L. and Carter, T.R. (1998) *Climate Impact and Adaptation Assessment: A Guide to the IPCC Approach,* London: Earthscan.

Parry, M.L., Canziani, O.F., Palutikof, J.P., van der Linden, P.J. and Hanson, C.E. (ed.) (2007) *Climate Change 2007: Impacts, Adaptation and Vulnerability, Contribution of Working Group II to the Fourth Assessment Report of the Intergovernmental Panel on Climate Change,* Cambridge, UK: Cambridge University Press.

PLANAT (2004) *Strategie Naturgefahren Schweiz,* Biel, Schweiz: PLANAT.

Preston, B.L., Yuen, E.J. and Warburton, R.M. (2011) 'Putting vulnerability to climate change on the map: a review of approaches, benefits and risks', *Sustainability Science* 6 (2), pp. 177–202.

Ribeiro, M.M., Losenno, C., Dworak, T., Massey, E., Swart, R., Benzie, M. and Laaser, C. (2009) *Design of Guidelines for the Elaboration of Regional Climate Change Adaptations Strategies*, Study for European Commission – DG Environment – Tender DG ENV. G.1/ETU/2008/0093r, Vienna, Austria: Ecologic Institute.

Scheraga, J.D. and Grambsch, A.E. (1998) 'Risks, opportunities, and adaptation to climate change', *Climate Research* 11 (1), pp. 85–95.

Schneiderbauer, S., Pedoth, L., Zhang, D. and Zebisch, M. (2011) 'Assessing adaptive capacity within regional climate change vulnerability studies – an Alpine example', *Natural Hazards*, pp. 1–15.

Swiss Confederation (2012a) Anpassung an den Klimawandel in der Schweiz. Ziele, Herausforderungen und Handlungsfelder. Erster Teil der Strategie des Bundesrates (www.bafu.admin.ch/klimaanpassung).

Swiss Confederation (2012b) Strategie Biodiversität Schweiz.

van der Linden, P. and Mitchell, J.F.B. (ed.) (2009) *ENSEMBLES: Climate Change and its Impacts: Summary of Research and Results from the ENSEMBLES Project*, Exeter, UK: Met Office Hadley Centre.

Wilby, R.L. and Dessai, S. (2010) 'Robust adaptation to climate change', *Weather* 65 (7), pp. 180–185.

Wilk, J., Hjerpe, M., Jonsson, A., Andre, K. and Glaas, E. (2013) *A Guidebook for Assessing and Managing Integrated Vulnerability to Climate Change*, Linköping, Sweden: Linköping University Press.

Willows, R. and Connell, R. (2003) *Climate Adaptation. Risk, Uncertainty and Decision-making*, Oxford, UK: UKCIP Technical Report.

6

INITIATE ADAPTATION, ENSURE COMMITMENT AND MANAGEMENT

Andrea Prutsch, Sabine McCallum, Torsten Grothmann, Inke Schauser and Rob Swart

Explanation of the guiding principle

National and international public authorities should provide a framework and an institutional, legal and socio-economic environment for adaptation (Frankhauser et al. 1999, Agrawala & Frankhauser 2008, Mysiak et al. 2010) that enables municipalities, organisations (e.g. NGOs, businesses) and citizens to adapt to the specific conditions at the sub-national scale (e.g. regional, local) (Amundsen et al. 2010).

When realising adaptation in a region, sector or within an organisation, a clear commitment from decision-makers (e.g. political leaders, business managers) to support the adaptation process is highly beneficial. This formal consent on the part of decision-makers – together with the engagement of stakeholders (cf. Chapter 8) – will often enhance the acceptance of adaptation and help to overcome bureaucratic resistance (Smith et al. 2009, Ribeiro et al. 2009, UKCIP 2005). To facilitate such commitment, it is often necessary to raise the awareness of climate change impacts and adaptation needs among decision-makers (cf. Chapter 7). Decision-makers should be made aware of the fact that climate change will proceed over the course of several decades, a length of time that (e.g. in the case of politicians) certainly extends beyond their mandates. Thus, they must take responsibility for building the long-term capacity of society and ecosystems to adapt to climate change.

To sustain the adaptation process, a clear mandate for the management of the process should be assigned to an organisation or an individual. The role of the managing unit can include:

- defining clear responsibilities, aims and rules for the adaptation process,
- facilitating the adaptation process by motivating stakeholders to participate (cf. Chapter 8),
- promoting dialogue and understanding among the various actors and stakeholders (cf. Chapter 8), and

- providing reliable knowledge and scientifically based results (cf. Chapter 7).

Ideally, the managing unit should be free of self-interests. In addition, for the purposes of promoting an effective working environment, the power and responsibilities of the managing unit should be transparent and agreed upon in advance. It is important to note that such an environment requires the trust of the stakeholders involved in the adaptation process. Negative experiences from previous processes may have resulted in lingering distrust, hindering the success of later cooperation (Smith et al. 2009). As adaptation can be a resource-intensive process, the long-term personnel and financial resources for the managing unit and for the adaptation process in general should be ensured.

Chapter overview

The three case studies in this chapter present practical experiences and reflect upon specific questions related to initiating adaptation and ensuring commitment and proper management. Governments worldwide have begun to develop governance innovations to support adaptation at various decision levels. The first case (Section 6.1) presents the results from a survey of 10 OECD countries that describes the policy instruments they have on hand to 'steer' society towards adaptation. Despite these tools, a variety of challenges could impede governments as they seek to facilitate climate change adaptation, as this case explains.

The second case (6.2) explores how the strong commitment to adaptation has evolved in the case of the United Kingdom. In particular, it considers the leadership that has been established in terms of the individuals and institutions involved in achieving the current level of climate change response. The chapter concludes with five key steps that have brought the UK to its current position and that can likely be transferred to other countries seeking to develop their adaptation response.

The third case, presented in Section 6.3, focuses on the specific situation of the private sector in relation to adaptation to climate change, pointing out what governments can do to support businesses. It draws conclusions based on three practical initiatives aimed at promoting adaptation in the private sector in the UK.

6.1 The role of governments in adaptation: A comparison across Europe

Reinhard Steurer and Anja Bauer

6.1.1 *Why governments are concerned with adaptation and the challenges they face*

Since adaptation to climate change – i.e. the "adjustment in natural or human systems in response to actual or expected climatic changes or their effects, which moderates harm or exploits beneficial opportunities" (IPCC 2001, 2007) – has been the rule rather than the exception throughout human history, why are governments increasingly getting involved in adaptation policy-making? As Cimato and Mullan explain, people are expected "to take autonomous action to adapt when it is in their interest and power to do so; that is, they will take measures where the private benefits outweigh the costs to them" (Cimato & Mullan 2010: 16). However, because the pace of anthropogenic climate change in recent decades has been unprecedented, it is likely that 'adaptation as usual' may no longer suffice (Berkhout 2005). Consequently, several adaptation scholars have emphasised that autonomous adaptation (i.e. adaptation by individuals, organisations and businesses without political intervention) is likely to be inadequate, in particular when (Berkhout 2005; Cimato & Mullan 2010):

- Those affected by climate change are unaware of the need to adapt (e.g. because future impacts are marginalised by everyday routines or are hard to foresee), or rational action is impeded by behavioural obstacles (such as difficulties with complexity and long-term thinking);
- Those affected by climate change are aware of the need to adapt but do not have the necessary capacity (such as financial resources, knowledge of what to do and how to do it, technical expertise); and/or when
- Affected individuals face socio-economic obstacles to adaptation, such as market failures (some of which may result from the obstacles described above), (new) external effects or public-good dilemmas connected to climate change impacts and the respective adaptation measures (e.g. farmers or fishermen downstream of a large-scale irrigation system that dries out the river under drier climate conditions).

Because of the numerous circumstances that can prevent individuals and businesses from autonomously adapting to climate change, and because of the negative macroeconomic consequences that will result from inadequate adaptation (Cimato & Mullan 2010), international organisations and governments around the world have recognised the need to facilitate adaptation to climate change with a wide variety of public policies (for the international level, cf. Burton et al. 2006; for the EU, cf. Rayner & Jordan 2010; for OECD and EU countries, cf. Bauer et al. 2012, Biesbroek et al. 2010, Keskitalo 2010).

When governments seek to facilitate climate change adaptation, they struggle with a plethora of challenges, as climate change is a complex issue that cannot be solved using quick fixes. Among the most important challenges for traditional forms of public governance are the following (Bauer et al. 2012):

- *Climate change impacts affect many sectors.* Governments are usually organised according to a sectoral logic (reflected by their ministerial/departmental structure), but they will have to integrate (or mainstream) adaptation policies horizontally across policy sectors.

- *Climate change impacts transcend political borders*, i.e. they materialise locally as well as globally. Governments are territorial entities that are embedded within a multi-level system; they must strive to integrate adaptation policies vertically across different governmental jurisdictional levels.
- *Future climate change and the respective potential impacts are riddled with uncertainties.* In order to avoid (costly, inefficient) maladaptation, governments are advised to strengthen the science-policy interface and to base their policies on sound knowledge.
- *Although local stakeholders often have valuable expertise regarding climate impacts, they may not be aware of the need to adapt to a changing climate or the possible options for adaptation.* Involving these stakeholders in adaptation policy-making can be supportive for both policy-makers and local stakeholders, but also implies new challenges (e.g. managing expectations and sustaining commitment).

As the following section shows, governments have become aware that they must play a role in facilitating adaptation to climate change by changing behaviours via public policy. The emergence of numerous governance innovations signifies that governments are also aware of the fact that climate change adaptation challenges the status quo of public governance, i.e. the traditional ways in which public policies are formulated and implemented.

6.1.2   Public policies and governance on adaptation in Europe

Public policies on climate change adaptation should be concerned with, inter alia, raising awareness of present and future impacts and vulnerabilities, building adequate capacities to cope with the impacts, putting already existing adaptation capacities into action (Adger et al. 2005, Nelson et al. 2007), resolving conflicts of interest and reducing any external effects that may be triggered or reinforced by climate change (Cimato & Mullan 2010). What policy instruments do governments have on hand to steer society towards better adaptation? As in other policy fields (cf. e.g. Hood 1986, Baldwin & Cave 1999), adaptation policy instruments include:

- *Informational instruments*, such as government-sponsored studies, websites, guidelines, handbooks, brochures and campaigns that seek to raise awareness of climate change impacts and the associated adaptation needs and options (cf. e.g. the UKCIP website);
- *Legal instruments* (i.e. laws and regulations), such as the Delta Act in the Netherlands that provides a legal basis for adaptation in the water sector (Bauer et al. 2012);
- *Fiscal or economic instruments*, such as tax breaks or subsidies that foster e.g. green-roof projects to combat the heat-island effect in cities (Mees et al. 2012);
- *Partnering instruments*, such as negotiated agreements, public–private partnerships, or regional adaptation partnerships (like those in the UK and in Canada; cf. Bauer & Steurer forthcoming);

- *Hybrid instruments*, such as comprehensive adaptation strategies/plans/ programmes that aim to mainstream and orchestrate other adaptation instruments and policies across various sectors (e.g. the National Adaptation Strategy in Austria and the National Adaptation Programme in the UK).

In addition to steering society (or parts thereof) towards better adaptation with the policy instruments mentioned above, adaptation policies are also concerned with building new or adapting existing public infrastructure (such as sewage or transportation systems) to address expected future climate change impacts (Cimato & Mullan 2010).

While some of the adaptation policy instruments listed above have been investigated by (relatively few) local case studies, broader comparative studies on efforts across Europe exist only for governance issues (cf. e.g. Biesbroek et al. 2010, Hulme et al. 2009, Keskitalo 2010, Massey & Bergsma 2008, Swart et al. 2009, Bauer et al. 2012). Consequently, we can only speculate that the latter have overshadowed the implementation of actual adaptation policies so far, and that many of the adaptation policies already underway are concerned with raising awareness via informational activities. Since uncertainties and a lack of awareness are key barriers to adaptation, these informational policies are a legitimate starting point.

What governance innovations can be found that seek to address the challenges of horizontal integration, vertical integration, uncertainty and stakeholder involvement? Based on desk research and a series of telephone interviews with public administrators from 10 OECD countries, Bauer et al. (2012) provide the following answers (for an overview, cf. Table 6.1.1).

Overall, governance of climate change adaptation includes several relatively narrowly defined approaches that address only one or two of the four challenges outlined above, and a few comprehensive approaches that seek to address three or all four challenges at once. The most prominent comprehensive approach, the national adaptation strategy (NAS; sometimes also referred to as national adaptation programmes or frameworks) has emerged in many EU Member States since 2005. In addition to being a comprehensive governance approach that facilitates and orchestrates several of the narrower governance approaches (cf. Table 6.1.1), a NAS is also a hybrid policy instrument that encompasses non-binding targets, adaptation options and measures. In other words, the ideal NAS is a policy document (concerned with the 'what?' of adaptation) and a governance process (concerned with the 'how' of adaptation policy-making) at the same time.

With regard to the challenge of integrating adaptation policies horizontally across various sectors, the process of developing a NAS marks the first systematic governance response in most countries. For this purpose, most countries have established either temporary or permanent inter-ministerial coordination bodies (the temporary bodies are generally dissolved once the NAS has been formulated). The horizontal mainstreaming of adaptation is also facilitated by sectoral strategies (e.g. in Denmark and Finland, on forestry, sustainable development and biodiversity) and by departmental adaptation plans (e.g. in the UK). While both the NAS and

TABLE 6.1.1 Overview of governance challenges and governance approaches

		Governance challenges			
		Horizontal integration	*Vertical integration*	*Knowledge integration*	*Participation*
Governance approaches	Policy frameworks (linked to several of the governance approaches summarised here)	+	+	+	~
	Temporary coordination and/or consultation for elaborating the NAS	+	+	+	+
	Institutionalised coordination bodies (horizontal, vertical or both jointly)	+	+	+	~
	Strategies apart from the NAS addressing adaptation	+			
	Monitoring, reporting (and evaluation) schemes		+	+	
	Networks and partnerships		+	~	+
	Status quo assessments and studies	~	~	+	~
	Research programmes	~	~	+	~
	Scientific advisory bodies and services	~	~	+	
	Institutionalised consultation bodies			~	+
	Temporary stand-alone consultation			~	+

Source: Bauer et al. 2012: 299

+   most governance approaches of the respective type address the challenge extensively
~   some governance approaches of the respective type address the challenge to some degree

the narrower governance approaches seek to establish a common national or federal approach to climate change adaptation across different ministries, their coordination achievements have been relatively moderate thus far. What most of these approaches have accomplished is the formulation of a common ground, a first impulse and an improved awareness of climate change adaptation across different ministries. In other words, they have put adaptation on the national policy agenda.

Although the horizontal and vertical integration of adaptation policies are clearly distinct challenges from an analytical point of view (see above), governments often address them jointly. This applies in particular to the temporary coordination during the NAS formulation, but also to some of the institutionalised coordination bodies in unitary states. Federal countries (such as Germany) tend to address the challenge of vertical integration not only by means of stand-alone approaches but also earlier in the process of formulating a NAS by involving sub-national actors in working groups. In addition, some countries address this challenge by supporting networks and partnerships concerned with adaptation planning and decision-making at sub-national levels, and a few use monitoring and assessment schemes as a means of vertical integration as well. For example, the Norwegian government requires municipalities to conduct risk and vulnerability analyses to ensure that they consider climate change adaptation in spatial planning.

Government actions aimed at addressing uncertainties usually start by commissioning impact and vulnerability assessments and launching research programmes that have a close link to adaptation policy-making. These first steps often precede the formulation of a NAS. Later, this is generally complemented by the enlargement of existing (or the establishment of new) scientific advisory bodies and service units that can provide information, guidance tools and advice to policy-makers (frequently also to non-state actors) on an ongoing basis. These 'knowledge brokerage institutions' develop, select and communicate scenarios, provide a range of assessment and decision-support tools and/or organise courses and seminars for policy-makers on a regular basis. In most countries, scientific expertise is also fed into the policy coordination bodies outlined above.

Many governments involve organised non-state stakeholders as fellow experts in temporary coordination bodies (e.g. in workshops) and consult the public in temporary consultation rounds (e.g. in public hearings or online consultation phases) during the formulation of their NAS. Less common approaches of stakeholder involvement include institutionalised forms of stakeholder involvement (e.g. in the Dutch Delta sub-programmes) and networks and partnerships that unite policy-makers from different levels of government and non-state stakeholders (e.g. the regional climate change partnerships in the UK).

6.1.3   Concluding thoughts

Adaptation to climate change is obviously a young and complex policy field that cuts across not only many sectors but also all levels of government. Since policy-makers at all levels (in particular at the local level of government) struggle, inter alia, with a plethora of uncertainties, we are currently witnessing a rather cautious starting phase in which issues of governance (in particular, the strengthening of relatively weak science-policy interfaces) play a dominant role. As vividly demonstrated by climate-related events that have resulted in negative economic consequences, things are likely to change at all levels of government as uncertainties shrink and expected climate change impacts and adaptation pressures materialise. One of the key challenges at that stage will be to shift from a primarily reactive policy response pattern to an anticipatory one, in particular in sectors that demonstrate little interest in climate change adaptation. A major challenge will be to mainstream adaptation in such a way that it will no longer be perceived as an environmental policy issue, but instead as an important sectoral concern. Thus, sectoral adaptation efforts such as the Dutch Delta Programme could provide interesting insights into how to advance adaptation in key sectors.

6.2 Building commitment for adaptation: The right place at the right time?

Julian Wright

6.2.1 Adaptation policy in the UK

The UK is now in a relatively strong position in terms of the country's response to climate change. The PEER report (Swart et al. 2009) comparing national adaptation strategies in Europe describes the UK as a "frontrunner country in many respects: a comprehensive approach, strong scientific and technical support, attention to legal framework, implementation and review". The UK's Climate Change Act 2008 sets out a legal duty for the government to conduct a climate change risk assessment every five years and to develop a national adaptation programme in response to those risks, in addition to endowing the government with the power to require key institutions to report on their climate change risks and adaptation responses. The first UK climate change risk assessment was published in January 2012, and the National Adaptation Programme for England was presented to Parliament in July 2013. Ninety-one key infrastructure providers covering water, flood defence, transport and energy distribution have submitted adaptation plans setting out their responses. A process of co-creation has allowed the development of a comprehensive response – i.e. one that includes action by businesses, NGOs and other institutions in addition to the government.

However, reaching this position has required time and sustained commitment to climate change (and specifically adaptation). This chapter explores the steps that have enabled the UK to achieve its current position, in particular the elements that have promoted the necessary leadership. The following question is addressed: How much of this achievement is based on luck and the specific conditions in the UK, and how much on conditions that may be replicated elsewhere?

6.2.3 Links between science and decision-makers

The British Isles have a relatively strong history of academic exploration of meteorology and climate ranging from John Tyndall's 1859 demonstration of the existence of greenhouse gases to the establishment of the Climatic Research Unit at the University of East Anglia in 1971, the UK Climate Impacts Programme in 1997 and the Tyndall Centre for Climate Change Research in 2000. As in many other countries, however, a scientific understanding of the risks of climate change did not automatically translate into mainstream public and political interest.

To get an idea of how the issue of climate change first started to gain political momentum in the UK, we must look back about 20 years. At that time, the UK's ambassador to the UN, Crispin Tickell, had become interested in the problem of climate change. While on sabbatical at Harvard, he wrote a book on the subject, *Climatic Change and World Affairs* (Tickell 1986). Sir Crispin, together with then-Prime Minister Margaret Thatcher, worked to raise the profile of climate change within the UN and domestically. Thatcher's interest in the issue (no doubt supported by her background in the physical sciences) helped kick-start a UK climate change response. This included backing Sir John Houghton (head of the UK Met Office) and others in the organisation of the UN Intergovernmental Panel on Climate Change and the foundation of the Hadley Centre for Climate Change in 1990 with the aim of developing state-of-the-art global climate models.

The key element in this first step was that several individuals in senior political positions had a sound scientific background. This understanding, combined with a willingness to take a leadership position on climate change in the international arena, automatically translated into a domestic response. Of course, an element of chance was involved: the UK happened to have these specific leaders just at the time when scientific certainty on the anthropogenic link to climate change was developing. Other countries could attempt to promote this type of leadership at an early stage by ensuring that key researchers, particularly those with good communication skills, have access to political leaders. For example, the UK now has a Chief Scientific Advisor for every government department, responsible for working with departmental boards and ministers to ensure that science and engineering are at the core of decision-making. It also helps if political leaders have a willingness to make evidence-based rather than populist policy decisions; unfortunately, this is not always the case.

Although the focus in these early days was on modelling the link between greenhouse gas emissions and global climate change and on the mitigation required, the evidence that was being developed at the Hadley Centre provided a robust basis for the communication of all elements of the climate change debate. Not least, this evidence was able to show that some climate change would occur even if there were dramatic cuts in the level of emissions, and that adaptation had to be considered alongside mitigation. It was fortunate that the Hadley Centre employed scientists skilled in communication. These scientists worked with the civil service to develop an interest in adaptation amongst policy-makers, politicians and research funders. The next key step was the creation of the UK Climate Impact Programme (UKCIP) at Oxford University.

The British government established the UKCIP in 1997 to facilitate a stakeholder-led integrated assessment of climate change impacts (Hedger et al. 2006), funding the programme until April 2012 (when the majority of government adaptation support in England was transferred to the Climate Ready Support Service at the Environment Agency). A key part of the early work of UKCIP was the close involvement of stakeholders in defining research needs and in joint efforts seeking to produce evidence directly applicable to user planning and action. This capacity-building has been useful across the sectors covered by UKCIP, and now similar organisations exist in a number of European countries (for example, KomPass in Germany and the Information Centre on Adaptation in Denmark).

The critical lesson learned is again that close interaction between research and decision-makers is essential for the development of commitment. In particular, the development of an understanding of climate risks, vulnerabilities and adaptation responses needs to be 'owned' by the end users to prevent the process from remaining a solely academic exercise. UKCIP's greatest successes have come in situations in which there has been a clear mandate from a senior level within businesses and other organisations to carry out the work. Unfortunately, the degree of interest from senior levels has differed widely between organisations, a variation that arguably has had little to do with the organisation's level of exposure to climate risk. In order to mainstream adaptation more broadly across society, a new approach was needed – the approach that is currently being delivered under the adaptation provisions of the Climate Change Act.

6.2.3   Binding national legislation for responding to climate change

The Climate Change Act 2008 is a ground-breaking piece of legislation. It is the first legislation in the world to legally obligate a national government to respond to climate change. However, including adaptation in the Climate Change Act was by no means straightforward. By the mid-2000s, public and political interest in climate change had increased dramatically, and there was consensus between the government and the opposition that new legislation was needed to deal with the emissions responsible for climate change. Initially, the draft legislation only covered targets for emission reductions and the mechanisms to achieve them, failing to include adaptation. A few key individuals from organisations with an interest in climate impacts (in particular, Defra, the Environment Agency for England and Wales and Natural England) pushed for the inclusion of simple provisions to legally bind the government to deliver an adaptation response. The debate on the climate change bill coincided with dramatic weather events in the UK, specifically the flooding in the summer of 2007; although these events were not directly attributable to climate change, they helped to maintain a focus on the kinds of risks the UK would increasingly face without an adequate adaptation response. The Climate Change Act that was eventually passed in 2008 includes the commitments described above (conducting a risk assessment every five years and developing a national adaptation programme), as well as the power to require public bodies and statutory undertakers (such as privatised water companies) to produce their own adaptation reports.

This legal duty imposed on the government, and particularly the scrutiny that will be provided by both Parliament and the independent Adaptation Sub-Committee, has arguably produced the greatest change in terms of the level of commitment to adaptation. It has led to a significant commitment to resource the development of National Adaptation Programmes for the countries of the UK. It is recommended that other European countries should also integrate legally binding adaptation duties into mitigation policy as it evolves. To this end, capitalising on the opportunities that events such as floods or droughts may present in order to gain commitments from politicians and senior policy-makers to deal with future climatic challenges can be an effective strategy.

6.2.4 Ownership of adaptation across society fosters implementation

England has recently completed the process of developing its National Adaptation Programme (NAP) under the Climate Change Act, with the publication of a document summarising commitments on 1st July 2013. A particularly important element of the NAP is that it seeks commitment to adaptation action from all parts of society, not only from the government. In order to achieve this, a process of 'co-creation' was carried out. This involved working with key stakeholders across five themes (agriculture and forestry, business, health and well-being, buildings and infrastructure, and natural environment, the sectors identified by the UK Climate Change Risk Assessment as encompassing the most significant risks) in order to build a comprehensive set of adaptation responses. The messaging associated with this co-creation approach was that ownership of the risks and responsibility for the adaptive response rests with the sector concerned. To the greatest extent possible, the role of the government is to provide the appropriate policy and regulatory framework to ensure that these sectors can implement the actions themselves. For example, the government and regulators expect that water companies will address climate change in their investment cycles, but leaves the details of planning to the companies themselves. The overall adaptation response is therefore a combination of sector-led and government-led actions. The co-creation process was supported by senior-level commitment (driven by the Climate Change Act) within the government, which has drawn in equally senior commitment from business and other leaders (for example, via a series of minister-hosted breakfast meetings). It is worth noting that this ministerial commitment exists in large part because of the duties set out in the Climate Change Act and the scrutiny by Parliament that the act entails.

TABLE 6.2.1 Summary of some of the key steps in the establishment of the adaptation response in England and the UK

Date	Significant step in developing adaptation response in the UK
1989	Margaret Thatcher speaks to the UN on climate change
1990	Hadley Centre for Climate Change established
1997	UK Climate Impacts Programme established
2008	Climate Change Act passed
2012	First UK Climate Change Risk Assessment published
2012	Climate Ready Support Service established
2013	National Adaptation Programme published

Source: Author's compilation

6.2.5 Concluding thoughts

The UK has a strong long-term political commitment to adaptation, driven by the legislation of the Climate Change Act. This commitment is beginning to translate into practical delivery of adaptation on the ground.

The steps that have brought the UK to its current position include, in order:

1 High-profile political leadership to initiate understanding of climate change
2 Development of climate change research capabilities
3 Research findings fed into policy-making on mitigation and adaptation
4 Adaptation provisions included in policies being developed for mitigation
5 Co-creation of a practical adaptation response that includes all parts of society.

An interesting consideration is whether this pathway represents a standard procedure that could be followed by other industrialised or perhaps even developing countries. It is worth noting that both chance and design have played a part in the UK's success. The UK has been fortunate in that senior political commitment to climate change was developed at an early stage, and that this level of commitment has been (to date) sustained. There is no doubt that the UK will continue to rely at all levels on pioneering individuals capable of assuming leadership roles. However, the country's response to climate change has been supported at all stages by strong links between scientific research and politics, and more recently by the legally binding commitments set out in the Climate Change Act. Focusing efforts on these elements – building the link to scientific evidence and long-term policy commitments – could help other countries to develop their responses to the challenges of climate change.

6.3 Initiating and sustaining adaptation in the private sector

Magnus Benzie and Oskar Wallgren

6.3.1 Background

The 'private sector' is made up of a wide spectrum of actors. It ranges from small firms with very few employees to large multi-national companies with balance sheets and staff numbers rivalling small countries. The private sector can also be understood as the set of activities in society that are primarily driven by market logic. In that sense, the housing market and the forestry sector both have private-sector characteristics. The functions, risks and needs of private-sector actors are diverse; a range of different governance measures will therefore be required to initiate and sustain adaptation within the private sector.

Who should initiate adaptation in business, and why?

It is not immediately clear why external actors should have an interest in or mandate for initiating adaptation in the private sector. Businesses constantly manage risk and are well placed to decide their own tolerance for risk – unless public safety or the environment is threatened. Competitive pressures from other firms should in theory ensure that businesses will mitigate climate risks and adapt in order to gain advantages over rivals. Adaptation would thus be in a business's self-interest.

The role of government may therefore simply be to ensure that businesses have the information they require to identify appropriate adaptation strategies. However, in cases in which private operations provide public goods or common resources (such as water services, transport, security, or even large-scale employment) or are partially financed through public funds, governments may wish to avoid societal costs from mal- or under-adapted business operations and will therefore take more proactive measures to ensure that adaptation occurs. Governments also have an interest in correcting market failures, i.e. when the allocation of goods, services and risks in the free market can be improved from a collective point of view. There is an ongoing debate regarding the appropriate division of financial responsibility for adaptation between private and public funds – e.g. should private homeowners cover the full cost of climate-proofing their properties or neighbourhoods, or should public money be used to support such work?

Other actors may also have an interest in initiating adaptation in businesses. For example, insurers are exposed to their clients' climate-related risks, and investors (especially long-term investors and pension funds) want their returns to be resilient to changes in the climate. Investors, particularly, will also benefit when companies are able to seize commercial opportunities presented by a changing climate; both groups thus have an interest in ensuring that their clients or companies adapt.

Practical experiences from the UK

Below, we consider three initiatives that seek to initiate adaptation in the private sector (limited here to firms):

1   Government as information providers (to business): UK Climate Projections and UKCIP
2   Government as information receivers (from business): the Adaptation Reporting Power
3   Businesses as information providers (to markets): the Carbon Disclosure Project.

6.3.2   Government as information providers to business

Autonomous adaptation in the private sector will not always take place due to the existence of various well-known barriers, including the long-term nature of climate risks versus shorter-term business planning cycles, uncertainty regarding the timing and scale of impacts, lack of information and data in useable formats, low awareness and capacity and a lack of clarity regarding the division of responsibilities between state and private actors. The main response from governments thus far has been to invest in the provision of information that will help actors (including private actors) to adapt.

There is an obvious role for public investment in climate science, which is a non-rivalrous, non-excludable good that will typically be under-provided by the market. The UK government has made significant investments to enable the production of freely availably climate change scenarios: UKCIP02 and UKCP09.[1] This information has been used by a number of businesses in the UK to assess future climate impacts and risks.

The government also established the UK Climate Impacts Programme (UKCIP), which from 1997 to 2012[2] provided government-funded adaptation support to decision-makers in the public and private sectors. UKCIP developed a range of services and tools for generic business contexts, including the CLARA resource for advisors working with small and medium-sized enterprises (SMEs) and the BACLIAT and speed-BACLIAT business tools, in addition to a range of other resources.[3] BACLIAT is a process-oriented tool that consists of a set of workshops in which businesses can assess possible climate change impacts and explore possible adaptation responses. Other governments offer similar resources that can be used by businesses: for example, the KomPass initiative[4] in Germany and Climate-ADAPT, the European Climate Adaptation Platform.[5] In this way, governments have sought to initiate adaptation by overcoming the barriers of insufficient information and low awareness.

However, not all of this information is appropriate for businesses attempting to plan adaptation. The formats may not be relevant, the information may not be detailed enough, and the content may not be sufficiently geared towards a particular sector or size of enterprise. In summary, awareness and capacity barriers remain in areas of the private sector (CDP 2012, OECD 2011).

Is information enough?

Among businesses, awareness of climate change and adaptation issues has increased, according to various surveys. A poll found that over 70% of larger companies in the UK have begun to consider adaptation, although on the whole private companies demonstrate lower awareness than local government (Ipsos Mori 2010). A different survey, drawing from a sample of more 'climate-aware' businesses in the UK, found that over 80% have identified climate risks, although a much lower proportion (less than half) have incorporated adaptation into their business strategy (CDP 2012). A global survey commissioned by UK Trade and Industry (UKTI) revealed lower levels of awareness, with only 30% of businesses planning and making changes (despite over 90% having experienced extreme weather-related disruptions). However, almost 40% of businesses in the

UKTI survey have seen competitors deriving advantages from adaptation, and close to half of the companies surveyed are carrying out research on the topic of adaptation (EIU 2011). The reasons behind the apparently higher awareness of climate change adaptation needs among UK businesses in comparison to their international peers have yet to be identified.

However, despite raising awareness, the Adaptation Sub-Committee has found no evidence that information provision is leading to "tangible action" on the ground in the UK (ASC 2010). This is in part a reflection of how difficult it can be to observe and characterise adaptation action, particularly within the private sector (for example, developing adaptation indicators for the business and economy sector is highly complex; see Chapter 14 of this manual). It could also be a result of the unwillingness of businesses to 'advertise' their risks and adaptation activities in the same way in which they publicise their mitigation efforts.[6]

Beyond information

It is perhaps too early to say how effective a strategy of information provision will be in spurring private-sector adaptation, but evidence seems to suggest that for action to occur, information provision must typically be paired with other types of activities that engage stakeholders and provide space for deliberation and learning. Governments will have more leverage to initiate processes that will lead to learning where there are already close links between private and public operations – for example, through government procurement, construction, critical national infrastructure (see below) or other sectors in which public regulation plays a vital role.

As with all adaptation, the extent to which change takes place in the private sector will be highly dependent on the processes by which information, guidance and tools are made available to decision-makers. Knowledge is a necessary but insufficient prerequisite for voluntary action, unless incentives (e.g. prices or insurance premiums) push action in a certain direction. Research has shown how arenas for meeting and learning (e.g. seminars, multi-stakeholder processes, public inquiries) have played key roles in integrating climate adaptation into policy, by providing spaces in which scientists and business practitioners can debate, share and co-create knowledge (Ulmanen et al. 2012). In the UK, there are only a few examples of this kind of sustained, close collaboration between scientists/ researchers, businesses and stakeholders. UKCIP has attempted this approach within its limited remit as a boundary organisation, via the Changing Climate for Business partnerships and other longer-term research programmes such as the ARCC project.[7] However, in these instances, businesses are usually involved only as 'risk sharers' (i.e. they have a direct stake in the risks presented by climate change) rather than innovators and solution providers. In a more active, problem-solving role, the private sector's contribution could be significant (EIU 2011) and may currently be undervalued.

6.3.3 Government as information receivers from business

A key innovation in the UK's adaptation strategy is the Adaptation Reporting Power[8] (ARP), which enables the government to request assessments of current and future climate risks and adaptation plans from critical infrastructure and service providers, many of which are private actors. The ARP enables the government to assess the level of preparedness for climate change among the enterprises covered. It also provides an overview of likely adaptation measures and insight into where further government action may be required.[9] This information may be of interest not only to the government itself, but also to other non-government actors (incl. the business community) and citizens at large.

In its first application during 2010/11, the ARP was a catalyst for many organisations to take adaptation seriously. A common finding was that the ARP process "led to greater visibility of climate change risks at the organisational and board level and has enabled management of these risks within corporate risk management processes" (Defra 2012). The ARP is an example of a government initiating adaptation in business operations that provide public benefits.

A parallel example from the United States is provided by the Securities and Exchange Commission, which issued revised guidance in 2010 requiring companies to include information on "material risks" related to climate change in their financial reports. The impact of the SEC's guidance on company reporting has been questioned (Riedel 2011), but it is an example of a minor regulatory adjustment seeking to initiate adaptation through improved risk disclosure.

6.3.4 Business as information providers to markets

The Carbon Disclosure Project (CDP) has grown to become the dominant method used by global businesses to report their greenhouse gas emissions and exposure to risks from physical climate change. CDP currently works with 655 institutional investors holding US$78 trillion in assets.[10] The CDP is built on investors' need for information on companies' exposure to future regulation or climate impacts.

The CDP initiative provides an example of a demand-driven response by 'the market'. It encourages companies to disclose their climate risks (and implicitly, their plans or measures taken to adapt), which drives awareness and initiates adaptation in businesses wishing to attract or maintain investment. As well as demonstrating to investors their awareness of the potential climate risks – and the existence of plans to mitigate and manage those risks – companies can (and some do) also use the CDP as a medium for communicating potential climate-driven commercial opportunities to their investors. Disclosure initiatives of this kind can be seen as the de facto mainstreaming of adaptation into existing financial market incentive structures and information streams (cf. Table 6.3.1).

TABLE 6.3.1 Strengths and weaknesses of the ARP and CDP schemes for initiating adaptation in the private sector

Adaptation Reporting Power	Carbon Disclosure Project
Very limited coverage: 91 reporting authorities addressed in 2011, mostly limited to infrastructure operators	3,715 respondents worldwide in 2011, including 81% of the Global 500 from a range of sectors
Reluctance among some reporting authorities to disclose commercially sensitive information (Benzie et al. 2010) and few mechanisms by which the government can verify results or force the disclosure of all risks	Respondents choose what they disclose and may exaggerate adaptation efforts or under-state levels of risk
A relatively thorough assessment of ARP reports was conducted by risk experts[a]	CDP process does not involve standards for climate risk assessment, although initiatives for global disclosure standards are underway[b]
ARP reports contain detailed information on risks, costs and adaptation options	Demand-driven: investors can improve the process if they are dissatisfied with the results
Report documents available to download, but access to data is not easy or user-friendly	Verification is 'open-source'; the database is freely available. The CDP produces several analytical reports of results with partner organisations. Third parties have developed products based on CDP data, e.g. the Acclimatization Index

Source: Authors' compilation

a   Carnfield University conducted an evaluation of risk assessments in the ARP reports, available at: http://archive.defra.gov.uk/environment/climate/documents/interim2/report-framework.pdf
b   For more information, see the Climate Disclosure Standards Board (an initiative seeded by the CDP) – for example, the Consistency Project. More information is available at: http://www.cdsb.net/ and https://www.globalreporting.org

6.3.5   Concluding thoughts

Governments should continue to support businesses by investing in scientific research and the effective communication of climate change, climate impact and adaptation science. Bridging the gap between science and on-the-ground decision-making remains a challenge, but there are plenty of successful experiences to draw on, not least from the UK. Information must be matched by efficient approaches for awareness-raising and capacity-building, including investments in the co-production of adaptation knowledge in collaboration with business decision-makers and researchers. Deliberate efforts have to be made to foster and sustain processes of learning. Climate change adaptation is a relatively 'new' issue for which governance regimes, regulation and cost-sharing models have only recently begun to evolve. Most jurisdictions lack principles governing cost-sharing between private and public actors, or such principles exist only in a rudimentary state. Regular dialogue between governments, business and civil society is thus crucial, both to clarify responsibilities and to maximise the public benefits of private-sector adaptation.

The exchange and disclosure of information on climate risks and adaptation between businesses and other actors – namely, governments and investors – has the potential to initiate adaptation. Both ARP- and CDP-style initiatives can be used to facilitate dialogue between relevant stakeholders on the importance of climate risks, how to address them and how to seize commercial opportunities. This will be important in order to sustain adaptation as a learning process in the private sector and thereby prevent adaptation from becoming a compliance or one-off issue for businesses to deal with. The main advantage of risk disclosure initiatives is that they mainstream adaptation into existing incentive structures within the market framework.

While there are reasons to believe that large parts of the private sector will adapt autonomously, governments should devote special attention to those parts of the private sector in which inadequate adaptation would result in societal costs and the loss of public goods. An important prerequisite is investment in the provision of climate change information, but governments should go further to offer support during the adaptation process. Institutionalising climate risk disclosure is a 'light touch' way to initiate adaptation in the private sector. The UK's approach offers one possible example to countries that have not yet considered the role of government (or civil society and investors) in private sector adaptation.

Acknowledgement

The authors wish to thank Mistra (the Swedish Foundation for Strategic Environmental Research) for its support through the Mistra-SWECIA Programme in the preparation of this article. All views expressed are those of the authors alone.

6.4 Lessons learned for the management and governance of adaptation

Andrea Prutsch

This chapter on initiating adaptation and ensuring commitment and proper management touches upon a number of important points concerning the start of an adaptation process as well as possible roles for governments and businesses in adaptation.

As shown by Steurer and Bauer (6.1), with climate change accelerating its pace in recent decades, a growing number of governments have grasped the urgency of the need to facilitate adaptation and have begun to use various policy instruments to steer society towards better adaptation. Thus far, governments seem to be primarily concerned with establishing a knowledge base and new governance approaches for climate change adaptation (including adaptation strategies) rather than with concrete adaptation measures, which are generally planned for later development. In many countries, more urgent issues related to the economic and financial crises have pushed climate concerns further down the political agenda. The relatively slow pace of adaptation policy-making may change when uncertainties (e.g. with regard to the frequency of extreme weather events) decrease and/or are better understood, or when expected climate change impacts materialise.

The contribution by Julian Wright (Section 6.2) revealed insights into the history of adaptation in the UK and the conditions that resulted in a strong commitment to adaptation in that country. A number of lessons learned from the UK can likely be generalised to other countries. These are:

- A strong (preferably formalised) link between scientific research and politics, particularly at higher levels, is important in kick-starting any climate change response.
- A strong response to climate change can be initiated by relatively few people, but these individuals must be in positions of real authority and have the dynamism to engage others more broadly.
- Building leadership on adaptation is easier if a country or region has already achieved political interest in mitigation, from which an understanding of the need for adaptation can more easily follow.
- Clear long-term policy commitments, particularly those that are legally binding for governments (and, where appropriate, for critical service providers in the public and private sectors), should be the aim of any country that seriously intends to increase and sustain its climate resilience.

The UK is certainly a frontrunner in adaptation, with a comparatively long history of more than 15 years of experience. However, in most countries across Europe, climate change adaptation as a policy field is still in its infancy.

Nevertheless, the ownership of adaptation across society – not just in the government – is a necessity in order to build a comprehensive adaptation response. The third case in this chapter, provided by Benzie and Wallgren (6.3), focuses on the governmental role of supporting the private sector in adaptation, as climate change will have a range of impacts on businesses. Impacts are expected to affect all kinds of businesses, both large firms and SMEs, by disrupting business operations, causing property damage, disrupting supply chains and infrastructure (leading to higher costs for maintenance and materials) and raising prices. In other cases, climate change may also offer new business opportunities (e.g. for products and services that will help people to adapt) in the form of expanding market share, wealth creation in communities (innovation and job creation) and access to new financial streams (increased public funding and financial products and services).

The current state of the art in adaptation in the private sector shows that proactive measures may be required from governments, researchers, boundary organisations and non-government actors in order to ensure progress in the initiation of adaptation. From the case on private-sector adaptation (6.3), we can derive a number of lessons learned:

- Governments can act as providers of climate information to businesses, but this alone is unlikely to be sufficient to initiate adaptation.
- Governments can kick-start adaptation processes by requesting information on risks and adaptation plans from businesses, particularly those that deliver public benefits.
- Businesses can be encouraged by investors and civil society to disclose information on their climate risks to the market, thus providing incentives to adapt (and thereby reducing risks).
- Information disclosure of this kind holds promise as a 'light touch' governance option, but neither information provision nor risk disclosure guarantees that effective or appropriate adaptation will take place. Sustained efforts to facilitate learning and adaptation in certain areas of the private sector are likely to be required.

Notes

1 As well as related statistical tools, such as the Weather Generator and a user interface for climate projections, see the UK Climate Projections website for more information: http://ukclimateprojections.defra.gov.uk/.
2 The UK Climate Impacts Programme ended in 2012 when the Environment Agency became the delivery body for adaptation in England (with different arrangements in the other devolved administrations). UKCIP, hosted at the Environmental Change Institute at Oxford University, continues to provide free tools and support.
3 For more information, see UKCIP's business website (http://www.ukcip.org.uk/business/) and tool information (http://www.ukcip.org.uk/clara/ and http://www.ukcip.org.uk/bacliat/).
4 For more information, see: http://www.anpassung.net/.
5 The Climate-ADAPT platform is available at: http://climate-adapt.eea.europa.eu/.
6 Action on mitigation is often associated with positive brand value and a corporate image of responsibility and sustainability. There appears to be less brand value in visible adaptation. In fact, alerting customers to the fact that a company has risks and therefore needs to adapt may have negative connotations for brand value, as well as for investors.
7 For more information, see: http://www.ukcip.org.uk/business-case-studies/.
8 The Adaptation Reporting Power is part of the Climate Change Act 2008. It addresses organisations with functions of a public nature and statutory 'undertakers' that are licensed to undertake statutory functions on behalf of the government, such as infrastructure operators (e.g. airports, ports, rail networks, etc.) and some infrastructure service providers, including utilities (e.g. water and energy). The ARP was first enacted with a deadline of 2011. ARP guidance encourages reporting authorities to base their assessments on their existing risk management processes and thereby aims to mainstream adaptation in business practices. For more information, see: http://www.defra.gov.uk/environment/climate/sectors/reporting-authorities/.
9 The results of the ARP process have been used to inform the first national climate change risk assessment, and they will be used to inform the design of the first National Adaptation Programme (due in 2013), including the strategy to facilitate adaptation in the private sector.
10 For more information, see https://www.cdproject.net.

References

Adaptation Sub-Committee (ASC) (2010) *How Well Prepared is the UK for climate change?* UK.

Adger, W.N., Arnell, N.W. and Tompkins, E.L. (2005) 'Successful adaptation to climate change across scales', *Global Environmental Change* 15 (2), pp. 77–86.

Agrawala, S. and Frankhauser, S. (ed.) (2008) *Economic Aspects of Adaptation to Climate Change. Costs, Benefits and Policy Instruments*, Paris: OECD.

Amundsen, H., Berglund, F. and Westskog, H. (2010) 'Overcoming barriers to climate change adaptation – a question of multilevel governance?', *Environment and Planning C: Government and Policy* 28 (2), pp. 276–289.

Baldwin, R. and Cave, M. (1999) *Understanding Regulation: Theory, Strategy, and Practice*, Oxford, UK: Oxford University Press.

Bauer, A. and Steurer, R. (forthcoming) 'Multi-level governance of climate change adaptation through regional partnerships in Canada and England', *Geoforum*.

Bauer, A., Feichtinger, J. and Steurer, R. (2011) *The Governance of Climate Change Adaptation in Ten OECD Countries: Challenges and Approaches*, InFER Discussion Paper 1/2011, Vienna.

Bauer, A., Feichtinger, J. and Steurer, R. (2012) 'The governance of climate change adaptation in ten OECD countries: challenges and approaches', *Journal of Environmental Policy and Planning* 14 (3), pp. 279–304.

Benzie, M., Pooley, M., Siddiqi, A. and Horrocks, L. (2010) *Engaging with Transport Authorities in the South East on the Adaptation Reporting Power*, Final Report to CSE and Defra, Oxfordshire, UK.

Berkhout, F. (2005) 'Rationales for adaptation in EU climate change policies', *Climate Policy* 5 (3), pp. 377–391.

Biesbroek, G.R., Swart, R.J., Carter, T.R., Cowan, C., Henrichs, T., Mela, H., Morecroft, M.D. and Rey, D. (2010) 'Europe adapts to climate change: comparing National Adaptation Strategies', *Global Environmental Change* 20 (3), pp. 440–450.

Burton, I., Diringer, E. and Smith, J. (2006) *Adaptation to Climate Change: International Policy Options*, Toronto, Canada: University of Toronto.

Carbon Disclosure Project (CDP) (2012) *Insights into Climate Change Adaptation by UK Companies*, a report for Defra by the Carbon Disclosure Project, London, UK.

Cimato, F. and Mullan, M. (2010) *Adapting to Climate Change: Analysing the Role of Government*, London, UK, Department for Environment, Food and Rural Affairs.

Defra (2012) *Adapting to Climate Change: Helping Key Sectors to Adapt to Climate Change*, Government Report for the Adaptation Reporting Power, UK.

Economist Intelligence Unit (EIU) (2011) *Adaptation to an Uncertain Climate: A World of Commercial Opportunities*, a report for UKTI, UK: Crown Copyright.

Frankhauser, S., Smith, J.B. and Tol, R.S.J. (1999) 'Weathering climate change: some simple rules to guide adaptation decisions', *Ecological Economics* 30 (1), pp. 67–78.

Hedger, M., Connell, R. and Bramwell, P. (2006) 'Bridging the gap: empowering decision-making for adaptation through the UK Climate Impacts Programme', *Climate Policy* 6 (2), pp. 201–215.

Hood, C. (1986) *The Tools of Government*, Chatham, UK: Chatham House Publishers.

Hulme, M., Neufeld, H., Colyer, H. and Ritchie, A. (2009) *Adaptation and Mitigation Strategies: Supporting European Climate Policy. The Final Report from the ADAM Project*, Norwich, UK: Tyndall Centre for Climate Change Research, University of East Anglia.

IPCC (2001) *Synthesis Report. A Contribution of Working Groups I, II, and III to the Third Assessment Report of the Intergovernmental Panel on Climate Change*, Cambridge: Cambridge University Press.

IPCC (2007) 'Summary for policy-makers', in Parry, M.L., Canziani, O.F., Palutikof, J.P., van der Linden, P.J. and Hansen, C.E. (ed.) *Climate Change: Impacts, Adaptation and Vulnerability. Contribution of Working Group II to the Fourth Assessment Report of the Intergovernmental Panel on Climate Change*, Cambridge: Cambridge University Press, pp. 7–22.

Ipsos Mori (2010) *Climate Change Adaptation: A Survey of Private, Public and Third Sector Organisations.*

Keskitalo, E.C.H. (2010) *Developing Adaptation Policy and Practice in Europe: Multi-level Governance of climate change*, Dordrecht: Springer.

Massey, E. and Bergsma, E. (2008) *Assessing Adaptation in 29 European Countries*, Report W-08/20, Amsterdam: Institute for Environmental Studies, Vrije Universiteit.

Mees, H., Driessen, P., Runhaar, H. and Stamatelos, J. (2012) *Governance Arrangements for Climate Adaptation: The Case of Green Roofs for Storm Water Retention in Urban Areas*, Symposium 'The Governance of Adaptation', Amsterdam.

Mysiak, J., Henrikson, H.J., Sullivan, C., Bromley, J. and Pahl-Wostl, C. (ed.) (2010) *The Adaptive Water Resource Management Handbook*, UK: Earthscan.

Nelson, D.R., Adger, W.N. and Brown, K. (2007) 'Adaptation to environmental change: contributions of a resilience framework', *Annual Review of Environment and Resources* 32, pp. 395–419.

OECD (2011) *Private Sector Engagement in Adaptation to Climate Change: Approaches to Managing Climate Risks*, OECD Environment Working Papers, No. 39, OECD Publishing.

Rayner, T. and Jordan, A. (2010) 'Adapting to a changing climate: an emerging European Union policy?', in Jordan, A. et al. (ed.) *Climate Change Policy in the European Union: Confronting the Dilemmas of Mitigation and Adaptation?*, Cambridge: Cambridge University Press, pp. 145–162.

Ribeiro, M.M., Losenno, C., Dworak, T., Massey, E., Swart, R., Benzie, M. and Laaser, C. (2009) *Design of Guidelines for the Elaboration of Regional Climate Change Adaptations Strategies*, Final report, Study for European Commission – DG Environment, Vienna: Ecologic Institute.

Riedel, A. (2011) *Companies Should Recalibrate Climate Risk Disclosure towards Adaptation Risk*, Acclimatise paper.

Smith, J.B., Vogel, J.M. and Cromwell III, J.E. (2009) 'An architecture for government action on adaptation to climate change, an editional comment', *Climatic Change* 95 (1), pp. 53–61.

Swart, R., Biesbroek, R., Binnerup, S., Carter, T.R., Cowan, C., Henrichs, T., Loquen, S., Mela, H., Morecroft, M., Reese, M. and Rey, D. (2009) *Europe Adapts to Climate Change, Comparing National Strategies*, Helsinki: Partnership for European Environmental Research.

Tickell, C. (1986) *Climatic Change and World Affairs*, 2nd edn, Lanham, MD: University Press of America Inc.

UKCIP (2005) *Identifying Adaptation Options*, UK Climate Impacts Programme, Oxford.

Ulmanen, J., Swartling, Å.G. and Wallgren, O. (2012) *Climate Change Adaptation in Swedish Forestry: A Historical Overview, 1990–2012*, SEI Working Paper Project Report (in press), Stockholm: SEI.

7

BUILD KNOWLEDGE AND AWARENESS

Torsten Grothmann, Andrea Prutsch,
Inke Schauser, Sabine McCallum and
Rob Swart

Explanation of the guiding principle

Due to the complexity of climate change and its potential impacts, building knowledge and updating information regularly based on new research results and practical experience require particular attention (Grothmann 2011). Open access to information and the communication of scientific and practical knowledge, among other elements, are essential preconditions for good adaptation. Building knowledge (cf. Chapters 5, 9 and 10) also raises awareness of climate change and adaptation needs and helps to support the implementation of adaptation actions (Smith et al. 2009).

The awareness of climate change and adaptation also has an emotional side. Emotion research shows that emotions are necessary for practical decision making and for the motivation to act (Roeser 2012). Both pleasant emotions (e.g. hoping for safety from climate impacts in the future) and unpleasant emotions (e.g. worrying about climate change impacts) can contribute to adaptation action. When creating concern over climate change impacts, one should always communicate adaptation options at the same time to minimise the impacts, so that recipients will react to their worry with adaptation action and not with fatalistic strategies or denial of the existence of the climate problem (Grothmann & Patt 2005).

Decision-makers, stakeholders and the general public should have access to reliable information about potential climate change impacts, vulnerabilities, uncertainties, adaptation options and tools, good-practice examples and the advantages of adaptation, potential trade-offs, synergies and conflicts with mitigation and existing policies, to the extent such information is available. Researchers, policy-makers and practitioners should jointly bundle information sources, improve access to understandable and usable information (Glicken 2000) and fill knowledge gaps by means of targeted assessments, especially for users at the local level.

It is important for adaptation actors to agree upon guidelines for communication and educational efforts that will:

- differentiate between stakeholder groups that need different information and communication instruments (Grothmann 2011) (cf. Chapter 8),
- develop a common language among stakeholders to support stakeholder cooperation (cf. Chapter 8),
- raise awareness among the general public (e.g. by integrating climate change and adaptation information into school curricula),
- provide information in the simplest, most understandable and most usable form possible but also maintain necessary complexity,
- include ways to communicate gaps in and uncertainties about the knowledge of climate change, climate change impacts and adaptation options without discouraging adaptation efforts (cf. Chapter 11), and
- organise communication, also in the form of mutual dialogue and learning between different stakeholders and scientists (Grothmann 2011, Lebel et al. 2010) (cf. Chapter 8).

Several communication formats are available. Personal consultations (e.g. public talks, small group discussions) can build trust, facilitate mutual learning and effectively stimulate adaptation action. Mass media and internet communication are well suited for eliciting a political debate and for knowledge transfer and raising awareness among a wide range of people (Grothmann 2011). Internet platforms allow easy access to information (e.g. climate change impact scenarios and good-practice examples of adaptation actions), but require dedicated long-term resources to keep them up-to-date; significantly, these platforms lack the face-to-face interactions that are often required to interpret the relevance of knowledge in a specific context.

Chapter overview

In the following chapter, four cases illustrate the practical realisation of the considerations described above on how to build knowledge and awareness through the communication of knowledge. The first case (Section 7.1) presents the European Climate Adaptation Platform (Climate-ADAPT), an internet platform that provides information on climate change, vulnerabilities and adaptation strategies in Europe. The second case (7.2) emphasises the importance of target group-specific communication, primarily in connection to supporting the adaptation actions of municipalities in Denmark by the national government. The third case (7.3) shows how climate change projections have been made useful for regional adaptation plans in France by means of an intense dialogue between climate scientists and regional users. The fourth case (7.4) also addresses the regional level, but stresses ways to stimulate adaptation action among business actors and the general public in northwest Germany by the use of target group-specific communication tools (including movies).

7.1   The European Climate Adaptation Platform (Climate-ADAPT)

André Jol and Stéphane Isoard

7.1.1   Short general background

Current status

The European Climate Adaptation Platform (Climate-ADAPT[1]) (EEA 2012a), launched on 23 March 2012, is a joint activity of the European Commission (DG CLIMA, DG Joint Research Centre and other DGs) and the European Environment Agency (supported by its European Topic Centre on climate change impacts, vulnerability and adaptation, ETC CCA[2]). Climate-ADAPT is a publicly accessible, web-based platform designed to support policy-makers at the EU, national, regional and local levels in the development of climate change adaptation measures and policies. As one of the first international information systems on adaptation managed by a public institution, it helps users to access and share information on:

- Expected climate change in Europe
- Current and future vulnerability of regions and sectors
- EU, national and transnational adaptation strategies and actions
- Adaptation case studies and potential adaptation options
- Tools that support adaptation planning.

Climate-ADAPT organises information under the following main entry points:

- Adaptation information (observations and scenarios, vulnerabilities and risks, adaptation measures, national adaptation strategies, research projects)
- EU sectoral policies (agriculture and forestry, biodiversity, coastal areas, disaster risk reduction, financing, health, infrastructure, marine issues and fisheries, water management)
- Transnational regions, countries and urban areas
- Tools (Adaptation Support Tool, Case Study Search Tool, Map Viewer).

Background

The European Commission's White Paper on climate change adaptation (EC 2009) presented the framework for adaptation measures and policies intended to reduce the EU's vulnerability to the impacts of climate change, focusing on:

- building a stronger knowledge base
- taking climate change impacts into consideration in key EU policies ('mainstreaming')

- financing climate change policy measures
- supporting international efforts on adaptation by helping, for example, non-EU countries to improve their resilience and capacity to adapt to climate change.

The White Paper emphasised that information on climate change impacts and vulnerability and on the costs and benefits of adaptation measures in Europe remains scarce and fragmented and that more spatially detailed information would be needed to develop adequate adaptation strategies. It called for a European Climate Change Impacts, Vulnerability and Adaptation Clearinghouse to be established by 2012. The name of this endeavour was changed in early 2012 to 'European Climate Adaptation Platform (Climate-ADAPT)', hereafter referred to as 'Climate-ADAPT' or 'the platform'.

Climate-ADAPT is intended to be a tool with added value in comparison to the national adaptation knowledge platforms currently in place or being developed in various countries across Europe. Climate-ADAPT aims to support cooperation across countries and regions (transnational, between neighbouring countries, interregional in areas with similar characteristics (such as mountainous regions) and/or interdependencies, e.g. transnational river catchments). Climate-ADAPT can also help to identify gaps in available information and thus support the potential improvement efforts of countries and/or at the EU level (e.g. through research).

The focus will initially be on governmental decision-makers (and organisations providing support, such as agencies, boundary organisations and researchers) working on the development and implementation of adaptation strategies or actions. The geographical scope of Climate-ADAPT is the EEA32 member countries.[3] In a later phase, the scope may be extended to include the seven EEA cooperating countries. Although the platform focuses on Europe, it provides access to information that is also likely to be useful for other regions.

7.1.2   Main content of Climate-ADAPT

Mainstreaming adaptation in EU sector policies

Mainstreaming climate change adaptation in EU policies is one of the pillars of the European Commission's 2009 White Paper. The Europe 2020 strategy for smart, sustainable and inclusive growth (EC 2011a) states: "We must also strengthen our economies' resilience to climate risks, and our capacity for disaster prevention and response". To achieve this goal, adaptation actions need to be integrated into the EU's sector policies and focus on the areas in Europe that are most vulnerable to climate change (e.g. mountains, coastal areas, flood-prone river areas, the Mediterranean and the Arctic). Climate-ADAPT provides an overview of existing mainstreaming in EU policies[4] in the following areas: water resources, the marine environment, coastal areas, biodiversity, agriculture, forestry, infrastructure, the urban environment, environmental assessment and disaster risk reduction.

Although mainstreaming is taking place, there is a need for further action across the EU, and the European Commission has developed a comprehensive adaptation strategy. The underlying impact assessment and the strategy were published in April 2013 (EC 2013, EEA 2013).

Mainstreaming is also very important within the multiannual financial framework (MFF), which includes a proposal from the Commission (EC 2011b) to increase the share of resources for climate change mitigation and adaptation combined to 20% of the EU budget. The MFF includes proposals for major funding policies and financial instruments, such as structural funds, the EU Common Agricultural Policy, Trans-European Networks (transport, energy), Horizon 2020 (the future research policy) and the legislative and non-legislative developments related to these instruments. Following consultations with the Member States and the European Parliament in 2012 and 2013, final decisions on the MFF were taken in 2013.

Adaptation strategies

EEA member countries are at different stages in preparing, developing and implementing adaptation strategies.[5] This development depends on the magnitude and nature of the observed impacts, the assessment of current and future vulnerability and the country's capacity to adapt. A number of countries have national strategies in place (Austria, Belgium, Denmark, Finland, France, Germany, Hungary, Ireland, Lithuania, Malta, the Netherlands, Portugal, Spain, Sweden, Switzerland and the United Kingdom, as of Spring 2013) and some of these have additionally adopted national action plans. Most EEA member countries have voluntarily submitted information on their national strategies and plans, assessments, climate services and priority actions to Climate-ADAPT.

Adaptation Support Tool

The aim of the Adaptation Support Tool[6] is to assist users in adaptation planning, based on the policy cycle (highlighting that climate change adaptation is an iterative process). The steps of this cycle can be re-considered periodically to ensure that adaptation decisions are grounded in up-to-date data, knowledge and policies. This iterative process will also allow monitoring and the timely assessment of successes and failures and encourage adaptive learning.

Adaptation measures

Adaptation measures[7] seek to reduce climate risk to an acceptable level, taking advantage of any positive opportunities that may arise. Adaptation measures can focus on:

- Accepting the impacts and bearing the losses that result from risks
- Offsetting losses by sharing or spreading the risks or losses
- Avoiding or reducing exposure to climate risks
- Exploiting new opportunities.

There are also other ways of considering adaptation in terms of the types of actions that can be undertaken, e.g. temporary, managerial, technical and strategic. In Climate-ADAPT, potential adaptation options can be explored by selecting a specific climate impact and/or adaptation sector of interest. Thus far, the primary focus is on adaptation in the sectors of water resource management and flood risk prevention; options in other sectors will be added in the future.

Indicators

A range of indicators[8, 9] are included for past and projected climate change and impacts. These were published by the EEA in 2012 in association with the report 'Climate change, impacts and vulnerability in Europe 2012' (EEA 2012b) and are regularly updated. The indicators include changes in the climate system and in the cryosphere (glaciers, snow and ice) and impacts on the following environmental systems: oceans and the marine environment, coastal zones, inland waters (water quantity and quality), terrestrial ecosystems, and biodiversity and soil. In addition, the EEA published a report on adaptation actions (EU, transnational, national, local) in April 2013 (EEA 2013).

Uncertainty guidance

This guidance[10] seeks to help decision-makers to understand the sources of uncertainty in climate information that are most relevant for adaptation planning. It also provides suggestions for dealing with uncertainty in adaptation planning and for the communication of uncertainty. The guidance is organised around three main topics and addresses a list of questions:

- *What is meant by uncertainty?* (What are the sources of uncertainty in adaptation planning? Why is there uncertainty in climate information? Which emission scenarios form the basis of most climate projections?)
- *How can uncertainty be factored into adaptation decision making?* (What are the possible ways to account for uncertainty in decision making? What are different types of adaptation options? What are no-regrets adaptation measures? How are uncertainties quantified? Which scenario should be used for adaptation planning?)
- *How can uncertainty be communicated?* (What are the lessons in communicating uncertainty? How can uncertainty be presented? How are uncertainties communicated in European national portals?)

Case study search tool

Case studies are reports of practical adaptation actions that have been implemented, with the results described and, where feasible, also analysed. Using the Case Study Search Tool,[11] the user determines the location of interest and the tool will display case studies, differentiating between case studies within and outside the biogeographical region. Thus far, Climate-ADAPT includes adaptation case studies from the EU project OURCOAST (EC 2012a) and from EU-funded Life+ projects (EC 2007). Several ongoing or (almost) finalised EU-funded and other projects could provide additional contributions, in particular projects on urban adaptation and ecosystem-based adaptation, and programmes and projects involving transnational cooperation.[12]

Database and map viewer

A variety of online interactive tools providing support for adaptation are included,[13] among them a Search and Discover function[14] for integrated searches of the Climate-ADAPT database (and the full website). The database contains quality-checked information on the following content items: publications and reports, information portals, guidance documents, tools, maps, graphs, datasets, indicators, research and knowledge projects, adaptation options, case studies and organisations.

The platform also includes a Map Viewer,[15] which provides projections of climate change impacts, vulnerability and risks from a range of projects and organisations, including ClimWatAdapt,[16] ESPON Climate,[17] JRC-IES[18] and ENSEMBLES.[19] It is expected that through the emerging national climate services in many countries and the EU programme Copernicus (Global Monitoring for Environment and Security),[20] further observations and re-analysis data with longer time series and improved climate change projections will become available in the future.

Furthermore, the platform includes several other tools, based in particular on impact assessment and human health.

7.1.3 Where to go next?

During the development of Climate-ADAPT, the Commission and the EEA put a great deal of effort into involving both the providers and the intended users of the information. This was challenging because of the many different types and perspectives of organisations involved. Moreover, some providers can also be users, e.g. countries that have provided their national information may also be using the adaptation support tool in Climate-ADAPT or comparing their national approaches to those of other countries.

Advice on the content was provided by a technical Working Group on Knowledge Base (WGKB), consisting of the European Commission, the EEA, the ETC CCA and country representatives (usually from environment agencies), researchers from a range of EU-funded FP projects and various other organisations (e.g. WHO Europe, ECDC). Guidance from a policy perspective was provided by the Adaptation Steering Group, chaired by DG CLIMA and consisting of representatives from the European Parliament, EU Member States (usually from ministries of the environment) and other stakeholders, e.g. environmental and sectoral and/or business NGOs. In addition, the network of Environmental Protection Agencies and the EEA member countries' national reference centres provided comments. Thus, the key stakeholders (with the possible exception of industry) were sufficiently involved, and most of the advice provided could be incorporated.

How and to what extent the business sector should be involved will be addressed in the future in further discussions of the platform with users and providers. Industry (e.g. manufacturing) may be (potentially) affected by climate change (see also Section 6.3), but businesses may also be able to provide certain types of adaptation solutions (e.g. engineering companies or the insurance sector). Businesses sometimes have their own networks and information systems, and whether Climate-ADAPT can and should provide specific support should be assessed.

Several countries have national information systems or portals in place to a varying extent covering types of information similar to Climate-ADAPT, but they are generally focused more on national issues and present information in the national languages. Further discussion will take place with these countries to enhance synergies with Climate-ADAPT and further clarify the added value of Climate-ADAPT, e.g. focusing on EU and transnational issues and common issues across countries.

Much work will be required to keep the platform continuously updated with the large amount of new scientific, methodological and policy information that is becoming available. The experience thus far shows that the approach of 'sharing information' works, but direct outreach to the providers of information is also needed.

Ongoing and new research can improve the various tools on the platform. Monitoring and the evaluation of the effectiveness and efficiency of adaptation strategies and actions will become increasingly important (see Chapter 14), and the platform can be improved in this respect. The case studies should be enhanced, in particular by including experiences from the increasing number of cities across Europe actively involved in adaptation.

Climate-ADAPT is amongst the most frequently visited thematic information pages on the EEA's website. Visitors most often enter using a direct link, then through search engines and third by means of the EEA website. Visitors come from countries across Europe and beyond, although eastern European countries are underrepresented among visitors. These statistics demonstrate a healthy interest in the platform, and further work will be done to assess which parts are of the greatest interest and how outreach to users can be improved.

After the publication of the EU adaptation strategy by the European Commission (EC 2013), further discussion will take place on the possible revision and updating of national information and on how information could be made more comparable across countries, including consideration of indicators. It is anticipated that countries will be requested by the European Commission to update their respective country pages once per year. Furthermore, the MFF (adopted in 2013) includes opportunities for EU funding for climate change adaptation, and an analysis will be conducted to determine how such information can be included in Climate-ADAPT.

Finally, regarding the governance of Climate-ADAPT, a steering group has been established (composed of DG CLIMA, the EEA and the JRC) to steer the platform's strategic direction, while the regular management and maintenance of the system is organised by the EEA (supported by its ETC CCA). It is expected that contributions will be provided by the JRC in the form of reference data sets, maps and information sources and by the Commission with regard to the relevant EU sector policy developments.

7.2   Awareness of climate change adaptation in Denmark: How to address a target group

Louise Grøndahl

7.2.1   Background: The Danish adaptation strategy and its information campaign

The national adaptation strategy of Denmark was adopted in 2008. The strategy includes a description of the vulnerability of those sectors in which climate change is expected to have significant consequences (The Danish government 2008). The focus is on adaptation measures already underway and what will be necessary to advance the process. The strategy emphasises that the design of adaptation efforts should take into consideration the consequences of climate change, the probability of their occurrence and the costs of prevention by means of autonomous adaptation (i.e. adaptation not supported by the government). This includes a focus on what will be attainable in the individual sectors within the next 10 years. The designed measures should be scientifically, technically and socio-economically appropriate for implementation within the given period. The strategy is based on the concept that adaptation is a long-term process, and that it is still uncertain what the consequences of climate change will be and how soon they will take effect.

The strategy features an information campaign aimed at various target groups (municipalities, citizens and business). This campaign includes the continued development of a web portal with the goal of ensuring that climate change will be incorporated into planning and development so that public authorities, businesses and citizens have the best possible basis for determining whether, how and when climate change should be taken into account. The web portal, launched in 2009, is a one-stop shop (single point of contact) for obtaining various information on climate change, its impacts and adaptation options. The portal provides relevant data as well as experiences with respect to how adaptation measures have been implemented in different sectors and by various target groups (i.e. citizens, municipalities and businesses). The aim of the web portal is to satisfy the users' needs for information, tools and guidelines on climate change adaptation. Among the main target groups of the information campaign are municipalities.

Following the adoption of the adaptation strategy in 2008, an action plan for a climate-proof Denmark was launched in 2012 (The Danish government 2012). This action plan describes a series of initiatives to establish sustainable, green and innovative climate change adaptation efforts throughout Denmark. One initiative by the government is intended to prevent future flooding in sewers and streams. As a part of this initiative, the municipalities have been made responsible for preparing local action plans for adaptation, including the mapping of flood-prone areas.

In order to provide guidance and facilitate the process of establishing municipal action plans, a mobile team has been established with the aim of providing assistance and facilitating collaboration between municipal authorities and other stakeholders.

Also, a campaign seeking to raise awareness among citizens was launched. This campaign provides inspiration and guidance to help prevent property damage during extreme precipitation events.

In order to better understand user needs for information and to ensure that the web portal was structured in a user-oriented fashion during the first two years, the following instruments were applied:

- Focus groups including several municipalities
- Questionnaire survey sent to all 98 municipalities in Denmark
- *Tour de klimatilpasning* – a tour designed to establish a dialogue with the municipalities.

In the following section, these instruments are described in greater detail.

7.2.2   Instruments for understanding information needs

As the web portal will be continuously improved, there was a need to review the usefulness of the information provided with the first release of the web portal. Consequently, a number of the Danish municipalities were contacted in 2009 in order to establish *focus groups*. Two focus groups were formed, consisting primarily of civil servants with responsibilities in planning and GIS. Approximately 20 people participated in each of the groups. The participants in the focus groups were asked to identify specific information needs for their municipality in relation to the data provided by the webGIS. The webGIS part of the web portal includes datasets (information on temperature, precipitation, cloud coverage, wind and groundwater) for the predicted future climate based on IPCC scenarios (A2, B2, A1B) and the EU 2-degree climate scenario. The focus groups were asked to find the information that their municipality would need to adapt to climate change and to evaluate the webGIS on that basis. In addition, a questionnaire examining how the municipalities were handling the challenges of a changing climate was prepared for the focus group meetings.

In general, the focus groups determined that the content of the web portal was usable for the municipalities; however, the groups did indicate needs for data and tools that were not yet part of the portal. As the webGIS content was in the early stages of development, only a few datasets for climate projections were available to the users, and their comments reflected the need for further development.

In 2010, after having had contact with the focus group, a *survey* was conducted among the Danish municipalities in order to obtain an overview of their implementation of adaptation measures. All 98 Danish municipalities were invited to participate in the survey, which was mailed to them as a questionnaire. A total of 73 municipalities took part, corresponding to a response rate of 74%. The primary objective of the survey was to improve the web portal and make it easier for local governments to exchange knowledge and experiences regarding climate change adaptation. The survey asked questions such as:

• How is climate change adaptation included in local government work?
• How far has your municipality progressed with adaptation strategies and concrete initiatives?
• To what extent does your municipality have the knowledge required to carry out adaptation activities?
• To what extent does your municipality know about and use the web portal for assistance in adaptating to climate change?

The survey results indicated that adaptation is high on Danish municipal agendas. Work is in progress to map the risk of flooding and establish adaptation strategies, as well as to establish measures to manage increased water volumes. Fewer municipalities are implementing measures in other sectors. Many municipalities expect more flooding in the future from precipitation and the sea. The majority of the municipalities therefore identify sewage/wastewater and planning as their areas of priority with regard to adaptation over the next few years. The results of the survey are summarised on the web portal and in three reports (Hellesen et al. 2010, Hedensted Lund 2010, Hedensted Lund & Nellemann 2012).

The focus on flooding issues in most municipalities is probably due to the fact that flooding is a concrete phenomenon for which municipalities and the public perceive an urgent need for solutions. Eighty-one per cent of the participating municipalities expect more floods from precipitation in the future to a great extent or to a moderate extent, while 65% of the coastal municipalities expect greater challenges from flooding events from the sea in the future. The municipalities have gone to great lengths to identify the scope of the problem and identify risk areas. Seventy-three per cent have (to a great or a moderate extent) mapped areas vulnerable to flooding from storm water, groundwater or sewage, and 67% of the coastal municipalities have mapped areas vulnerable to flooding from the sea (Hellesen et al. 2010).

The survey also revealed that many municipalities lack knowledge and tools. Only 7% of the municipalities believe that to a great extent they have the knowledge and tools necessary to make decisions about adaptation to climate change. Forty-one per cent reported having the knowledge and tools necessary to a moderate extent, while half of the municipalities believe that they do not have, or have only to a limited extent, the necessary knowledge and tools. The majority of municipalities lack knowledge regarding the consequences of climate change in different sectors and local areas, as well as knowledge regarding the concrete options available for climate change adaptation in municipalities. They also express a need for tools to perform socio-economic analyses and risk analyses, and they lack both decision-making tools to prioritise efforts and tools to map flood-threatened areas.

Collaboration with other municipalities on climate change adaptation is increasingly taking place. More than half of the municipalities (52%) reported that they collaborate with other municipalities on climate change adaptation. Examples of these collaborative projects include concrete projects on storm water/surface water retention, sewage/wastewater management and establishing dikes; in addition, the exchange of experiences via municipal and regional climate networks was reported.

Quite a few of the municipalities have involved the public in their work in one way or another. A total of 41% have to a great or a moderate extent involved the public in climate adaptation, while 28% have informed the public and enterprises about what they can do to adapt to climate change (e.g. through information folders, information boards and websites).

Furthermore, the survey revealed that a majority of municipal plans state that climate change must be considered in planning, and two-thirds of the municipalities assert that they intend to prepare a municipal adaptation plan within the next few years.

Following the survey, in the summer of 2011 the *Tour de klimatilpasning* was held to strengthen the dialogue with the municipalities. The tour consisted of six regional seminars attended by a total of 82 municipalities. The seminars included demonstrations of the web portal and presentations by municipalities about their work on climate change adaptation. Some of the municipalities presented their efforts on adaptation action plans; others had been asked to test the tools on the web portal and reflect on how the tools could be used and further developed. The main challenge at present for municipalities in Denmark turned out to be water management, e.g. water from extreme precipitation, water from flooding and groundwater. A number of ways in which municipalities could cooperate were identified, e.g. cooperation within catchment areas, the use of the same climate scenarios for future planning and the organisation of groups in which experiences in handling adaptation could be exchanged. In general, a need for increased cooperation between municipalities and regions was identified.

FIGURE 7.2.1 Group work at the *Tour de klimatilpasning* (source: author's photo)

7.2.3 Including the formulated user needs in the web portal

For the majority of the municipalities, adaptation is about establishing protection against increased water volumes from precipitation – which leads to greater pressure on sewage and wastewater management processes – and protection against rises in groundwater and sea levels.

The focus groups pointed to the need for interactive tools to identify flooding from the sea and from precipitation. In addition, the need for data on extreme precipitation was identified. In line with these recommendations, an interactive tool to identify flooding from the sea was developed and launched on the web portal in the beginning of 2012. To facilitate adaptation in the municipalities, new datasets (e.g. on future groundwater levels and extreme precipitation) were developed and provided on the web portal in 2012. Furthermore, a dataset has been developed to identify areas prone to flooding due to extreme precipitation (so-called 'blue spots'). The content of the web portal is continuously being improved and updated in order to include the latest information on climate change impacts and adaptation, additionally educating users on the uncertainties related to the climate scenarios.

7.2.4 Concluding thoughts

Including members of the target groups in the development of the web portal has proven to be of great value in terms of the portal's quality and usefulness. The focus groups, established immediately after the release of the web portal, clearly indicated that its content still needed development. The survey, conducted approximately one year after the focus groups, showed that the web portal was being used by the municipalities in their adaptation efforts, but that there was still a need for specific adaptation tools, e.g. to identify the location of flooding from the sea. The *Tour de klimatilpasning* focused on intensifying the dialogue with the municipalities and helped to train them on how to use the tools provided by the web portal. The users thus profited directly from participation in the focus groups, the survey and the tour: they learned about the data content of the portal, which has been improved over the past few years, and about adaptation tools that can be used by municipalities.

Awareness of climate change impacts has been on the agenda of planners and other actors in Denmark for several years. The public – in the author's personal experience – has become aware that extreme precipitation can have severe impacts on property. Therefore, providing information on how individuals can act to help their property better withstand more extreme events is an important part of information campaigns on climate change impacts and adaptation. Good-practice examples for adaptation are also important in order to provide inspiration for solutions. The results of the survey of the municipalities indicate that the public is interested in climate change and in possibilities for adaptation. The next step in the information campaign of the Danish adaptation strategy is to further increase the awareness of the public regarding the consequences of a changing climate and potential adaptation options.

7.3 Making climate change scenarios useful for regional adaptation plans in France

Christian Pagé

7.3.1 Background: Adaptation plans at the regional level

In France, several administrative regional entities (*départements*) are preparing adaptation plans related to climate change. This is in response to the French Climate National Plan, which prescribes the production of plans to lower greenhouse gas emissions and to adapt to climate change at the regional level. In this context, several studies are taking place in the French *départements*, sometimes through collaboration between specialised consultancies (in the case presented here: TEC-Conseil), climate data producers (here: CERFACS) and stakeholders (here: the governmental administration of one *département*[21] in France). The region considered in the case presented here is *département* 13, Bouches-du-Rhône (5087 km²), situated on the Mediterranean coast (latitude 43.5°) and encompassing the city of Marseille, some mountains and very low-lying terrain along the coast near the Rhône delta. The sectors involved are extremely diverse, since all the services from the *département* had to participate: agriculture, education, health, transportation and infrastructure, tourism, energy and industry, housing, employment, forestry, water management, urban affairs and others.

7.3.2   Practical experiences from the Bouches-du-Rhône département

The main objective of the case presented here was to inform public policy of the potential climate change impacts in Bouches-du-Rhône over short-term (2030) and mid-term (2050) periods and to support adaptation to these impacts. This was a challenging task, given the number of people involved in the stakeholder administrations (i.e. the many sectoral administrations with an interest in information on climate change). The General Council of Bouches-du-Rhône launched a specific call for project proposals to fulfil the obligations set out in the French National Adaptation Plan; going beyond these limited obligations, an in-depth analysis of climate change impacts, vulnerabilities and possible adaptation measures (IVA) was also included. The project proposed by the specialised consultancy TEC-Conseil in cooperation with CERFACS was approved. TEC has extensive experience in translating climate science jargon for end users and in working with people to determine their needs for information on IVA. CERFACS, a research organisation with experts in climate science, completed the consortium, providing data, interpretation, processing and analysis.

Addressing stakeholder needs by meetings and interviews

In order to tailor climate change information to user needs, the project consisted of two main parts: stakeholder meetings and stakeholder interviews. The project started with a first meeting (a kick-off) with all stakeholder administration heads present. At this meeting, the aim was to present available options for regional climate change scenarios[22] (such as statistical or dynamic downscaling (Goodess et al. 2011) and regional climate datasets like ENSEMBLES (Goodess et al. 2009)) and the course of actions to be undertaken during the project. The stakeholder involvement design was developed by TEC based on previous experiences, e.g. in the tourism sector and in an IVA study for the Wallonie region in Belgium.

Subsequent meetings were held regularly to inform the stakeholders about the project's progress. This involved a process of 'vulgarising' the methodology and the climate science background that the consultancy and the climate experts envisioned using in two ways. First, the scientific methodology was adapted to the stakeholders' needs and to the availability of suitable climate scenarios. This was done using a stakeholder-driven selection of scenarios from a broad range of scenarios with respect to temperature and precipitation changes for Bouches-du-Rhône, using material such as that presented in Figure 7.3.1, which shows the mean change in precipitation (%) and temperature (degrees Celsius) for the Bouches-du-Rhône region in winter (December-January-February) for the period 2030–2050 in comparison to 1961–1990 for several regional climate models. This made the climate data more useful for the stakeholders, as the methodology allowed the selection of a median-response[23] scenario (used as

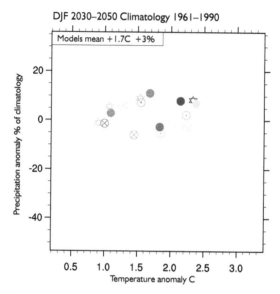

FIGURE 7.3.1 Change in precipitation (%) and temperature (degrees Celsius) for the Bouches-du-Rhône *département* in winter (DJF: December-January-February) for the period 2030–2050 in comparison to 1961–1990 for several regional climate models (symbols) (source: author's compilation)

a 'reference') and four additional scenarios to cover the uncertainties. The process of 'vulgarising' the methodology and the climate science background was essential for establishing a real dialogue and 'bridging the gap' between the stakeholders, the consultancy and the climate experts, since each group had its own specific expertise. It also helped in the selection of climate scenarios that best met stakeholders' needs. For example, depending on the climate variables that have the strongest impacts on local vulnerabilities, the selection of climate scenarios can change because the climate models do not react in the same way for all variables. The methodology for selecting scenarios and using climate data for assessing climate change impacts was heavily based on IPCC (2007). In addition, it was necessary to clarify the specific vocabulary used in climate science, as well as the implicit limitations, uncertainties and hypotheses behind the scientific methodologies and the data that the researchers planned to use. Climate change information was also communicated to the stakeholders in synthetic reports following each meeting that summarised the material presented during the meetings, the minutes of the meetings' discussions and the conclusions reached. The stakeholder meetings helped the project to achieve several of the recommendations listed in the guiding principle 'Build knowledge and awareness', notably the construction of a common knowledge base and the use of common language, as well as a simple formulation of the information. Furthermore, the meetings fulfilled many of the recommendations made by Swart and Avelar (2011).

The second core section of the project (tailoring climate change information to user needs) involved 25 individual interviews by the consultancy with each head of the stakeholder administrations. The interview questions were related to the identification of suitable climate indices, the knowledge of the stakeholders about extreme climatic events, their impacts, how they have reacted and other aspects. This step was crucial in order to accurately characterise the vulnerabilities and to identify common climate indices that could be used within the study across all sectors. Based on the interviews, 27 climate indices were selected, including the number of frost days, heating degree days (number of degrees below 18°C for a day's average temperature), number of rainy days, maximum number of dry days and the length of the vegetation growth period. This step also required the consultancy to explain to the stakeholders (during the interviews and meetings) which climate indices were possible and how they would need to be tailored, given data availability and regional characteristics. Tailoring in this case implied adjusting the thresholds of the indices given the spatial scale of available data (e.g. extreme values for daily total precipitation) and evaluating uncertainties with respect to the climate-internal variability (i.e. focusing on model uncertainties rather than uncertainties related to emission scenarios for the short-term time range (Hawkins & Sutton 2009), since the largest uncertainties come from the models for periods before 2050). Thus, the interviews can be described as dialogues – two-sided interactions in which the stakeholders explained their data expectations and the consultancy explained which of these expectations could be fulfilled and which could not.

Furthermore, a parallel dialogue took place between the consultancy and the climate experts, since the consultancy needed support from the climate experts during the interview period to consider the requests of the stakeholders (given the limitations and uncertainties of the climate data). One example involves the limitations due to the spatial scale of the data, especially for precipitation data. At a spatial resolution of 25 km, the extremes that can be sampled are an average over a surface of 125 km^2; this makes the possible extremes smaller than what can be observed at a single point (such as an observing station).

All of these interactions – in meetings and in interviews – helped to better tailor the climate information to the stakeholders' needs, taking into consideration the data limitations and uncertainties. This interactive process, which required frequent contact between the stakeholders and the consultancy experts as well as between the climate experts and the consultancy, was therefore essential to achieve the objectives of the study. This demonstrates that good and frequent communication is crucial when dealing directly with stakeholders; in addition, the active participation of all actors increases a project's chances for success. This experience from our study is very much consistent with the recommendation of the guiding principle 'Build knowledge and awareness' to differentiate between different stakeholder groups in providing climate change information.

Communication of uncertainties and central concepts of climate change science

The communication regarding uncertainties and the central concepts of climate change science was prepared using carefully chosen material presented during the general stakeholder meetings. For example, a report on 'Pitfalls when using climate data' was created, describing in detail the following concepts in a simple and comprehensible fashion: internal climate variability versus climate change signal, decadal oscillations versus oceanic cycles, Special Report on Emissions Scenarios (SRES) and spatial scales of downscaled climate data versus observation stations for historical data. These concepts are closely related to the uncertainties in climate change, primarily the types of uncertainties that are important for a given time range and spatial scale. The study also included the use and selection of multiple climate change scenarios that portray the ranges of uncertainty for a region of interest in terms of potential precipitation and temperature changes.

Build knowledge and awareness of climate change impacts, vulnerabilities and possible adaptation measures

Once the information requirements of the stakeholders were defined, the consultancy could start the data analysis to assess climate change impacts, vulnerabilities and possible adaptation measures based on initial discussions with the climate experts. In parallel to the start of the requirements analysis, more in-depth interviews were conducted with each sectoral stakeholder with respect to their sectoral resources, in addition to a transverse analysis in sectors and strategic priorities definition. Data from these interviews were the basis for the vulnerability analysis on transverse themes (such as water, health, transportation, biodiversity, etc.) and the adaptation plan. This required close interaction between the stakeholders and the consultancy, but not much involvement of the climate experts. This led to the IVA report for the Bouches-du-Rhône *département* and to the presentation of the results to the stakeholders. Because these results were developed in close collaboration with the stakeholders, they can easily be used by the stakeholders to define suitable adaptation policies.

7.3.3 Concluding thoughts

This case highlights the fact that experienced consultants and communicators are needed in order to translate climate science knowledge to stakeholders, establishing a continuous dialogue between the stakeholders and the climate experts. The characterisation of stakeholder needs as well as the clarification of climate science concepts to stakeholders and consultants can also be identified as necessary steps. This initiative helped to improve the methodology for selecting appropriate climate change scenarios that are better tailored to stakeholder needs. The challenge now is to convince other stakeholders to consider several climate change scenarios rather than just one, as is frequently the case (ignoring scientific uncertainties). With a rapidly increasing database of climate model outputs (cf. CMIP5), improved and transparent methodologies for the selection of scenarios will be required. This will increase the challenge of selecting appropriate scenarios because of the large amount of data processing needed, which in turn will require sizable data downloads from the data archive to the local analysis and data-processing servers.

This study has made it clear that the spatial resolution of climate products will have to be improved to make them more usable for stakeholders at the local level. However, this will pose some problems regarding increased uncertainties in the data, as the significant uncertainties at coarser spatial resolutions translate into much larger uncertainties at finer resolution.

The methodology used here was very successful, notably due to the significant time investments of the stakeholders as well as the close and frequent interactions between the climate experts, the consultancy and the stakeholders. The careful planning of the communication of uncertainties was certainly beneficial to both the stakeholders and the climate experts, as the stakeholders obtained a much more useful climate change IVA report to inform their decisions on upcoming policies, and the climate experts will benefit from the successful translation of this important information to the end users.

7.4 Adaptation communication with business actors and the general public at the metropolitan level: Experiences from northwestern Germany

Claudia Körner and Andreas Lieberum

7.4.1 Northwest Germany and the 'Roadmap of Change'

The regional and local consequences of climate change and options for adaptation are only slowly finding their way into the public awareness. While there is significant awareness of climate change and mitigation of climate change, the topic of adaptation to climate change has yet to find a place among decision-makers in industry, politics and administration, or among the media and the general public in Germany.

The research project *nordwest2050*, including its various adaptation communication and sectoral stakeholder involvement activities, was launched in 2009 with support from the German Federal Ministry for Education and Research. Its goal is to ensure that a vision of a climate-adapted and resilient society is established in the metropolitan area of Bremen-Oldenburg and to clearly communicate that there are solutions to the challenges of climate change.

The metropolitan area of Bremen-Oldenburg is a coastal and port region in Germany's northwest with almost three million inhabitants. *nordwest2050* examines the institutional structures and key economic sectors in this region that are predicted to be significantly affected by climate change and at the same time play an important role in the economy of the region. These sectors include energy and food production as well as ports and logistics. The energy industry in the northwest will have to overcome a number of direct challenges resulting from climate change. These include the limited supply of cooling water for thermal power plants – for example, due to more frequent and longer heat waves – and the growing risk of disruptions in the supply of commodities. In addition, the energy sector faces a rising demand for energy to run air-conditioning systems in summer and a dwindling demand for heat in winter. The food sector is exposed to the growing risk of delays or interruptions in deliveries, including those resulting from extreme weather events, and to cost increases that are due in part to an increased need for cooling in hotter summers (especially for fattening houses, food warehouses and food production facilities). Ports are particularly affected by extreme weather that can impede the workflow. Thus, there is a need for significant investments to protect existing infrastructures. The usability of shipping routes or connections of harbour logistics with the hinterland may also be put at risk. On top of this, the consequences of climate change might also force changes in global transport chains elsewhere in the world and thus affect the regional flow of goods, affecting the status of individual ports in the process. Outside of economic sectors, the region's citizens are also threatened by climate change, especially by extreme weather events such as heat waves, heavy rain and storm surges (Schuchardt & Wittig 2012).

The main product of the project will be the 'Roadmap of Change', a strategy for adaptation to climate change, which should provide the region with guidance up to 2050.

7.4.2 Principles and examples for adaptation communication in the nordwest2050 project

In order to strongly establish the concept of climate adaptation in the region, extensive public relations work is needed. *nordwest2050* has made use of its funding opportunity to try different approaches to communication in order to evaluate the experiences and to discuss them with other research projects. Key principles for the successful communication of adaptation needs and options have been developed in *nordwest2050* and now serve as guidelines for the development of various tools and communication formats (Born et al. 2012).

These communication principles are based on the findings of the adaptation to climate change project 'Klimawandel Unterweser' funded by the Federal Ministry for Education and Research, which was conducted from 2007 to 2009 in the Unterweser region in northwestern Germany. These communication principles were further developed in *nordwest2050* and put into practice in various ways for the respective target groups (cf. Table 7.4.1).

The use of understandable language is essential for communication with various stakeholders. To this end, *nordwest2050* has published a series of publications ('short+sharp') in which the research results of the project are presented in simple

TABLE 7.4.1 Communication principles in *nordwest2050* (translated and adapted from Born et al. 2012)

Phase 'Perception of the problem'
- Explain terms using common language
- Differentiate between adaptation to climate change and mitigation of climate change
- Convey risks and opportunities of climate change and adaptation options that can reduce risks and take advantage of opportunities
- Use windows of opportunity when the public attention for climate change issues is high
- Address the recipients' emotions
- Use target-group specific communication
- Reduce complexity.

Phase 'Decision-making'
- Convey ways to deal with uncertainties
- Communicate the goals of adaptation to climate change
- Create role models; use good-practice examples; involve reliable climate change ambassadors
- Use practical knowledge and create compatibility with practical experiences.

Phase 'Realisation'
- Ensure the participation of practitioners and stakeholders
- Use planned and ongoing processes of change
- Address conflicts when developing and implementing climate change measures.

Source: Authors' compilation

language. Each issue of 'short+sharp' is visually well designed and consists of four to six pages in A4 format. The series is available both in a printed version and online as a PDF download on the project website. Because the phrases 'climate adaptation' and 'adaptation to climate change' are still not very familiar to the general public, it is also necessary to create a common understanding of them. The expressions 'mitigation of climate change' and 'adaptation to climate change' must be clearly defined and differentiated from each other in order to create a better understanding of the necessity for strategies and measures in both fields. In order to generate a sense of concern, recipients must be approached through their own sphere of activities. The use of images and examples from the relevant fields provide the recipients with points of reference that they can relate to, facilitating an emotional link to the subject addressed. Following this principle, *nordwest2050* has produced a series of video clips[24] that show the results of the regional vulnerability analysis. In ten episodes, individual examples of regional vulnerabilities are presented, together with commentary from practitioners from the northwest region who have developed solutions to reduce their specific vulnerabilities. In the video clips, the practitioners explain their vulnerability and their solutions themselves. Thus, the vulnerabilities are described concretely and their importance for the 'everyday world' of the viewers of the clips is clearly communicated.

Furthermore, the clips fulfil another requirement of successful communication in climate adaptation: to stimulate adaptation, messages and images of potential climate change impacts should not be presented 'catastrophically' (e.g. represented by climate disasters), as this often leads to reactions such as denial and fatalism in the recipient (Grothmann 2005: 214). However, findings from prospect theory (Tversky & Kahneman 1992) demonstrate that losses motivate people more to act than gains, particularly if the subjective judgement of the probability of loss occurrence is low. Messages should thus include the presentation of risks but should also show possible courses of action to reduce the risks – a principle that is upheld admirably by the clips.

7.4.3 The right tool for each target group: Example – Businesses

nordwest2050 places a particular focus on adaptation to climate measures in the economic sector. Businesses in northwest Germany currently feel very little concern about climate change. Adaptation to climate is predominantly seen as an issue for the future and, as the topic itself is burdened with great uncertainty, it remains largely unclear which adaptation actions might be necessary and useful. In order to establish the topic in this context, the project sets up sector-specific networks, in particular for the energy, food and port/logistics sectors.

Dialogue with business representatives is sought in workshops in order to permit reciprocal learning and to increase knowledge about the regional consequences of climate change and the resulting opportunities and risks (cf. Chapter 8 in this manual on the identification of and cooperation with stakeholders). This requires a foundation of trust upon which an atmosphere of continuity and listening can be created.

nordwest2050 also seeks out dialogue with intermediary organisations, such as chambers of commerce and industry and economic development agencies. These bodies enjoy a high level of credibility among businesses and are therefore well placed to promote the topic of climate adaptation in the economic sector. Because of the uncertainties surrounding the topic, the credibility of the sources of information is particularly highly valued.

Since small and medium-sized enterprises generally do not have sufficient resources at their disposal to review complex scientific information, they turn to the sources they trust most. Decision-makers feel safer when they are copying the adaptation measures of other businesses. Therefore, *nordwest2050* makes an effort to include good-practice examples of adaptation measures in its adaptation communication. For example, in 2012, the project organised its first competition for businesses. Businesses were asked to submit projects or ideas about adaptation to climate change. The winners were rewarded publicly and thereby became role models and served as good examples for others.

The 'Quick Check for Businesses', also developed in *nordwest2050*, is a tool to provide guidance to businesses in the complex field of climate change. The quick check consists of ten questions in an online survey assessing whether a business is affected by climate change. This first assessment may generate an initial interest in adaptation and create contact between enterprises and the project.

7.4.4 Creating a buzz: Addressing the general public

Addressing the general public is important in order to familiarise people with adaptation to climate change and to establish the topic widely. A key measure used to achieve this goal is media relations. For its regional and local media efforts, *nordwest2050* primarily uses events hosted by the project itself (e.g. conferences, workshops) or 'regional windows of opportunity'. These include extreme weather events in the region, which are used as a 'hook' when approaching the media. Extreme weather events that illustrate certain risks of climate change are often widely publicised in the media and thus can be used to inform the public about ongoing adaptation activities and projects in the region.

Easily accessible, informative and entertaining online information can also be used to sensitise the public to the consequences of climate change and adaptation options. *nordwest2050* uses the medium of scrap-paper animation clips[25] (cf. Figure 7.4.1) in this regard. This method has the advantage that even abstract topics can be conveyed in simple and easily understandable images. Through the simplicity of the images and the relatively light-hearted imagery and text, the clips attract the viewer's attention.

FIGURE 7.4.1 Example screenshot from scrap-paper animation clip with off-voice explaining the story of *nordwest2050* (see www.nordwest2050.de, Mediathek) showing four main adaptation sectors in *nordwest2050*: energy, agriculture/food, harbour/logistics, governance (from left to right by line) (source: authors' screenshot)

FIGURE 7.4.2 User interface for *nordwest2050* online climate game (see www. nordwest2050.de, Mediathek) showing symbols of different climate change impacts (e.g. first left in the first line shows a flooded house; if the user clicks on the symbol a second window opens with regional climate projections for the next decades. The user can then choose to like or dislike the projected climate situation. After clicking on ten symbols and like/dislike votings the user gets a short personalised climate change impact assessment) (source: authors' screenshot)

Furthermore, an online climate game was developed as another tool to convey the potential impacts of climate change in a playful manner (cf. Figure 7.4.2). *nordwest2050* also uses elements of web 2.0, such as a YouTube channel and a Facebook profile. These serve to spread knowledge on climate change and adaptation and to raise the profile of the project as a whole. These channels also offer the project the opportunity to interact directly with target groups and to receive feedback. Here, it is important to follow the rules of social media communication and above all to be authentic. People want to communicate with people, and abstract project talk or commercial PR talk will be dismissed by the Facebook community.

7.4.5   Concluding thoughts

The responses to an online questionnaire filled out by 29 users of *nordwest2050* have shown that the 'short+sharp' series and the scrap-paper animation films were particularly well received and were evaluated as helpful and informative. The series of video clips showing the results of the regional vulnerability analysis was also evaluated as informative, but these videos have significantly lower viewer numbers on YouTube than the scrap-paper animation films. Further discussions with the recipients should provide more information about the reasons for this discrepancy. One assumption is that for good distribution, not only are the topic and style of the films important, but also their embedding on the project website. Films present a visualisation of the text and enable a more emotional appeal but, particularly for the professional public, these are perceived primarily as support for the text. Therefore, films and clips that are meant for a targeted audience should not stand alone without a contextual link to further online textual information. Films and clips probably play to their strengths more effectively when they are integrated into the presentation of results on climate change vulnerabilities and adaptation options and when their function is to emphasise, complement or highlight these results.

The fact that many audiences currently do not feel affected by climate change and have difficulty dealing with the uncertainties of climate change projections remains a challenge. Greater interest is often shown for the topic of climate change mitigation, which many actors cannot distinguish from adaptation – and, in fact, there is a strong link between the two issues. Thus, to reach certain target groups (especially in the business sector and civil society), it can be very helpful to approach them through their mitigation activities and add aspects of adaptation step by step. Linking these aspects of adaptation with current activities (such as risk management in the business sector or sustainable consumption in private households) could also help to stimulate attention.

Collaboration with certain economic sectors is particularly difficult. In general, the time horizon of 2050 – the time horizon chosen in *nordwest2050* – goes far beyond the planning horizon of many businesses; the focus of business strategies and questions about the future generally lies within the range of a few years, or at most a decade. Because of the lifespan of their infrastructure, ports, transportation networks and the energy sector are among the very few economic sectors that think beyond this time frame.

The collaboration with intermediary organisations such as chambers of industry, commerce or agriculture and economic development agencies and their liaison offices has proven helpful. They have a fundamental interest in providing new opportunities for innovation in which the adaptation technologies initiated by *nordwest2050* can play a role. Integrating climate adaptation into their long-term action plans through interactive involvement will be an important factor in ensuring the success of an adaptation strategy to be implemented in economic policies.

Civil society remains difficult to mobilise. Environmental organisations have mainly taken up the issue of climate adaptation where it touches upon their traditional issues (e.g. nature conservation); other aspects such as 'social vulnerability' currently have no champions. Here, new communication approaches and tools need to be developed.

Overall, we find that for the successful communication of climate adaptation, it is essential to establish an appropriate target group approach and a dialogue with the stakeholders. Climate adaptation is an individual process that depends on specific vulnerabilities and subjective risk assessments. Communication measures must consider these in order to reach the target groups and to evoke a profound understanding of appropriate climate adaptation measures. Therefore, an intensive dialogue between science, specific target groups, politics and society is required.

7.5 Lessons learned for the communication of impacts, vulnerabilities and adaptation options

Torsten Grothmann

All of the cases presented in this chapter on building knowledge and awareness for adaptation to climate change stress the importance of a target or user group-specific approach. This involves the clear definition of the target groups, the analysis of their user needs (in terms of climate change and adaptation information, but also their preferred communication channels) and the design of tailor-made communication formats to reach the target groups and to provide them with useful information.

User needs in relation to information content change over time, in particular because there are different stages in the adaptation process. Often, at the beginning of an adaptation process, users primarily need information on potential climate change impacts relevant to them. Subsequently, they mainly need information on adaptation options and experiences. Therefore, information tools (e.g. internet platforms) have to be updated not only to include new scientific knowledge on climate change and adaptation but also to provide appropriate information for the adaptation stage of the users. This illustrates that building knowledge and awareness is a long-term process and requires long-term resources.

All of the cases also stress the importance of a dialogue with the user groups (e.g. in workshops or personal consultations) in order to investigate their needs (including their current stage of adaptation, to provide them with the necessary knowledge to move on to the next stage) and to learn from their practical experiences with weather-related impacts and adaptation measures that have proved effective in the past. In particular, when potential climate change impacts and useful adaptation options are very unclear, this type of dialogue and mutual learning between experts on climate change and adaptation and local experts seems to be beneficial. Consultants who combine climate expertise and communication skills can play a very effective role.

The communication formats used in the cases presented in this chapter include not only unidirectional formats for providing information (e.g. online information platforms, flyers, brochures, movies) but also unidirectional formats for gathering information from users (e.g. surveys) and bidirectional formats for mutual learning (e.g. workshops, focus groups, training and counselling). This demonstrates that practices concerned with building knowledge and awareness (depicted in this chapter) are closely related to practices that focus on involving and cooperating with stakeholders (cf. Chapter 8).

In all of the cases, three groups are involved in the process of building knowledge and awareness:

- The target/user group: in most cases, decision-makers in policy, but also decision-makers and stakeholders in business and the general public
- The data/information providers: mostly climate change scientists and adaptation experts
- The communicators: often bridging organisations (e.g. environmental protection agencies or consultants) that try to organise the information transfer between the data/information providers and the target/user group.

Pagé (Section 7.3) stresses that the active participation of climate experts in meetings with the users of the climate change scenarios can build confidence in the scenarios among the users. Both Pagé and Körner and Lieberum (7.4) emphasise the importance of using understandable language rather than scientific jargon.

This underlines that there is much more to effective adaptation communication than simply providing scientifically accurate knowledge – which is already a challenge in itself, as Jol and Isoard (7.1) point out. Körner and Lieberum (7.4) recommend the use of extreme weather events in adaptation communication and stress the importance of addressing recipients' emotions and creating role models for adaptation. Körner and Lieberum, and indirectly also Grøndahl (7.2), underline the importance of communicating not only information on climate change and its potential impacts but also information on adaptation options, so that recipients will get an idea of what they can do to adapt to climate change. More specifically, Grøndahl identifies the provision of good-practice examples for adaptation as particularly effective at stimulating adaptation.

One main challenge of adaptation communication seems to be the uncertainties in scientific climate change and impact scenarios. Körner and Lieberum explicitly cite the uncertainties as a barrier for adaptation action, and Pagé stresses that careful preparation for the communication of uncertainties in climate change scenarios is crucial. How uncertainties can be communicated without demotivating adaptation action is addressed in detail in Chapter 11.3.

Thus, we can conclude the subject of building knowledge and awareness with the following lessons learned:

- Building knowledge and awareness for adaptation to climate change requires a target or user group-specific approach.
- Building knowledge and awareness also requires a dialogue with the user groups to investigate their user needs and to learn from their practical experiences, using common language as much as possible.
- There is much more to effective adaptation communication than simply providing scientifically accurate knowledge. Other practices, such as building trust in the knowledge, communicating existing scientific uncertainties well, addressing emotions for adaptation motivation and providing role models and good-practice examples for adaptation, are also essential.
- In most cases, three groups are involved in the process of building knowledge and awareness: the target/user group, the data/information providers and the communicators.

Notes

1 http://climate-adapt.eea.europa.eu/
2 http://cca.eionet.europa.eu/
3 http://www.eea.europa.eu/about-us/countries-and-eionet
4 http://climate-adapt.eea.europa.eu/web/guest/eu-sector-policy/general
5 http://climate-adapt.eea.europa.eu/web/guest/countries
6 http://climate-adapt.eea.europa.eu/web/guest/adaptation-support-tool/step-1
7 http://climate-adapt.eea.europa.eu/web/guest/adaptation-measures
8 http://climate-adapt.eea.europa.eu/web/guest/observations-and-scenarios
9 http://climate-adapt.eea.europa.eu/web/guest/vulnerabilities-and-risks
10 http://climate-adapt.eea.europa.eu/web/guest/uncertainty-guidance
11 http://climate-adapt.eea.europa.eu/sat
12 http://climate-adapt.eea.europa.eu/web/guest/transnational-regions
13 http://climate-adapt.eea.europa.eu/web/guest/tools
14 http://climate-adapt.eea.europa.eu/web/guest/data-and-downloads
15 http://climate-adapt.eea.europa.eu/map-viewer
16 http://www.climwatadapt.eu/
17 http://www.espon.eu/main/Menu_Projects/Menu_AppliedResearch/climate.html
18 http://ies.jrc.ec.europa.eu/
19 http://ensembles-eu.metoffice.com/
20 http://copernicus.eu/
21 France is divided into 96 *départements*.
22 A climate scenario is understood here as a climate projection using a specific climate model and a GHG emission scenario.
23 The 'median-response' scenario is chosen by calculating (for the area and time period of interest) the average change in temperature and precipitation for each climate scenario (a scenario being defined as a climate projection using a specific climate model and a GHG emission scheme), and by then choosing the median scenario. This is called a 'response' scenario because here we are looking at the change (anomalies) in temperature and precipitation of the future climate (2030–2050 average) compared to the present climate (1961–1990 average).
24 See www.nordwest2050.de (Mediathek).
25 See www.nordwest2050.de (Mediathek); clips are available in English.

References

Born, M., Lieberum, A. and Winkelseth, C. (2012) *Prinzipien der Anpassungskommunikation im Projekt 'nordwest2050'*, nordwest2050-Werkstattbericht, nr. 15., Bremen. www.nordwest2050.de.

EC – European Commission (2007) 'Regulation No 614/2007 of the European Parliament and of the Council of 23 May 2007 concerning the Financial Instrument for the Environment (LIFE+)', *Official Journal*, Brussels. http://ec.europa.eu/environment/life/funding/lifeplus.htm.

EC – European Commission (2009) *Towards a European Framework for Adapting to Climate Change Action*, Brussels.

EC – European Commission (2011a) *Europe 2020 – A Strategy for Smart, Sustainable and Inclusive Growth*, Brussels. http://ec.europa.eu/europe2020/index_en.htm.

EC – European Commission (2011b) *EU Budget Proposal – Multiannual Financial Framework 2014–2020*, Brussels. http://ec.europa.eu/budget/reform/.

EC – European Commission (accessed) (2012) *OURCOAST Project and Database Supporting the Exchange of Experiences and Best Practices in Coastal Planning and Management*, Brussels. http://ec.europa.eu/ourcoast/index.cfm?menuID=18.

EC – European Commission (2013) *An EU Strategy on Adaptation to Climate Change*, Brussels. http://ec.europa.eu/clima/policies/adaptation/what/documentation_en.htm.

EEA – European Environment Agency (2012a) *Major New Website to Assist with Climate Change Adaptation in Europe*, Copenhagen. http://www.eea.europa.eu/highlights/major-new-website-to-assist.

EEA – European Environment Agency (2012b) *Climate Change, Impacts and Vulnerability in Europe 2012*, EEA Report nr. 12, Copenhagen. http://www.eea.europa.eu/publications/climate-impacts-and-vulnerability-2012/.

EEA – European Environment Agency (2013) *Adaptation in Europe – Addressing Risks and Opportunities from Climate Change in the Context of Socio-economic Developments*, EEA Report nr. 3/2013, Copenhagen. http://www.eea.europa.eu/publications/adaptation-in-europe.

Glicken, J. (2000) 'Getting stakeholder participation "right": a discussion of participatory processes and possible pitfalls', *Environmental Science & Policy* 3 (6), pp. 305–310.

Goodess, C.M., Jacob, D., Deque, M., Guttierrez, J.M., Huth, R., Kendon, E., Leckebusch, G.C., Lorenz, P. and Pavan, V. (2009) 'Downscaling methods, data and tools for input to impacts assessments', in van der Linden, P. and Mitchell, J.F.B. (ed.) *ENSEMBLES: Climate Change and its Impacts: Summary of Research and Results from the ENSEMBLES project*, Exeter: Met Office Hadley Centre, pp. 59–78.

Goodess C.M., Anagnostopoulou, C., Bárdossy, A., Frei, C., Harpham, C., Haylock, M.R., Hundecha, Y., Maheras, P., Ribalaygua, J., Schmidli, J., Schmith, T., Tolika, K., Tomozeiu, R. and Wilby, R.L. (2011) *An Intercomparison of Statistical Downscaling Methods for Europe and European Regions – Assessing their Performance with Respect to Extreme Temperature and Precipitation Events*. Climate Research Unit (University of East Anglia, UK) Research Report no 11. http://www.cru.uea.ac.uk/cru/pubs/crurp/CRU_RP11.pdf.

Grothmann, T. (2005) *Klimawandel, Wetterextreme und private Schadensprävention – Entwicklung, Überprüfung und praktische Anwendbarkeit der Theorie privater proaktiver Wetterextrem-Vorsorge*, (Dissertation), Magdeburg: Universitätsbibliothek Magdeburg. http://diglib.uni-magdeburg.de/Dissertationen/2005/torgrothmann.pdf.

Grothmann, T. (2011) 'Governance recommendations for adaptation in European urban regions: results from five case studies and a European expert survey', in Otto-Zimmermann, K. (ed.) *Resilient Cities – Cities and Adaptation to Climate Change – Proceedings of the Global Forum* 2010, Hamburg: Springer, pp. 167–175.

Grothmann, T. and Patt, A. (2005) 'Adaptive capacity and human cognition: the process of individual adaptation to climate change', *Global Environmental Change* 15 (3), pp. 199–213.

Hawkins, E. and Sutton, R. (2009) 'The potential to narrow uncertainty in regional climate predictions', *Bulletin of the American Meteorological Society* 90 (8), pp. 1095–1107.

Hedensted Lund, D. (2010) *Baggrundsrapport til: Klimatilpasning i de danske kommuner – et overblik. Kvalitativ analyse af udvalgte kommuners klimatilpasningsstrategier*, Arbejdsrapport nr. 120, Skov & Landskab, Københavns Universitet, Frederiksberg, pp. 1–45. http://www.klimatilpasning.dk/media/5342/ktpstrategier_slutvers151110.pdf.

Hedensted Lund, D. and Nellemann, V. (2012) *Klimatilpasning i de danske kommuner – det siger politikerne*, Arbejdsrapport nr. 146, Skov & Landskab, Københavns Universitet, Frederiksberg, pp. 1–45. http://kft.au.dk/fileadmin/www.kft.au.dk/Publikationer/KTP_Kommuner_Politikere.pdf.

Hellesen, T., Hedensted Lund, D., Nellemann, V. and Sehested, K. (2010) *Klimatilpasning i danske kommuner – et overblik*, Arbejdsrapport nr. 121, Skov & Landskab, Københavns Universitet, Frederiksberg, pp. 1–85. http://kft.au.dk/fileadmin/www.kft.au.dk/Publikationer/Rapport_Overblik_revAug11_slutvers_web.pdf.

IPCC – Intergovernmental Panel on Climate Change (2007) *General Guidelines on the Use of Scenario Data for Climate Impact and Adaptation Assessment*. http://www.ipcc-data.org/guidelines/TGICA_guidance_sdciaa_v2_final.pdf.

Lebel, L., Grothmann, T. and Siebenhüner, B. (2010) 'The role of social learning in adaptiveness: insights from water management', *International Environmental Agreements: Politics, Law and Economics* 10 (4), pp. 333–353.

Roeser, S. (2012) 'Risk communication, public engagement, and climate change: a role for emotions', *Risk Analysis* 32 (6), pp. 1033–1040.

Schuchardt, B. and Wittig, S. (ed.) (2012) *Vulnerabilität der Metropolregion Bremen-Oldenburg gegenüber dem Klimawandel (Synthesebericht)*, nordwest2050-Berichte, Heft 2, Bremen/Oldenburg: Projektkonsortium 'nordwest2050'.

Smith, J.B., Vogel, J.M. and Cromwell III, J.E. (2009) 'An architecture for government action on adaptation to climate change. An editorial comment', *Climatic Change* 95 (1), pp. 53–61.

Swart, R.J. and Avelar, D. (ed.) (2011) *Bridging Climate Research Data and the Needs of the Impact Community* – Proceedings of IS-ENES/EEA/CIRCLE-2 Workshop, 11–12 January 2011, EEA, Copenhagen.

The Danish government (2008) *Danish Strategy for Adaptation to a Changing Climate*, pp. 1–52. http://www.klimatilpasning.dk/media/5322/klimatilpasningsstrategi_uk_web.pdf.

The Danish government (2012) *How to Manage Cloudburst and Rain Water*, pp. 1–32. http://en.klimatilpasning.dk/media/590075/action_plan.pdf.

Tversky, A. and Kahneman, D. (1992) 'Advances in prospect theory: cumulative representation of uncertainty', *Journal of Risk and Uncertainty* 5 (4), pp. 297–323.

8

IDENTIFY AND COOPERATE WITH RELEVANT STAKEHOLDERS

Torsten Grothmann, Andrea Prutsch,
Inke Schauser, Sabine McCallum and Rob Swart

Explanation of the guiding principle

Because of its complex nature, adaptation to climate change should be a joint effort of the affected stakeholders from civil society, business and government. This is often referred to as 'governance' (Mayntz 1993), which includes close interactions not only between the different levels of government but also private-sector actors and actors from civil society.

Adaptation is a process of social learning (Paavola & Adger 2006, Nilsson & Swartling 2009, Lebel et al. 2010) and should utilise stakeholder knowledge in discussions of potential climate change impacts, vulnerabilities and adaptation options in different sectors and regions, by different actors and at different policy levels (Mukheibir & Ziervogel 2009) (cf. Chapter 7). Stakeholders should also be incorporated in processes of prioritising climate change impacts, vulnerabilities and adaptation actions and in discussions on acceptable risks (Grothmann 2011) (cf. Chapters 5 and 10). Rather than stakeholders merely being 'involved', they should be 'engaged', i.e. encouraged to play an active role.

Furthermore, Lebel et al. (2010) stress that stakeholder engagement helps to build consensus on the criteria used in monitoring and evaluation, which are essential elements of adaptive management and the adaptive governance schemes often used to adapt to climate change (cf. Chapter 14). Engagement in which actors share knowledge and responsibility in participatory processes can empower stakeholders to influence adaptation and to take appropriate actions themselves. Furthermore, it can reduce conflicts and identify synergies between adaptation activities and with the other activities of various stakeholders, thus improving the overall chances of success. Perhaps most importantly, stakeholder engagement can improve the likely fairness, social justice and legitimacy of adaptation decisions and actions by addressing the concerns of all relevant stakeholders (Lebel et al. 2010).

Relevant stakeholders from civil society, different business sectors and governmental decision levels should be invited to take part in the adaptation process. One way to identify relevant stakeholders is by asking the following questions (based on Lim et al. 2004):

- Who will likely be affected by the impacts of climate change or the outcomes of decisions related to adaptation?
- Who is responsible for realising the potential adaptation actions?
- Who can increase the effectiveness of the adaptation actions through participation or decrease its effectiveness by non-participation?

To address equity and social justice, particular attention should be devoted to including groups (such as low-income individuals or minority groups) that are often marginally represented in participatory processes and that are often the most vulnerable (Christoplos et al. 2009, Grothmann 2011).

It is important to note that successful stakeholder cooperation is a time- and resource-intensive process. At the beginning of the process, a common understanding about adaptation and its role must be developed among the stakeholders. To motivate participants, it is important to set a clear vision or goal for their engagement. The stakeholders' extent of influence over the adaptation process should be defined to avoid unrealistic expectations (Few et al. 2006, Larsen & Gunnarsson-Östling 2008, Reed 2008). The cooperative process needs to be efficiently structured (cf. Chapter 6) and well integrated into the larger decision context to ensure policy relevance (cf. Chapter 13).

Cooperation with the relevant stakeholders can be established at different levels of involvement, e.g. access to information, consultation or participation in decision-making (Lim et al., 2004). The appropriate approach to cooperation is case-specific and should be decided after taking into account factors such as the stage of the adaptation process, the group of stakeholders involved and the available resources.

Chapter overview

In the following chapter, three cases illustrate the practical realisation of the previously discussed considerations of how to identify and cooperate with relevant stakeholders. The first case (Section 8.1) describes the comprehensive stakeholder involvement in the process of the development of the national adaptation strategy in Austria. The second case (8.2) addresses stakeholder involvement in adaptation to climate change at the regional level in a largely rural area of Berlin-Brandenburg (Germany), focusing on the agricultural sector. The third case (8.3) shows how stakeholder engagement has contributed to building resilience to climate change at the city level in Ancona (Italy). This case also includes an overview of methods for stakeholder engagement.

8.1 Lessons learned from stakeholder involvement in the development of the Austrian National Adaptation Strategy

Andrea Prutsch and Therese Stickler

8.1.1 Introduction

In general, there is an assumption that for complex and socially relevant issues – such as climate change and adaptation – an extensive discourse with stakeholders is important for the policy-making process, in order to enhance the quality of the content as well as the acceptance of the policy and its chances of successful implementation. In the case of Austria, the development of the National Adaptation Strategy (NAS) was accompanied by a process engaging a wide range of stakeholders (defined here as representatives from the government, civil society and business who have a stake or special interest in adaptation), scientists and the general public. This broad involvement in the development of the NAS was required by the respective chapter of the Austrian governmental programme 2008–2013. Table 8.1.1 provides an overview of the involvement formats implemented, structured according to the level of intensity.

In the following case study, we focus on one specific format, namely the active stakeholder involvement (shown in bold in Table 8.1.1). We will draw lessons learned from an examination of the process and outcomes on the basis of the goals originally set and communicated by the process management/owners (Environment Agency Austria):

a   Raise awareness of climate change and the need for adaptation among affected stakeholders
b   Provide a platform to discuss opinions, experiences, needs, preferences and (possible) conflicts and to develop balanced solutions (or at least recommendations)
c   Enhance the quality of the NAS by including stakeholder knowledge and experience (particularly with regard to adaptation measures)
d   Increase the acceptance of the NAS as well as the strength of the commitment to implementation.

The analysis is based on the results of desk research (review of policy documents, scientific and grey literature) and on the personal experiences of the authors, who were involved in the development of the Austrian NAS and were responsible for the design and implementation of various involvement activities, especially in the active stakeholder involvement process. In addition, the stakeholders involved in this process were asked to evaluate it in a written questionnaire. Thus, the lessons learned also include the stakeholders' perspectives.

TABLE 8.1.1 Overview of involvement formats in Austria, structured according to the level of intensity

	Level of involvement	Time frame			
		2007/2008	2009	2010	2011
Informative	Information giving (provide information to assist stakeholders' understanding of potential climate change impacts and adaptation options)	6 informal workshops for stakeholders	Website on climate change and adaptation in Austria for broad public		
			Internet database on existing adaptation measures in Austria for broad public		
					Newsletter on climate change adaptation for broad public
Consultative	Information gathering (collect opinions and preferences that will assist decision-making)	E-mail survey of scientists from AustroClim	Online survey of broad public		
	Consultation (obtain feedback on specific policies or proposals)			3 stakeholder consultation rounds on drafts of the NAS	
Decisional	Active involvement (involve people actively and allow some influence on ownership of decision)		13 workshops with scientists to develop lists of adaptation options		
			16 stakeholder workshops to discuss adaptation options identified by scientists		
	Partnership (redistribute power through negotiations between powerholders and stakeholders)	(no involvement at this level)			
	Empowerment (place final decision-making in the hands of the stakeholders)	(no involvement at this level)			

Source: Authors' compilation; structure based on Wilcox in Dialogue by design 2010, Arnstein 1996, Green & Hunton-Clarke 2003 and Standards of Public Participation 2008

8.1.2   Background: The Austrian NAS

Before focusing on the active involvement process, we first briefly describe the process of developing the Austrian NAS in order to establish the link between the strategy and the involvement process. In 2007, the Federal Ministry of Agriculture, Forestry, Environment and Water Management assumed the responsibility for developing an Austrian NAS. The strategy, adopted in October 2012, aims to reduce the negative impacts of climate change, make use of positive impacts and increase resilience to climate change (Lebensministerium 2012). It intends to create a national framework to ensure coordination and harmonisation of the various climate change adaptation activities in all areas and at all levels of decision-making. The strategy consists of two parts: the framework document and the action plan for 14 sectors. Due to the intensive stakeholder involvement and the cooperative identification of adaptation options as part of the action plan, involving the scientific community and various stakeholders, the development of the NAS took several years.

8.1.3   Practical experiences from the active stakeholder involvement process

In all, 16 stakeholder workshops were organised in the active stakeholder involvement process to discuss the adaptation options identified by scientists. Within this process, the *stakeholders' role* can be described as that of *practical experts* whose knowledge complemented the adaptation recommendations provided by the scientists. The process consisted of two phases:

- Phase I, Summer 2008 to Autumn 2010: Discussion of the sectors perceived as being the most vulnerable (agriculture, forestry, electricity, tourism and water)
- Phase II, Autumn 2010 to Summer 2011: Examination of other important sectors (natural hazards, health, building/construction, biodiversity/ ecosystems and transport infrastructure).

Both phases were initiated and carried out by the Environment Agency Austria (EAA) with financial support from the Austrian Climate and Energy Fund. The EAA was supported by two professional external moderators, who provided expertise for the process design and chaired the meetings. Due to the fact that the process was closely related to other activities carried out by the responsible ministry, a steering board was established, consisting of two representatives from the responsible ministry, one representative from each of the nine provinces, three representatives of the EAA and the two moderators.

Identification of stakeholders

At the beginning of the stakeholder involvement process, a stakeholder analysis and selection was carried out by the EAA in close cooperation with the steering board to ensure that all stakeholders from organisations likely to be affected by the impacts of climate change and/or relevant for the implementation of adaptation activities were invited to participate. Relevant organisations were identified through existing contacts from other processes, via brainstorming in the steering board and through internet research guided by questions such as:

- Who is potentially affected by climate change?
- Who can influence decisions with regard to adaptation?
- Who can contribute to the quality of the NAS?
- Who is responsible for implementing (and financing) the potential adaptation actions?
- Who can increase the effectiveness of the adaptation action through involvement or decrease its effectiveness by non-participation?

Via e-mail, the identified organisations were informed about the NAS development process and invited to participate. When there was no reply, a personal contact was established by phone. Although this approach was time-consuming, the personal nature of the conversation increased the motivation to participate in the process significantly. Thus, only a few of the organisations contacted declined to join the process due to a lack of interest in or experience with climate change adaptation. Climate change (adaptation) is largely perceived as a technical or an environmental topic, and stakeholders representing other issues do not always perceive that their interests may be affected. In the case of some environmental NGOs, a lack of resources was the main reason for non-participation.

To guarantee continuity in the group of participants, the involved organisations were invited to nominate two people – a representative and a substitute – who would be authorised to speak on behalf of their organisation. These nominated representatives were invited to all workshops in the active stakeholder involvement process and had the possibility of nominating further experts for specific sectoral workshops.

Involved stakeholders

The involved stakeholders can be divided into three roughly equal groups: one-third federal and provincial ministries or related organisations (e.g. Railway Austria, Austrian Federal Forests), one-third interest groups (e.g. Chamber of Agriculture, Austrian Automobile, Motorcycle and Touring Club) and one-third social/environmental NGOs (e.g. CARE, Austrian Alpine Club) and other organisations (e.g. companies such as Austrian Hail Insurance). In Phase I, 50 organisations were involved; Phase II included 100 organisations.

Background documents and written consultation process

In general, the discussions in the workshops were based on input documents containing adaptation recommendations developed by scientists (Haas et al. 2008, 2010). In most sectors (except health, transport and ecosystem/biodiversity), the recommendations were integrated into a policy paper (i.e. working papers, which later developed into the Austria NAS) and sent for a first consultation to the stakeholders before the beginning of the active involvement process. In the workshops, each adaptation recommendation was presented in the format of a table filled with information obtained from scientists and the consultation process; this table was printed out on A1 paper and hung on the wall of the meeting room. Within the discussion, these recommendations were clustered and/or restructured; some were deleted and others added. The discussion was guided by questions on existing directives, strategies or processes for mainstreaming adaptation, responsibilities and the financial resources required for implementation, potential conflicts and synergies between different adaptation options, knowledge gaps and the next steps necessary for implementation.

Design of stakeholder workshops

The format of the active stakeholder involvement process changed over time. In Phase I, five workshops (a kick-off, a workshop for the prioritisation of adaptation options to be discussed in two follow-up workshops and a workshop including stakeholders and scientists) were held; these were full-day events at which stakeholders from all of the sectors addressed in Phase 1 (agriculture, forestry, electricity, tourism and water) were present. The workshops alternated between plenary sessions and parallel sessions in smaller groups. In the plenary sessions, issues of importance for all sectors were discussed, while the parallel sessions mostly dealt with sector-specific issues. The last workshop was used to discuss open research questions. To foster the interaction between science and stakeholders, the group of participants was expanded at this workshop to include scientists from universities and research institutes.

In Phase II of the process, after a kick-off event including stakeholders from all of the sectors addressed in this phase (natural hazards, health, building/construction, biodiversity/ecosystems and transport infrastructure), a series of two half-day workshops for most of the five sectors was organised (exception: natural hazards, where stakeholders felt that due to the high quality of the input document no second meeting was needed). Due to the fact these half-day workshops took place on different dates, stakeholders had the opportunity to participate in workshops addressing different sectors. To encourage the exchange of the results obtained in the sectoral workshops and to discuss cross-sectoral issues, a full-day final conference including stakeholders from all five sectors was held at the end of this workshop series.

For the design of the workshops, the external moderators suggested a set of methods that were discussed with the steering board. The methods implemented were varied, ranging from expert presentations and expert discussions to world cafés, moderated discussions in small working groups and fishbowl discussions.

Results of the stakeholder involvement

The stakeholders' comments, experience and knowledge regarding concrete adaptation measures were included in the input documents for the policy paper. In addition, the results from the general discussion in the workshops were summarised in the form of minutes. Both documents – the supplemented input document and the minutes – were sent to the stakeholders no later than two weeks after the workshop. The stakeholders had the opportunity to provide corrections to statements in the documents. The final versions of the input documents for each sector were forwarded to the responsible ministry. In many cases, the ministry included the suggestions from the stakeholders verbatim in the NAS. In addition, highlights from the active involvement process were published on the website www.klimawandelanpassung.at in order to make them widely accessible to the interested public.

8.1.3   Lessons learned

We will draw the lessons learned from an examination of the process and outcomes on the basis of the goals originally set out and communicated. Because the NAS has only recently been politically adopted, the fourth aim regarding the increase in acceptance and the commitment for implementation cannot be addressed at the current stage.

a) Raise awareness of climate change and the need for adaptation among affected stakeholders

When the process began in 2008, climate change mitigation had clearly reached the stakeholders' awareness, but adaptation to climate change seemed to be a difficult concept to grasp. At the beginning of the process, the two concepts were often confused by the stakeholders or used synonymously, but clarification was achieved over the course of the involvement process. Thus, the first aim of the active stakeholder involvement – raising stakeholders' awareness of climate change and adaptation – was clearly achieved within the process. This impression was also shared by the stakeholders.

With regard to the second aspect of the aim – raising awareness of the need for adaptation – the situation was different. At the beginning of the process, most stakeholders involved had not put adaptation to climate change on their organisational or personal agenda and adaptation was therefore not a *priority*. As a result, participation in the active involvement process was mostly an *add-on* to their everyday business and was thus often constrained by a lack of time and resources.

Nevertheless, the interest in the topic and process was significantly high, although the level of engagement varied – depending on available resources – ranging from merely receiving information via mail up to attending all workshops and making a significant contribution. In addition, a number of the stakeholders involved are now starting climate change adaptation processes within their own organisations or are undertaking activities focused on adaptation (e.g. financing research projects, conducting workshops, developing their own adaptation strategies).

b) Provide a platform to discuss opinions, experiences, needs, preferences and (possible) conflicts and to develop balanced solutions (or at least recommendations)

A positive and constructive working atmosphere prevailed throughout the process, especially among the stakeholders who attended more than one workshop. In contrast, new stakeholders often acted more as lobbyists, focusing on their personal position and interests. In some sectors and for certain adaptation recommendations provided by scientists, stakeholders made very strong statements in an attempt to avoid any possible negative consequence for their clientele. Thus, in some cases, compromises or minimum solutions were developed in the hope of avoiding conflicts in the later stage of implementation.

Difficulties arose in the discussion of aspects that went beyond a sectoral view or beyond one specific discipline. This was particularly evident when attempts were made to discuss synergies and conflicts between adaptation measures in different sectors.

c) Enhance the quality of the NAS by including stakeholders' knowledge and experience (particularly with regard to adaptation measures)

The quality of the NAS was definitely improved through the active stakeholder involvement process. In general, quality depends on stakeholders' willingness to participate in the process on a voluntary basis. In this regard, it is extremely important to include the appropriate people from the start (Few et al. 2006, Glicken 2000). In the active involvement process, this was challenging in some cases due to the fact that adaptation is not commonly a priority for organisations (although this changed over the course of the process). As a result, some organisations sent representatives with limited mandates – they were to attend the process as listeners rather than as active participants capable of reacting in cases of conflicting interests. These representatives with a weak mandate could not speak on behalf of their organisations without consulting with their superiors.

During the stakeholder workshops, the social impacts of climate change (adaptation) might have attracted more attention if more stakeholders representing social topics had participated. The goal of stakeholder continuity could not be achieved in all cases, and climate change scenarios and complex topics such as uncertainty had to be presented and discussed again for newcomers. It was also evident that power inequalities between stakeholders with enough resources to participate in all processes and others that lacked such extensive resources (e.g. environmental NGOs) could not be counterbalanced. One solution to this dilemma might be the reimbursement of stakeholders for their participation in workshops.

8.1.4 Concluding thoughts

The active stakeholder involvement process in Austria confirms the assumption that a cross-cutting and complex theme such as adaptation to climate change calls for inter- and transdisciplinary approaches and the cooperation of all affected stakeholders. Nevertheless, the active involvement process showed that this new culture of cooperation on a specific issue that cuts across sectors and extends beyond organisational competencies and hierarchies challenges existing structures. Entrenched viewpoints and lobbyism, the fear of losing authority and influence, as well as negative experiences with similar processes obviously presented problems, especially at the beginning of the process.

Interestingly, even though there was a similar setting in every workshop, the results and the group dynamics within each individual workshop varied. This shows the limits of methods for inclusion processes: even when a method has been used successfully in previous workshops, additional methods/approaches may be needed in others. The human factor made every workshop different, and a diversified skill set was required to deal with the various possible developments in the phases of involvement.

To summarise our experiences: active stakeholder involvement processes are a challenging and resource-intensive task for all participating parties, stakeholders, the responsible ministry and organisers. They are characterised by mutual social learning on different levels, taking place within a framework of group dynamics that is also dependent on external (political) factors and the stakeholders' histories with one another (and the organiser team). Nevertheless, the effort was worthwhile, considering the results in the context of the goals of the process.

8.2 Stakeholder involvement for developing adaptation innovations in rural areas: Examples from Berlin-Brandenburg

Andrea Knierim

8.2.1 The Innovation Network on Climate Change Adaptation in Brandenburg Berlin

The Innovation Network on Climate Change Adaptation in Brandenburg Berlin (INKA BB) is a nationally funded, large-scale research project in two federal states of Germany – Berlin and Brandenburg – that connects scientists with practitioners for the development and implementation of adaptation innovations. The main aim of stakeholder involvement in this science-driven project is the identification of appropriate adaptation measures to climate change. In line with Brandenburg's major challenges in terms of the natural system, namely sparse rainfall and sandy soil in combination with periodic droughts, the network concentrates on innovations in land-use and water-management practices that will facilitate adaptation in order to overcome climate change challenges (MLUV 2008, Knierim et al. 2009).

In 2007, the German Federal Ministry of Education and Research (BMBF) issued a call for projects on regional networks for climate change adaptation that would increase the economic competitiveness of the network partners through innovation development in science-practice cooperation. In a two-year phase of stepwise refined proposal preparation, members from 12 different research bodies and various stakeholders from local, regional and state-level governance systems in Brandenburg and Berlin became involved and engaged in a working process of problem formulation and the development of adaptation measures. In 2009, the transdisciplinary network INKA BB was officially established, consisting of 24 modules in the activity fields of land-use, agriculture, water management, health management and network development. The publicly funded network project will run until April 2014, but the regional network cooperation is intended to attain long-term stability.

8.2.2 Insights from INKA BB's stakeholder involvement

The involvement of practice actors in research and development cooperation with science partners takes place in 20 out of the 24 INKA BB modules in the form of transdisciplinary group work (e.g. on adapted integrated and organic farming, adapted forest management, management instruments for watersheds, water-quality management for lakes, communal water management, etc.).[1] The actors involved in INKA BB are numerous and diverse (cf. Table 8.2.1). In addition to a large number of scientists, they include representatives from public agencies at the state and regional levels and from various professional associations, interest organisations and the private sector (of which a large number are farmers).

The process of stakeholder involvement and especially the identification of the practice partners has varied across the different modules. However, there was one core principle guiding the selection of who was contacted in all modules: problem orientation. During the preparation phase, scientists initiated the first contacts, addressing potential cooperation partners based on their (the scientists') 'problem assumptions' regarding the need for climate change adaptation measures. These assumptions were discussed and specified with interested practitioners before a research proposal was submitted to an internal module selection board. In the following section, stakeholder identification and cooperation is described, using the example of agricultural land-use management.

TABLE 8.2.1 Types of actors in INKA BB (close and occasional cooperation)

	Close cooperation (number)	In ... modules (out of 24)	Occasional cooperation (number)	In ... modules (out of 24)
Scientific institutions, departments, etc.	120	24	150	24
Public agencies	13	15	65	18
Associations, lobbying groups	15	15	50	18
Enterprises, entrepreneurs	30	17	70	21

Source: Author's compilation

Who are the involved stakeholders?

For agriculture, the rules for stakeholder involvement were not prescribed by the project coordinators, but rather evolved during the project's preparation in accordance with the specific problem orientations and foci of the scientists involved. The practitioners addressed were representatives from the sector (farmers, leaders of farmers' organisations, heads of public agencies and private consultancies) at the state and district levels. Consequently, agricultural stakeholders and actors form an important group of partners in INKA BB (cf. Table 8.2.2):

- With regard to the overall network coordination, the leader of the largest farmers' association in Brandenburg (*Landesbauernverband Brandenburg*) is one of the six members of the coordinating team.
- In the modules related to agriculture, nearly 50 farmers are partners, most of whom conduct on-farm research.
- In 7 modules, specialised farmers' associations (e.g. for organic farming, for livestock or plant breeding, etc.) are partners.
- In 6 modules, agricultural experts from public research and administration agencies are additionally involved.
- In some modules, consultancies or civil society organisations are included.

Thus, there is a great deal of diversity among the actors engaged in these modules although not every agricultural practice partner is a stakeholder (= representing larger groups' interests).

Empirical research over the course of the project has revealed that the involved farmers are mostly managers of large-scale farms who have high-level professional qualifications. With these socio-economic characteristics, the practice partners involved in INKA BB are far above the German average and even the relatively high standards in Brandenburg (Bundschuh & Knierim 2013). Whether and how this 'asymmetric' stakeholder involvement might represent an

TABLE 8.2.2 Overview of the agricultural stakeholders involved in INKA BB

	Integrated agriculture	Organic agriculture	Adapted varieties	Horticulture and tree-lined paths	Climate risk insurance	Pasture management	Wet grasslands	Agro-forestry	Irrigation systems
Research	1	2	1	2	1	2	2	5	2
Farmers	11	4	3	11		6	7	4	2
Farmers' organizations	2	3	1	3	1	4		2	
Public agencies	1		1	7		2	3	1	1
Companies, consultancies	2	3			1		1	3	2
Education, training, other NGOs	1		1	1		2			

Source: Author's compilation

obstacle for the project's achievement of its objectives requires more attention; this can only be answered through further thorough investigation.

How did the initial involvement take place?

The involvement of the agricultural stakeholders in INKA BB was guided by the general orientation of the network towards the solution of climate change adaptation challenges in agricultural land-use. Examples of the topics addressed include (i) climate-proof varieties in agriculture, horticulture and agroforestry, (ii) technologies and measurement techniques for water-conserving land-use practices, (iii) climate-adapted cropping systems and (iv) technological options for irrigation needs. The actual initial involvement took place at several levels of decision-making:

- The network initiators (social scientists) addressed and pro-actively invited the leader of the largest farmers' association to become a partner at the consortium's coordination level.
- Agronomists and livestock-husbandry specialists as well as horticultural and agroforestry researchers renewed ties with farmers and professional organisations they had previously collaborated with, developing bilateral working relationships for the cooperation in INKA BB.
- Additionally, over the course of the project, more farmers became involved in the modules – some directly contacted by scientists, some on their own initiative and others through intermediary parties (cf. Bundschuh & Knierim 2013).

In all cases, the cooperation of agricultural practice partners in the network took place on a voluntary basis, dependent on their specific research interests and questions and on their willingness to become an active participant. In all cases, these partnerships were formalised by a cooperation contract that was signed immediately after the project's start. Conceptually, this approach was based on the principle that it is not the representativeness but rather the motivation and engagement of farmers that will be decisive for successful changes in behaviour and the adoption of innovations related to climate change adaptation (Knierim & Hirte 2011).

In practice, the initial involvement took place in two different ways. (i) A few stakeholders, including the leader of the Brandenburg farmers' association, actively participated in the project's design and in the definition of research questions at the module level, supporting the elaboration of the proposal from 2007 to 2009. (ii) Most other farmers became involved once the project had started, when scientists at the module level organised the first transdisciplinary workshop. At this point, there was a certain margin for negotiation regarding the actual topics that would be addressed. For example, in the case of the organic agriculture module, although some farmers had already been involved in the module's conceptual design, the actual analysis of the problem situation and the identification of the partners' strengths and weaknesses and the opportunities and risks of climate change in Brandenburg took place in the context of the first transdisciplinary workshop. Here, the objectives of the module were revised and updated in a science-practice dialogue (Siart et al. 2012).

In summary, the identification of the INKA BB practice partners in the field of agriculture basically followed informal pathways. The process can be described as a *relationship-driven approach* to stakeholder involvement that followed one fundamental principle of networks: the idea that mutual trust and voluntary cooperation are key determinants of success (Willke 2001).

How is stakeholder involvement practised in INKA BB?

The involvement of stakeholders over the course of the project happens at two levels: in the project's overall coordination and in each of the modules. Within the modules, involvement takes place according to two parallel logics. (i) There is a general and formalised procedure that structures the overall project's working process into six phases (initial phase, implementation I, synthesis I, implementation II, synthesis II and consolidation) and prescribes certain tasks for all modules. In each of these phases, stakeholders are to be involved. (ii) There is also self-organised and flexible cooperation between the different partners at the module level, according to their specific needs, interests and habits.

Re (i): The conceptual framework of INKA BB prescribes a repeated transdisciplinary working process of situation analysis, planning, implementation and evaluation that is to be realised at the module level. From reports and from participant observations, it has become obvious that the actual involvement of the stakeholders varies considerably between the modules. In some modules, regular and intense exchange and practical cooperation are found, including on-farm research by practice partners. In other modules, contact and collaboration with practitioners are less frequent or less close in nature, consisting only of consultations. For example, in the interim evaluation of INKA BB, it was found that some scientists tend to avoid transdisciplinary group meetings, relying instead on bilateral communication (Siart & Knierim 2013). Here, the time-consuming burden of planning, organising, conducting and evaluating group events is mentioned, and some respondents report that practitioners are not readily available for such meetings or that the prescribed group meetings in the synthesis phases do not match the module's internal working logic.

Re (ii): As a complement to the formal stakeholder involvement in INKA BB described above, diverse forms of interaction and cooperation can be found throughout the modules. The following examples from the agricultural modules illustrate the variety of these interactions.

- In the organic agriculture module, field trials are designed, implemented and evaluated on four farms, with direct and repeated transdisciplinary communication and interaction (Siart et al. 2012).

- Similarly, several land-use modules have established a number of on-farm experiments related to their specific research topics (e.g. soil conservation techniques, adapted crop varieties, irrigation technologies). Of course, not all on-farm trials have been successful, and in some cases, farmers have decided to end their collaboration with INKA BB. However, other farmers would like to extend the scope of their contact and exchange with scientists.
- Several land-use related modules have jointly organised a demonstration event at a research station in which 70 to 100 farmers have participated; throughout the presentation of the various experiments, farmers are invited to comment on the results, exchange experiences and propose future research endeavours.

The involvement of the Brandenburg farmers' association in the network's internal coordination is ensured by the regular participation of its leader in the INKA BB coordination group. The coordination group is made up of the two network coordinators and four representatives from the different activity fields. It meets four to five times a year and discusses and deliberates upon all essential decisions to be taken in the network. By actively representing farmers' interests within this group, the leader of the farmers' association has gradually increased his influence on the network's coordination, making proposals for the design of events and for the potential outputs of the network. In the third project year, he started a science-practice dialogue in Berlin and Brandenburg in order to institutionalise the partnerships that had been initiated between farmers and research organisations and to prioritise farmers' questions and issues on official research agendas.

Overall, among the current agricultural practice partners in INKA BB, motivation to participate is generally quite high. However, the degree of involvement is also determined by the partners' interests, capacities and resources.

8.2.3 Concluding thoughts

Stakeholder involvement is an iterative process

In a network such as INKA BB that focuses on the development, implementation and evaluation of innovations, the identification and involvement of stakeholders must necessarily begin informally: real cooperation is heavily dependent on existing contacts, positive personal experiences and on the willingness of those concerned to engage. In the case of INKA BB, this involvement generally started with a research issue raised by a scientist that was recognised by one or more practitioners as corresponding to a real-life problem. Cooperation was then initiated, but frequently also led to a redefinition of the problem and to a reformulation or refinement of the research question such that it became more applicable to stakeholders' interests, thus fostering their involvement.

This type of iterative research and cooperation process has pros and cons. One strength is the (increased) influence on research that practitioners can obtain through cooperation, and thus the ability to augment the relevance of the research for practice. The challenges of such cooperation are clearly greater on the side of the scientists: they must be open and flexible with regard to research topics and must invest time in initial communication and interactions.

Identification of stakeholders can be driven by problem orientation

The science-practice cooperation in INKA BB shows that a problem-driven identification of stakeholders is an appropriate approach for involving people who perceive a problem, have an intense interest in the issue and a (certain) willingness to invest time and resources in the cooperation. However, there are certain limitations on the outcomes of this approach. Not all those involved are truly 'stakeholders', i.e. represent a larger group of people with common interests. Those involved neither speak for the average member of their professional group, nor do they stand for a representative selection of it. Especially with regard to the farmers engaging in INKA BB, they can be considered a 'positively biased sample' of the better-educated and more intellectually curious farmers from large farms.

It is possible that this asymmetric involvement is a reflection of a typical phenomenon of bottom-up stakeholder involvement: those who have the means (time, knowledge, resources) to participate do so, while other (less well-equipped) people tend not to get involved. Thus, before designing and implementing an involvement process with practitioners, scientists must be very clear about the objectives of the cooperation. Is *any* cooperating farmer appropriate, or should there be, for example, a certain range of farm sizes or socio-economic characteristics represented in the sample? A systematic and sound argumentation must be developed before practitioners are approached.

Personal and financial resources are crucial for successful cooperation

Finally, the experiences of INKA BB reveal that transdisciplinary cooperation is challenging for practitioners in terms of resources. In general, the amount of time required is initially underestimated, especially for communication and decision-making processes. Both practice partners and scientists have to learn the other group's 'language' first before they can become successful collaborators. Such personal engagement will only be maintained by practice partners if their interests are (at least partially) addressed. Some of the partners also contribute material resources, e.g. in the form of the fields and machinery used for on-farm trials. In INKA BB, there are considerable differences within and among the modules with regard to farmers' compensation: some farmers are financially compensated by INKA BB for cooperation, while others are not. In both of these groups (compensated and non-compensated), not all farmers contribute material resources. In addition, the motivation of researchers can be constrained by their primary goal (namely, developing scientific papers, not spending time working with stakeholders); this must be compensated by means of other incentives.

Resumé

In INKA BB, the main aim of stakeholder involvement was and is to identify appropriate climate change adaptation measures through cooperation with practitioners. For this objective, the iterative relationship-driven and problem-oriented approach to stakeholder identification and involvement is appropriate, even though in some cases it has meant changing the plans and prescribed stages for stakeholder involvement. Although the results of the transdisciplinary cooperation in the modules are valid, they are not 'representative results' for, e.g. farmers in Brandenburg. Both more systematic studies and exchange and dissemination activities will be necessary to assess the transferability of the developed adaptation measures.

8.3 Ancona is getting ready! How the city of Ancona is building resilience using a participatory process

Marco Cardinaletti

8.3.1 Background information: Ancona is getting ready for climate change

The city of Ancona (Italy) is the capital of the Marche Region and home to about 100,000 people (cf. Figure 8.3.1). It is the site of one of the most important ports in the Adriatic region for passengers, cargo and fishing. The area is characterised by a relatively low population density (814.97 people/km²) but features a rapid and land-use-intensive building development. The city borders are defined by the Adriatic Sea and by a hilly area in close proximity to the Monte Conero promontory.

On 13 December 1982, a large landslide affected the northern area of Ancona, specifically the 'Montagnolo hill'. This hill started sliding towards the sea, destroying private houses, public buildings and important infrastructure. The dramatic event involved about 180 million cubic meters of soil and rock. The landslide covered an area of 220 hectares, representing around 11% of the urban area of Ancona (Cardellini 2008). The event affected the entire community and produced severe social, environmental and economic impacts. For a number of days, the city remained without necessary services. The two main transport routes, the national railway (Bologna–Lecce) and the access road SS19 (Flaminia), were interrupted, as were gas and water supplies. As a result, the entire city system was non-functional for a long time. Important buildings (the faculty of medicine, hospitals, public offices) were destroyed and entire residential districts were completely evacuated. In total, more than 3,000 people were evacuated from the area.

After this disaster, the local community (institutions, NGOs, universities, experts, civil society) cooperated to rebuild the area that was destroyed. As a consequence of the experience of this disaster, the municipality has been heavily involved in implementing a Sustainable Urban Management Strategy. This strategy seeks

FIGURE 8.3.1   Geographical context of the city of Ancona

to increase the resilience of the city by implementing mitigation and adaptation projects and actions in order to reduce CO_2 emissions and disaster risk exposure connected to potential climate change impacts. According to the baseline scenario (ISPRA 2011) produced in 2011,[2] the City of Ancona will be affected by a progressively increasing trend in mean temperatures as well as greater uncertainty and more irregular behaviour in precipitation (high variability, non-seasonality) at the end of this century. These projected changes in climate trends could produce many impacts affecting the safety and the quality of life of the local community. Some important economic sectors such as tourism, fisheries and commercial and public services could be impacted as well. For example, the projected increase in the intensity of precipitation could result in an increase in the percentage of citizens directly affected by landslide risk, rising from the 10% of citizens currently exposed to a high level of risk to 14% by 2050.

In the following list, the most important steps that Ancona has taken towards increasing resilience over the last five years are described.

2006–2008:   Implementation of the early warning system for reduction of landslide risks.

2008:   The municipality of Ancona became a partner in the Sustainable Energy Campaign; in the same year, it signed the Covenant of Mayors, implementing its first Energy Master Plan.

2009:   The municipality of Ancona, within the framework of the White Paper 'Adapting to Climate Change: Toward a European

Framework for Action' (COM 2009), launched a new co-funded European project entitled 'ACT – Adapting to Climate Change in Time' (www.actlife.eu).

2010:   The municipality signed the World Disaster Reduction Campaign during the international conference 'ACT! Strategies and Experiences to Increase Urban Resilience to Climate Change', identifying the '10 Essentials for Making Cities Resilient' (UNISDR 2012) as priority goals for Ancona. Adaptation has been mainstreamed into all local policies, heavily involving the community throughout the process (Hoornweg et al. 2009).

8.3.2   Practical experiences from building a resilient community

The creation and the development of a resilient community must be fundamentally based on the active participation of all key stakeholders (UNEP 2005). This requires the selection of a physical place for discussions and meetings in which the participation process can become more concrete, grow and consolidate over time. In such a place, it will be possible to gradually build the collective consensus and the values upon which the community can move forward, thus creating a cultural identity. For this reason, the participation process must be well organised in order to be effective (Hemmati 2001). It must create organisational and managerial solutions that will support and assist the cooperation between all parties. In the city of Ancona, the process of building resilience has been fully embedded into a participatory process from the very beginning. This commitment came directly from the mayor and from the city council, who decided to approach climate change using a participatory planning process. The involvement process started in 2010 with the establishment of:

- *An internal 'core group'*: a team of administration-internal technicians and experts directly related to the adaptation issue: the head of the Urban Planning and Environmental Department, the individuals responsible for the Early Warning System and for civil protection, engineers and experts from the GIS systems, geologists and economists. Within this group, all roles and responsibilities are clearly defined. The internal core group was formalised by an official document signed by the city council.
- *A Local Adaptation Board (LAB)*, with a restricted group size. The LAB, which has been involved since the beginning of the process, was formed using the Direct/Indirect Influence Matrix (NRTEE 1993). This matrix was crucial for the identification of the key actors to be involved, and for assessing the level of influence and interest that they would have in policies and climate change strategies. Furthermore, the matrix is an effective instrument for defining the communication and involvement strategies that must be adopted to reinforce dialogue and partnerships during the various phases of the planning process.

Table 8.3.1 shows the stakeholder categories that were determined to be useful for the city of Ancona in the ACT Project. The Ancona LAB was established in March 2011 to ensure the effective participation of all local stakeholders involved in a particular sector, guaranteeing that these actors would have the possibility to discuss and decide on a common strategy and common actions to tackle the local impacts of climate change. The number of participants can vary according to the size of the city, but it should not be too high (in order to ensure effective collaboration and teamwork between members). The Ancona LAB consists of 10 members representing all stakeholder categories with a direct influence on the adaptation process in Ancona. As members of the board, they have signed a formal agreement of cooperation with the municipality. This group includes managers, experts, researchers and representatives from local institutions who operate in the sectors in which the city is attempting to build resilience.

- *Focus Teams*, formed by stakeholders directly interested in the specific impacts and issues analysed. These teams include some members of the LAB and certain other stakeholders who have specific knowledge of, but only indirect influence on, the issue assessed. The Focus Teams can be consulted at key points during the process. In the city of Ancona, four Focus Teams are currently working actively on the following issues: (i) landslide management, (ii) historical heritage conservation, (iii) coastline protection and (iv) infrastructure of transportation and city connections. These teams are consistent with the stakeholder categories presented in Table 8.3.1. Each of these groups, after developing a climate change impact and risk analysis for each sector, works to define feasible adaptation actions and solutions in order to improve the resilience of the city in the most vulnerable sectors.
- *A Local Climate Change Forum*, open to all citizens and the entire community. In this Forum, all community representatives have a voice. The Forum is usually organised annually in order to inform the entire community about the results achieved, discuss new projects, evaluate strategies and potential actions, propose ideas, suggest solutions and build consensus.

It should be noted that during the process of building a resilient community, the methodologies and the levels of involvement have varied depending on the specific activities to be carried out, the requirements that had to be satisfied and the knowledge that each stakeholder could provide. There were times when wider participation was required, and other times when a more restricted group of people was needed. There were also situations in which the level of involvement was limited to the simple transfer of information, and other circumstances in which there were debates and discussions or there was a necessity to build partnerships and collaborations without which it would not have been possible to implement projects and put concrete ideas into action (Dodds & Benson 2011). Even the instruments and the techniques that had to be adopted differed based on the level of involvement that had to be achieved. Table 8.3.2 systemises these various dimensions.

TABLE 8.3.1 Direct/indirect influence matrix for the Ancona Local Adaptation Board

Issue	Direct stakeholder	Indirect stakeholder	Wider community
Transport and infrastructure	Regional/Provincial Department of Transportation Rail sector Municipal Department of Mobility/Environmental/Urban Planning Civil protection University of Engineering	Private associations for mobility Society of Highway Management Port authority Private companies for logistics and freight transportation Airport authority Chamber of Commerce	Citizens Customers and city users Media Schools and universities (students)
Landslides	Civil protection University 'La Sapienza' – Faculty of Geology Early warning system ICT companies and software firms Regional Department	Families living close to the landslide area (exposed to high risk) Technical experts Foreign universities and technical expertise	Citizens Media International partners
Coastal erosion	Regional Department for ICZM University of the Sea Municipal Department of Urban Planning	Port authority Coast guard Fishermen	Citizens Media Schools
Historical heritage	Ministry Department for Cultural and Historical Heritage ISPRA – Institute for Environmental Protection Urban Planning Department	Institute of research for the vulnerability assessment Church institutions	Citizens Media Schools

Source: Author's compilation

TABLE 8.3.2 Overview of stakeholder involvement instruments used by the Ancona LAB

Stakeholder involvement	Low		Medium	High	
	Inform, train and educate	Consult	Participate	Collaborate	Form a partnership
What?	Inform, train and educate	Consult	Participate	Collaborate	Form a partnership
When?	Factual information is needed to describe a policy, programme or process; a decision has already been made (no decision is required) The public needs to know the results of a process There is no opportunity to influence the final outcome There is need for the acceptance of a proposal before a decision may be made An emergency or crisis requires immediate action Information is necessary to abate concerns or prepare for involvement The issue is relatively simple	The purpose is primarily to listen and gather information Policy decisions are still being shaped and discretion is required There may not be a firm commitment to do anything with the views collected – in this case, advise participants from the outset	Two-way information exchange is needed Individuals and groups have an interest in the issue and will likely be affected by the outcome There is an opportunity to influence the final outcome The organiser wishes to encourage discussion among and with stakeholders Input may shape policy directions and programme delivery	It is necessary for stakeholders to talk to one another regarding complex, value-laden decisions There is the capacity for stakeholders to shape policies that affect them There is an opportunity for shared agenda-setting and open time frames for deliberation on issues Options generated together will be respected	Institutions want to empower stakeholders to manage the process Stakeholders have accepted the challenge of developing solutions themselves Institutions are ready to assume the role of enabler There is an agreement to implement solutions generated by stakeholders
How?	Website Leaflets, flyers Educational material Documents and training books Newsletters Technical papers Energy days	Public comments Focus groups Seminars Surveys Online consultation	Multi-actor policy workshops Deliberate polling Seminars Online consultation Forum	Participatory decision making Citizen advisory committees Work/focus groups	Partnerships Project co-management PPP – Public/Private Partnership

Source: Author's compilation

TABLE 8.3.3 Advantages and disadvantages of stakeholder involvement tools

Activity	Advantages	Disadvantages	Comments
Workshops, conventions, seminars, focus groups	Encourages involvement and expression of views People feel their views are valued Can be very creative and flexible Targeted debate, possibly less confrontational, involves those who are interested and well informed Helps promote a common outlook	May arouse expectations that cannot be met Requires careful management, continuity and follow-up; dependent on quality of facilitation Does not necessarily represent a balanced point of view	Results depend heavily on participants Useful in a range of contexts, e.g. smaller community sessions, to break up larger meetings (sometimes known as scenario workshops)
Newsletters, technical papers, etc.	Sets scene for a dialogue opportunity for all contributors to enhance the image of the project Can be coordinated with a website	Open-ended commitment Can suffer from 'fatigue' if the process is extended May use too much professional terminology	A useful tool for communication, but must be attractive, relevant, accessible and clear
Exhibitions and info days	Can be seen by the entire community Opportunity to present contexts and issues to a large number of people in a very easy way Useful for distributing newsletters, leaflets, questionnaires, educational material or as background for a meeting Staff can directly answer questions and attract interest	May be poorly attended Not all venues are equally attractive Runs risk of dullness	A useful resource when combined with a wider information programme
Info points and urban centres	Permanent communications instrument in the region Enhances not only the transfer of information but also the rapport between the public administration and the community	Costly to develop, both in terms of time and resources Not very flexible Requires management and coordination	These are actual offices that have organisational duties with regard to the planning process, as well as communication duties. Also responsible for stakeholder involvement. They must actively listen and be able to negotiate

	Advantages	Disadvantages	Remarks
Press office activities and local media relations	Large potential audience Relatively cheap Good for public relations Raises awareness	Uncertainty over how media will use material Material may not be used at all, or the media may get the story wrong or stress conflicts	Good PR skills required; results may still be disappointing
Flyers, leaflets	Useful to identify key issues Easy to produce Useful public relations technique Wide coverage	Requires time and money to produce May over-simplify issues May encourage unjustified claims	Probably works best with targeted groups on specific issues, otherwise too expensive; stakeholders are likely to want more comprehensive documents
Website and internet tools for networking	Large potential audience Raises awareness and provides open access to data Positive image Possibility to use social networks or to form thematic groups	Intimidating medium for some sectors of the population Requires constant updating to remain relevant Can be expensive and impersonal	A web presence is the best instrument as long as it is sponsored, well known and user-friendly, permitting the easy acquisition of information and content
Information and training	Possibility of examining specific themes and issues Raises awareness Documents and papers are easily distributed	Difficult to involve some types of stakeholders (e.g. politicians) Very focused Activities are costly and consuming	Training courses are excellent instruments to share in-depth knowledge with specific parties (students, employees)
Gadgets (t-shirts, magnets, pens, etc.)	Have attraction potential; capture people's attention easily Easy to distribute and effective in transmitting a message Information consolidation; usually they are used and reused (t-shirts, for example)	Provides 'spot' messages and information Does not delve into problems, but merely raises awareness Can be expensive and impersonal	Powerful instrument to transfer brief messages or claims. Usually used with the objective of educating and increasing community awareness about a certain issue

Source: Author's compilation

This table was used by Ancona's LAB to develop its stakeholder engagement strategy (Cardinaletti & Di Perna 2009). It is a simple and very useful 'information transfer tool' that can be employed to determine a strategy based on the level of involvement that must be achieved.

8.3.3 Communication that informs, trains, educates and therefore involves

'Inform, train and educate' are three important actions that require focused attention in order to reinforce participation and involvement and to create a resilient community. Consequently, the available tools need to be utilised in the most effective way possible. Along with stakeholder mapping (cf. Table 8.3.1), a table of communication and stakeholder involvement tools can be useful. Table 8.3.3 presents a summary of the main tools utilised by the Ancona LAB, including their advantages and disadvantages for encouraging interactions and relationships.

8.3.4 Concluding thoughts: Difficulties and lessons learned

Many difficulties and obstacles were encountered in the course of the city of Ancona's participatory process; many lessons were learned as well. Most of the difficulties could be solved, but others were only partially mitigated. The lack of financial resources was the first difficulty. The implementation of the adaptation strategy required a huge amount of financial resources, which in most cases was difficult to secure. For this reason, a concerted strategy is required at both the government level and the community level as well. For instance, after the landslide event, in the reconstruction phase and during the full implementation of the early warning system, strong financial support was provided by the Marche Region. Overall, the fund-raising issue remains the most pressing problem; it must be resolved to ensure continuous investments in the new technologies and ameliorative solutions that are often required in the implementation of an adaptation strategy. Fostering Public/Private Partnerships (PPP) could be an innovative way of facilitating the implementation of urban adaptation projects. The involvement of the private sector and the implementation of new financial engineering mechanisms such as JESSICA[3] could be an interesting tool to exploit. On this issue, the city of Ancona has recently joined the CSI Europe project, networking with other European cities. This project, led by the city of Manchester and funded by the URBACT Program, focuses on the involvement of cities in Urban Development Fund (UDF) structures and on how these instruments can be more effectively embedded in future city planning and governance.

Another common difficulty, also seen in the case of Ancona, is the weak involvement of local communities and institutions, especially at the very beginning of the adaptation process. When they are not well organised and effectively led, the key local actors will not be able to work together on shared and concerted strategies. Over the last 10 years, the municipality of Ancona has invested a great deal of effort to strengthen community involvement by organising public events, workshops and public forums dedicated to the landslide issue as well as to the issue of climate change. The establishment of a permanent Local Adaptation Board (LAB) and the city's investment in a comprehensive stakeholder involvement strategy has allowed the main local stakeholders to be involved in a bottom-up adaptation planning process. The strong commitment of the municipal administration was the main success factor stimulating the entire community to actively take part in implementing local adaptation and mitigation strategies.

The last problem that should be highlighted is that there are not many consolidated experiences, knowledge or skills regarding adaptation to climate change; consequently, a learning-by-doing approach in which best practices, ideas and positive experiences are exchanged is the only road to follow. In fact, the climate change issue requires new expertise and highly skilled experts with multidisciplinary knowledge, which are often unavailable within municipal administrations. As a result, training courses on climate change and climate change adaptation are often necessary. For instance, the city of Ancona is preparing a dedicated project aimed at creating new professional profiles capable of managing the city's early warning system and the ICT applications. Of course, the 'adaptation issue' could also be an interesting driver for the creation of new jobs in Ancona.

8.4   Lessons learned for stakeholder engagement

Torsten Grothmann

The cases presented in this chapter highlight different positive outcomes of stakeholder involvement processes. Prutsch and Stickler (Section 8.1) and Knierim (8.2) showcase mutual learning and awareness-raising for climate change and adaptation options as well as the possibility of discussing adaptation requirements. Cardinaletti (8.3) stresses the importance of stakeholder involvement as a driver for the implementation of adaptation measures.

All of the cases presented in this chapter on stakeholder involvement in adaptation processes illustrate the importance of the differentiation of two main phases that require different methods: stakeholder identification (e.g. by stakeholder mapping) and stakeholder engagement (e.g. by various workshop formats).

8.4.1   Stakeholder identification

Whereas the stakeholder identification described by Knierim (Section 8.2) is an informal process, the stakeholder identification described by Cardinaletti (8.3) is based on systematic stakeholder mapping.

In their science-driven adaptation project, Knierim and her colleagues identified 'relevant' stakeholders, strongly relying on existing contacts, positive personal experiences and on the willingness of those concerned to engage in order to enable real cooperation. The stakeholder identification was primarily driven by *problem orientation*, a term that refers to the perception of stakeholders with regard to the need for climate change adaptation measures. The case presented by Knierim shows that the problem-driven identification of stakeholders is an appropriate approach to involve people who perceive a problem, have a deep interest in the issue and a (certain) willingness to invest time and resources in cooperation. However, there are certain limitations to the outcome of this approach. Not all those who become involved will truly be 'stakeholders', i.e. will represent a larger group of people with common interests. Instead, the people involved neither speak for the general spectrum of their professional group, nor are they a representative selection of it. Furthermore, only those who have the means (time, knowledge and resources) to participate will do so, while the less well equipped tend not to get involved. Thus, before designing and initiating a stakeholder involvement process, the objectives of the involvement should be clear: is 'any' stakeholder appropriate, or should there be a 'representative' sample of stakeholders present, e.g. to guarantee the involvement of representatives from civil society, business and government? Is mutual learning about climate change impacts and adaptation options the main aim (resulting in the need to primarily identify experts from practice in the stakeholder identification process), or is the main aim the discussion and implementation of adaptation measures (resulting in the need to primarily identify powerful stakeholders)?

Cardinaletti (8.3) identifies stakeholders based on their direct or indirect influence on adaptation measures, applying a systematic stakeholder mapping method (the direct/indirect influence matrix). As a result, here the method is much more systematic and policy-oriented than in the informal procedure described by Knierim, which focuses on involving stakeholders in transdisciplinary learning processes between science and practice. Cardinaletti's focus is on identifying stakeholders that are 'powerful players' and including them in the implementation of adaptation measures. Nevertheless, Cardinaletti also stresses that the aim was also to include stakeholders who have the expertise and knowledge relevant to realising adaptation. Hence, also in this case, learning was one of the aims of the stakeholder engagement process.

Although stakeholder identification precedes the actual stakeholder involvement process, Prutsch and Stickler (8.1) stress that it is sometimes necessary to go back to identifying relevant stakeholders during an involvement process. The goal should be the continuity of stakeholders throughout the process (to build trust between stakeholders and allow systematic learning), but this cannot be achieved in all cases. New stakeholders must sometimes be included, and climate change scenarios, their uncertainties and the difference between climate change adaptation and mitigation will have to be presented and discussed again for these newcomers.

8.4.2   Stakeholder engagement

All of the cases presented in this chapter primarily applied workshop methods to realise the stakeholder involvement. Nevertheless, the specific workshop methods differed between and within the cases. Particularly, Cardinaletti (Section 8.3) stresses the diversity of useful stakeholder involvement methods; their various application should depend on the activities to be carried out, the requirements that must be satisfied and the knowledge that each stakeholder can provide. In his case study, there were times when wider participation was required, and other times when a more restricted group of people was needed. There were also situations in which the level of involvement was limited to a simple transfer of information, and other circumstances in which there were debates and discussions or the need to build partnerships and collaborations without which it would not have been possible to implement projects and put concrete ideas into action. Consequently, the instruments and the techniques that were adopted differed depending on the level of involvement that had to be achieved.

Thus, stakeholder engagement methods – like the methods for building knowledge and awareness (cf. Chapter 7) – must be tailor-made for the group(s) of stakeholders involved. A specific workshop method suitable for one stakeholder group may not be suitable for another. Therefore, the stakeholder involvement process should be well planned and the advantages and disadvantages of particular engagement methods (cf. the overview in Section 8.3) should be systematically considered. Nevertheless, as can be seen in the case presented by Prutsch and Stickler (8.1), the suitability of specific workshop methods for the participating individuals can only be assessed in advance to a limited extent. During the course of a workshop, methods sometimes have to be changed and adapted to the participating stakeholders. This illustrates the need for experienced and flexible workshop moderators and their diverse skill sets to successfully manage the wide variety of possible developments in stakeholder engagement activities.

8.4.3   Challenges in the stakeholder engagement process

The authors of the cases describe various challenges and problems in the stakeholder engagement process. Many of these challenges and problems are known from stakeholder engagement processes in fields other than climate change adaptation.

All three cases presented in this chapter discuss the lack of financial and time resources of particular stakeholder groups as a major barrier to balanced stakeholder involvement. In some cases, this lack of resources has led to an 'asymmetric' stakeholder involvement; for example, underrepresenting environmental NGOs (cf. Section 8.1) or small-scale farmers (cf. 8.2). One solution to this dilemma is to reimburse certain stakeholders for their participation in workshops. If this is not possible, one should attempt to guarantee that the participating stakeholders perceive a personally beneficial output from their participation (e.g. seeing that their interests have been at least partially addressed).

One further challenge in the stakeholder engagement described in the cases was the limited perspectives of the stakeholders, especially at the beginning of the engagement process. Stakeholders acted as lobbyists and focused on the positions of their interest group without being able or willing to consider the interests of other stakeholders. In many of the engagement processes described, a positive and more constructive working atmosphere was achieved after some time, but when new stakeholders joined the group, they again entered as lobbyists. The perspectives of stakeholders are constrained not only by their limited interests but also by their limited expertise. Difficulties arose when aspects were discussed that went beyond the sectoral views or the disciplinary backgrounds of the stakeholders. This was particularly evident in discussions of synergies and conflicts between adaptation measures in different sectors.

Additional challenges – also known from other stakeholder engagement processes – included the fear of some stakeholders that they might lose authority, negative experiences with previous stakeholder involvement processes and group dynamics in the workshops, which can also depend on external (political) factors and the histories that stakeholders have with one another (and the organiser team).

A challenge that is perhaps specific to the issue of adaptation is the often low initial stakeholder interest in the topic, sometimes resulting in a small stakeholder involvement at the beginning of the adaptation process. Stakeholders either do not perceive their stakes in the adaptation topic or they do not view the issue as a priority for their organisation. As a result, in the case described in Section 8.1, some organisations sent representatives with limited mandates to attend the process more in the role of listeners than active participants capable of reacting in cases of conflicting interests (cf. 8.1). These representatives with a weak mandate could not speak on behalf of their organisations but instead had to consult with their superiors, slowing down the engagement process.

One way to address this challenge is to keep trying to motivate stakeholders to participate. This requires a strong commitment to a participatory approach among the organisers of the stakeholder involvement process. In the case of Ancona (cf. 8.3), the intense commitment of the municipal administration was the critical factor that successfully stimulated the entire community to actively take part in implementing local adaptation and mitigation strategies.

Another way to deal with the challenge of a low initial stakeholder interest in the adaptation topic is to raise stakeholders' awareness of potential climate change impacts and the potential benefits of implementing precautionary adaptation measures. This illustrates the close link between the methods for involving stakeholders presented in this chapter and the methods for building knowledge and awareness presented in Chapter 7. At times, before stakeholders are willing to get involved, they need to be told how they could be affected by climate change impacts and/or adaptation measures so that they become more aware of their 'stakes' in the issue.

This relates back to the objectives of stakeholder engagement processes discussed at the beginning of this section on lessons learned. Often, stakeholder engagement is aiming at mutual learning about potential climate change impacts and suitable adaptation options. As Cardinaletti (cf. 8.3) points out, there are not many consolidated experiences, knowledge or skills with respect to adaptation to climate change, such that a learning-by-doing approach in which best practices, ideas and positive experiences are exchanged is the only pathway to follow.

Thus, we can conclude with the following lessons learned:

- It is important to differentiate between two main phases, as they require different methods: stakeholder identification (e.g. by stakeholder mapping) and stakeholder engagement (e.g. through various workshop formats).
- Before designing and initiating a stakeholder engagement process, the objectives of the involvement should be clear: is 'any' stakeholder appropriate, or should there be a 'representative' sample of stakeholders present? Is mutual learning about climate change impacts and adaptation options the main aim, or is it the discussion and implementation of adaptation measures?
- All of the cases presented in this chapter primarily applied workshop methods to realise the stakeholder involvement. Nevertheless, the specific workshop methods differed between and within the cases. Stakeholder engagement methods must be tailor-made for the group(s) of stakeholders to be involved. Since workshop dynamics can only be predicted to a limited extent, there is a need for experienced and flexible workshop moderators who can manage the various possible workshops dynamics.
- The lack of financial and time resources of particular stakeholder groups can represent a major barrier to balanced stakeholder involvement. This lack of resources can lead to an 'asymmetric' stakeholder involvement – for example, the underrepresentation of environmental NGOs. One way to resolve this dilemma would be to reimburse certain stakeholders for their participation in workshops.
- A challenge that is perhaps specific to the adaptation topic is the often low initial stakeholder interest in the topic, sometimes resulting in limited stakeholder involvement at the beginning of the adaptation process. Before stakeholders are willing to get involved, they often need to be told how they could be affected by climate change impacts and/or adaptation measures, such that they become more aware of their 'stakes' in the issue.

Notes

1 The four remaining modules are concerned with: (i) methodologies for network cooperation, (ii) regional data availability on climate change, (iii) methodologies for knowledge exchange and dissemination and (iv) monodisciplinary research.
2 The climate change scenario projects a warming trend, stronger in summer and weaker in winter, with a progressively increasing trend in mean temperature reaching between 3.4 °C (SMHIRCA) and 3.7 °C (RM5.1) by the end of the century. Greater uncertainty and more irregular behaviour in precipitation is expected as well. The model projects a reduction in the annual cumulated precipitation during the last 10

years of the century ranging between −1.8% (SMHIRCA) and −17.0% (RM5.1), with an increase in the intensity of single precipitation events.
3 Joint European Support for Sustainable Investment in City Areas: http://ec.europa. eu/regional_policy/thefunds/instruments/jessica_en.cfm .

References

Arnstein, S.R. (1969) 'A ladder of citizen participation', *JAIP* 35 (4), pp. 216–224.

Bundschuh, A. and Knierim, A. (2013) 'Partizipation von Praxispartnern: wer repräsentiert die Landwirtschaft in INKA BB?', in *Partizipation und Klimawandel – Ansprüche, Umsetzung und Stand der Forschung*, München: Oekom Verlag.

Cardellini, S. (2008) *Living with Landslide: The Ancona Case History and Early Warning System*, first World Landslide Forum, Tokyo.

Cardinaletti, M. and Di Perna, C. (2009) *Tools and Concepts for the Local Energy Planning*, EASY Project.

Christoplos, I., Anderson, S., Arnold, M., Galaz, V., Hedger, M., Klein, R.J.T. and Le Goulven, K. (2009) *The Human Dimension of Climate Adaptation: The Importance of Local and Institutional Issues*, Commission on climate change and development report, Stockholm.

COM (2009) *Adapting to Climate Change: Toward a European Framework for Action*, Commission of the European Communities, Brussels.

Dialogue by Design (2010) *A Handbook of Public & Stakeholder Engagement*, London: Dialogue by Design.

Dodds, F. and Benson, E. (2011) *Multi-Stakeholder Dialogue*. http://pgexchange.org/images/ toolkits/PGX_D_Multistakeholder%20Dialogue.pdf.

Few, R., Brown, K. and Tompkins, E.L. (2006) *Public Participation and Climate Change Adaptation*, Working paper 95, Tyndall Centre for Climate Change Research.

Glicken, J. (2000) 'Getting stakeholder participation "right": a discussion of participatory processes and possible pitfalls', *Environmental Science & Policy* 3 (6), pp. 305–310.

Green, A.O. and Hunton-Clarke, O. (2003) 'A typology of stakeholder participation for company enviromental decision-making', *Business Strategy and the Environment* 12 (5), pp. 292–299.

Grothmann, T. (2011) 'Governance recommendations for adaptation in European urban regions: results from five case studies and a European expert survey', in Otto-Zimmermann, K. (ed.) *Resilient Cities – Cities and Adaptation to Climate Change – Proceedings of the Global Forum 2010*, pp. 167–175, Hamburg: Springer.

Haas, W., Weisz, U., Balas, M., McCallum, S., Lexer, W. et al. (2008) *Identifikation von Handlungsempfehlungen zur Anpassung an den Klimawandel in Österreich: 1. Phase 2008* (Study on the identification of necessary first actions and measures for climate change adaptation in Austria: 1st Phase), Report on behalf of the Federal Ministry of Agriculture, Forestry, Environment and Water Management, Vienna.

Haas, W., Weisz, U., Pallua, I., Hutter, H.P., Essl, F., Knoflacher, H., Formayer, H. and Gerersdorfer, T. (2010) *Handlungsempfehlungen zur Anpassung an den Klimawandel in Österreich, Aktivitätsfelder: Gesundheit, Natürliche Ökosysteme/Biodiversität und Verkehrsinfrastruktur*, im Auftrag des Klima- und Energiefonds, Klimaforschungsinitiative AustroClim.

Hemmati, M. (2001) *Multi-Stakeholder Processes – A Methodological Framework*, A UNED Forum (Draft) Report. London, UNED Forum. http://www.earthsummit2002.org/msp/ MSP%20Report%20Exec%20Summary%20April%202001.pdf.

Hoornweg, D., Freire, M., Lee, M.J., Bhada-Tata, P. and Yuen, B. (ed.) (2009) *Cities and Climate Changes – Responding to an Urgent Agenda*, Urban Development Series, Washington,

DC: The World Bank. https://openknowledge.worldbank.org/bitstream/handle/1098
6/2312/626960PUB0Citi000public00BOX361489B.pdf?sequence=1.

ISPRA (2011) *Baseline Climate Scenario – Climate Trends and Projections*, ISPRA, Rome.

Knierim, A. and Hirte, K. (2011) 'Aktionsforschung – ein Weg zum Design institutioneller Neuerungen zur regionalen Anpassung an den Klimawandel', in *Anpassung an den Klimawandel – regional umsetzen!*, München: Oekom Verlag, pp. 156–174.

Knierim, A., Toussaint, V., Müller, K., Wiggering, H., Bachinger, J., Kaden, S., Scherfke, W., Steinhardt, U., Aenis, T. and Wechsung, F. (2009) *Innovationsnetzwerk Klimaanpassung Region Brandenburg Berlin – INKA BB. Rahmenplan*, gekürzte Version, Leibniz-Zentrum für Agrarlandschaftsforschung (ZALF) e.V., Müncheberg.

Larsen, K. and Gunnarsson-Östling, U. (2008) 'Climate change scenarios and citizen-participation: mitigation and adaptation perspectives in constructing sustainable futures', *Habitat International* 33 (3), pp. 260–266.

Lebel, L., Grothmann, T. and Siebenhüner, B. (2010) 'The role of social learning in adaptiveness: insights from water management', *International Environmental Agreements: Politics, Law and Economics*, 10 (4), pp. 333–353.

Lebensministerium (Ministry of Agriculture, Forestry, Environment and Water Management Austria) (2012) *Österreichische Strategie zur Anpassung an den Klimawandel*, Vienna.

Lim, B., Spanger-Siegfried, E., Burton, I., Malone, E. and Huq, S. (ed.) (2004) *Adaptation Policy Frameworks for Climate Change. Developing Strategies, Policies and Measures*, UNDP, Cambridge: Cambridge University Press.

Mayntz, R. (1993) 'Governing failures and the problem of governability: some comments on an emerging paradigm', in Kooiman, J. (ed.) *Modern governance. New Government–Society Interactions*, London: SAGE Publications, pp. 9–20.

MLUV (2008) *Maßnahmenkatalog zum Klimaschutz und zur Anpassung an die Folgen des Klimawandels*, Ministerium für Ländliche Entwicklung, Umwelt und Verbraucherschutz Brandenburg. Potsdam.

Mukheibir, P. and Ziervogel, G. (2009) 'Developing a Municipal Adaptation Plan (MA) for climate change: the city of Cape Town', in Bicknell, J., Dodman, D. and Satterthwaite, D. (ed.) *Adapting Cities to Climate Change – Understanding and Addressing the Development Challenges*, London: Routledge, pp. 270–289.

Nilsson, A.E. and Swartling, A.G. (2009) *Social Learning about Climate Adaptation: Global and Local Perspectives*, Stockholm Environment Institute, Working Paper.

NRTEE – National Round Table on the Environment & the Economy (1993). *Building Consensus for a Sustainable Future: Guiding Principles: An Initiative Undertaken by Canadian Round Tables*, Canada.

Paavola, J. and Adger, W.N. (2006) 'Fair adaptation to climate change', *Ecological Economics* 56 (4), pp. 594–609.

Reed, M.S. (2008) 'Stakeholder participation for environmental management: a literature review', *Biological conservation* 141 (10), pp. 2417–2431.

Siart, S. and Knierim, A. (2013) 'Partizipative Planungs- und Entscheidungsprozesse zur Entwicklung von Klimaanpassungsstrategien in INKA BB', in *Partizipation und Klimawandel – Ansprüche, Umsetzung und Stand der Forschung*, München: OEKOM Verlag.

Siart, S., Bloch, R., Knierim, A. and Bachinger, J. (2012) 'Development of agricultural innovations in organic agriculture to adapt to climate change: results from a transdisciplinary R&D project in north-eastern Germany', in *Producing and reproducing farming systems: new modes of organisation for sustainable food systems of tomorrow*, Proceedings of the 10th European IFSA Symposium, 1–4 July 2012 in Aarhus, Denmark. http://ifsa.boku.ac.at/cms/fileadmin/Proceeding2012/IFSA2012_WS3.1_Siart.pdf.

Standards of Public Participation (2008) adopted by the Austrian Council of Ministers on 2 July 2008.

UNEP (2005) *From Words to Action, The Stakeholder Engagement Manual, Volume 2: The Practitioner's Handbook on Stakeholder Engagement*, AccountAbility, United Nations Environment Programme, Stakeholder Research Associates Canada Inc. http://www.accountability.org/images/content/2/0/208.pdf.

UNISDR (2012) *How to Make Cities More Resilient: A Handbook for Local Government Leaders*, UNISDR. www.unisdr.org/campaign/resilientcities/toolkit/handbook.

Willke, H. (2001) *Systemtheorie III: Steuerungstheorie*, 3rd edn, Stuttgart: UTB.

9

EXPLORE A WIDE SPECTRUM OF ADAPTATION OPTIONS

Inke Schauser, Andrea Prutsch,
Torsten Grothmann, Sabine McCallum
and Rob Swart

Explanation of the guiding principle

The complete portfolio of adaptation options should be used to reduce vulnerability and increase robustness and resilience to potential climate change impacts. After prioritising the climate change impacts to be addressed (cf. Chapter 5), a wide spectrum of adaptation options should be investigated at the appropriate (temporal and spatial) scales. Adaptation can have different forms – e.g. technological, financial, legal and informational (Smit et al. 2000). Past and current adaptation experiences and adaptation initiatives from a variety of regions, sectors or organisations can provide a valuable knowledge base in the assessment of adaptation options. A catalogue of adaptation options can be developed based on literature reviews (e.g. databases, such as the Climate-ADAPT platform) and stakeholder consultation (on experiences and future needs, cf. Chapter 8).

In order to enable decision-makers to make well-informed choices, all possible adaptation options should be characterised in as much detail as possible (van Ierland et al. 2007), including the following information:

- General aim of the adaptation option
- Direct and indirect effects in terms of reduction in vulnerability and increase in resilience
- Formal competences and supporting actors in the implementation of the adaptation option
- Spatial scope
- Social, economic and ecological context
- Existing policies, structures and processes that might be linked with adaptation
- Necessary steps of implementation and maintenance
- Financial resources required

- Time frame for planning and implementation
- Direct and indirect effects on climate change mitigation
- Interplay between other options and sectors (e.g. potential conflicts and synergies).

A detailed characterisation of possible options provides a good basis for prioritisation (cf. Chapter 10) and will help to determine how the adaptation actions can be best implemented.

Chapter overview

Three cases describe practical experiences with this guiding principle. The first case (Section 9.1) gives an overview of the different forms of adaptation options, demonstrating how a categorisation of options can be used to identify a wide range of adaptation options, establishing the theoretical adaptation potential and enabling a gap analysis. To collect information about different adaptation options and good-practice examples beyond existing databases, other sources can be explored or a competition between adaptation planners can be organised. The second case (Section 9.2) reflects on the adaptation options collected in the *Adaptation Inspiration Book*. This book provides practical examples that illustrate how diverse adaptation can be and how adaptation aspects can be incorporated into plans and projects primarily developed for reasons other than climate change. The third case (Section 9.3) highlights the importance of stakeholder consultation and the use of brainstorming techniques in collecting a range of adaptation options. It also clarifies that the identification of adaptation options is only one part of the overall adaptation process, and that it must be tailored to the needs of the organisations confronted with projected climate impacts.

9.1 Identifying and sorting adaptation options

Inke Schauser and Andreas Vetter

9.1.1   Collecting adaptation options: Where can they be found?

After the most significant or most urgent impacts a system faces due to climate change have been identified, information about suitable adaptation options must be collected. As much information as possible should be gathered on the available adaptation options, including information about their specific objectives, potential costs and benefits, socio-economic, ecological and institutional context, side-effects, the interplay between different options, cross-sectoral interactions, responsible actors, proponents and opponents, and spatial and temporal scopes.

In recent years, compilations of adaptation options have become available in the literature; these are in part sectoral compilations, i.e. for the water sector (e.g. Laaser et al. 2009), but also include multi-sectoral efforts (e.g. UKCIP 2005,

Van Ierland et al. 2007, Bizikova et al. 2008). These lists have been created to support decision-makers in the framework of research projects at the national or international level, usually based on literature reviews. In many lists, options are only superficially described – without reference to implemented examples and actual experiences – to give decision-makers an idea of the kind of adaptation measures that are generally available. However, examples of already implemented adaptation measures that have been realised under a similar general framework would be more convincing to decision-makers. When possible, best-practice examples should be identified to create adaptation awareness, motivate adaptation agents and disseminate adaptation possibilities widely. To this end, in recent years, databases of implemented adaptation options and good-practice examples have been developed on different scales (e.g. Austrian database for adaptation to climate change,[1] ccAlps,[2] European Climate Adaptation Platform Climate-ADAPT,[3] ClimWatAdapt Inventory of Measures,[4] ADAM Digital Adaptation Compendium,[5] Global and Regional Adaptation Support Platform[6]). These collections are based on literature reviews, surveys or on database entries by responsible actors. Because adaptation is still in its infancy, good-practice examples of actually implemented adaptation actions are relatively scarce. However, projects that primarily have non-climate objectives often incorporate climate adaptation features or can be relabelled as adaptation projects (e.g. urban green infrastructure). Such projects are just as relevant as (the few) projects that have been realised solely for climate adaptation reasons. To motivate actors to make the successes and failures of their projects publicly available, incentives must be provided.

To collect good-practice examples of adaptation in Germany, the 'Tatenbank'[7] database was established. The database was launched by a public competition: every institution or person who was conducting or had conducted an adaptation measure was invited to submit their measure to the Tatenbank using a specific form with informational requirements. Measures that were only in the planning stages were not considered in the competition. Eligible measures were those attempting to reduce the negative impacts of climate change, decrease the vulnerability of society, increase its adaptive capacity or take advantage of the opportunities presented by climate change. In particular, measures with co-benefits for mitigation or other policy goals were requested. Measures with either sectoral or cross-sectoral objectives at the local or regional level could be submitted. The submission had to include a general description of the measure, as well as structured information about the addressed climate change impacts and sectors, the instruments used to implement the measure, the measure's aim with regard to climate change and other policies, barriers to and the spatial location of implementation and the responsible actors. The measures were evaluated in the competition according to the following criteria:

- effectiveness and cost-benefit ratio
- co-benefits for other (ecologic, economic or social) goals

- participation and acceptance
- feasibility and transferability and
- innovativeness.

Experts from policy, administration, science and the economy selected four winners of the competition from around 50 participants. As a further incentive, prices were given to the winners at a ceremony during an adaptation conference focusing on public-relation activities (media relations, short films, good-practice brochures).

In general, because the available lists and databases are offered as guidelines, they have been structured according to different aspects in order to assist the decision-maker in the selection of potential measures. However, the creation of a robust portfolio as a basis for prioritising different adaptation options or for selecting a consistent package of options with regard to the specific impacts must be accomplished by the decision-makers themselves, often in a participatory process. The available databases can be used in the preparation of such a decision process. This includes networking with other adaptation actors and creating new ideas, as well as selecting and sorting the available and suitable options.

9.1.2 Sorting adaptation options: What kinds of options are available?

When searching for suitable adaptation options to create a robust portfolio, it is helpful to sort the available options into different categories to ensure that a wide spectrum of adaptation options is explored. Sorting enables a preliminary description of the options for a further in-depth assessment. Furthermore, a sorting procedure can facilitate the identification of useful combinations of adaptation options (e.g. technical and informational) or the development of new ones. A 'gap analysis' can be used to determine what impacts, vulnerabilities, adaptation approaches or types of options have not been considered, raising the question of why they have been neglected.

In sorting adaptation options, different categories with different objectives can be used, some of which overlap in part. The selection of the sorting categories determines the aspects of adaptation that will be investigated. For the first general description of adaptation options, the following categories can be useful (modified from Smit et al. 2000, Burton 2008):

- Intent (in relation to climatic stimuli): autonomous or planned
- Action: reactive, concurrent or anticipatory
- Temporal scope: short to long term
- Spatial scope: local to widespread
- Sector: sectoral or cross-sectoral
- Actor: individual or community, public or private
- Adaptation planning process: top-down or bottom-up.

For a more specific sorting of options, the most important categories would seem to be:

1 Climate change impacts and sectors addressed
2 Aims and effects of adaptation
3 Approaches and types of adaptation.

I. Adaptation options can be distinguished by the impacts and sectors addressed

For example, the Routeplanner (Van Ierland et al. 2007) sorts adaptation options into categories based on the impact addressed: sea-level rise, changes in river discharge, groundwater levels, storms, heat stress, drought stress and plant growth stress. In the German Federal Environmental Agency's Tatenbank and in the European Platform Climate-ADAPT, adaptation case studies can be sorted by users along addressed impacts and sectors. The Digital Adaptation Compendium produced by the European research project ADAM differentiates between adaptation options for various sectors and landscape types in order to consider the spatial and functional aspects of the impacts of climate change.

II. A frequently used method of sorting adaptation options is based on the types of effects or the strategic objectives of adaptation

Because of the uncertainty connected with projections of future climate change impacts, UKCIP (2005) distinguishes between (1) options that build adaptive capacity so that actors will be able to react adequately in the future when climate change impacts occur and (2) options that deliver adaptation actions to counteract specific climate change impacts. This dichotomy can be further divided, for example, into the seven classes described by Burton (1996): (i) bear the loss, (ii) share the loss, (iii) modify the threat, (iv) prevent effects, (v) change use, (vi) research and (vii) educate, inform and encourage behavioural change. This approach has also been proposed in modified forms by UKCIP (2005) and the European Commission (EC 2009). In the German Adaptation Action Plan,[8] planned and implemented adaptation options at the federal level were collected and split into four classes by differentiating (i) options building adaptive capacity, (ii) options creating a supportive (legislative or financial) framework for adaptation actions in Germany, (iii) options supporting international adaptation and (iv) direct adaptation actions under federal responsibility.

III. Adaptation options can also be sorted by the approaches and types of adaptation

Adaptation options may take a number of forms: Watkiss et al. 2009 differentiate between (i) infrastructure investment, (ii) financial incentives (autonomous and regulated price adjustments), (iii) voluntary behavioural change and (iv) the use of management decision-making processes that incorporate climate change risks. Adaptation options can also be divided into (i) policy/planning, (ii) capacity-building/awareness, (iii) information management, (iv) investment decisions and (v) practice/livelihood/resource management (McGray et al. 2007). The Digital Adaptation Compendium differentiates between (i) technological, (ii) managerial, (iii) financial, (iv) institutional and (v) other interventions. The PROVIA Guidance[9] (PROVIA 2013) differentiates between private and public options. The latter also includes policy instruments that encourage individual or collective actors to act themselves, e.g. by providing information or economic incentives or by enforcing regulations. An alternative and rather simplistic method of organising different approaches to adaptation has been applied by the European Commission (EC 2009), which differentiates between 'grey' or hard (technological), soft (non-structural) and 'green' approaches. Green approaches focus on increasing ecosystem resilience by using ecosystem functions and services.

9.1.3 Analysing the adaptation options: Are there gaps in the portfolio?

Describing and sorting the options into different categories can help create an overview of the theoretically available adaptation potential with regard to the objectives of the options, but without consideration of their effectiveness. By comparing this (theoretical) adaptation potential with the adaptation needs identified in an impact and vulnerability assessment, gaps can be determined and new ideas can also be generated. The different ways of categorising adaptation options are complements, not substitutes. Therefore, an analysis could include the following steps:

1   *Collection and general description of available options,* providing information regarding their objectives, potential benefits and side-effects, temporal and spatial scopes, whether they are sectoral or cross-sectoral in nature, which actors are responsible, which actors need to be involved and the financial resources required.

2   *Allocating available options to different impacts* (also with regard to the spatial and temporal scopes of the options), combined with an initial gap analysis determining which impacts (in which sectors and with which spatial and temporal scopes) can be counteracted using which options.

3   *Sorting the options based on their aims and intentions;* this permits a first impression of the potential effects of the options and a second gap analysis of the (missing) options for adaptation available for various impacts.

4   *Determining the approaches of the adaptation options,* in connection with a third gap analysis to ensure that sufficient types and/or approaches have been considered (e.g. green, grey, economic, behavioural, informational or other 'soft' options).

5   *Final gap analysis to ensure that a robust portfolio has been developed* that matches the selected impacts and vulnerabilities of concern and the aims of the adaptation process.

The sorting process and the criteria used must be adapted to the system at risk and the aims of the adaptation process. Thus, examining the adaptation possibilities of a system that will face a variety of climate change impacts and has many overlaps with other policy fields (e.g. nature conservation areas) will require wider screening than a system that only faces a few impacts and generally lacks interfaces with other systems (e.g. storm water drainage). In addition, in some circumstances, it will be of primary interest to determine which options are the responsibility of which actors at which administrative level or in which sector. Actors' areas of responsibility are currently only implicitly considered in the sorting process when the spatial and functional scopes of the options are considered. Consequently, it may be useful to explicitly assess the responsibilities for the adaptation options after the sorting process takes place.

9.2   Adaptation: What could it look like? Examples from the *Adaptation Inspiration Book*

Marjolein Pijnappels

9.2.1   Adaptation databases

To cope with the already occurring and potential future effects of climate change, several European countries have already developed (or are in the process of developing) national adaptation strategies, action plans or programmes, including the Scandinavian countries, the United Kingdom, France, Belgium, the Netherlands, Germany, Portugal and Spain. In these plans, adaptation needs are articulated and adaptation options are proposed. Publications and websites providing guidance or examples of good practice in adaptation are also becoming more common (Brown et al. 2011; see also examples in Section 9.1). The available collections and databases offer decision- and policy-makers an idea of the possibilities and the range of options. Many online databases provide short descriptions of projects, but lack detailed information, e.g. regarding budgets, the climate effects addressed and the costs and benefits. Unfortunately, a closer examination of the examples of good practice found in many of these online databases shows that they are often theoretical case studies of planned adaptation measures or small-scale pilot experiments rather than actual, fully implemented, ongoing adaptation measures. A reason often cited for this is that adaptation implementation is not yet very advanced in Europe – or anywhere else in the world, for that matter. Another reason for the apparently limited number of implemented adaptation measures in Europe is that a large portion of them are not labelled 'adaptation'. Rather, they are simply referred to by their direct effects, i.e. 'water safety measures', or measures to improve 'city climate'. Here we see a true gap between the domains of research, decision- and policy-makers and project developers. They do not speak the same language, which may lead to less-than-optimal implementation by developers and to adaptation measures that are hard to track, compare and analyse. All this adds to the perception that adaptation, unlike mitigation, remains somewhat elusive, difficult to describe and intangible.

In order to provide scientists, policy-makers and developers with a sense of what climate change adaptation is and what lessons can be derived from current implementations, the CIRCLE-2 network of European national climate impacts and adaptation research programmes conducted a search for currently implemented adaptation measures.

All three groups – policy-makers, developers and scientists – could benefit from an overview of implemented adaptation measures and the wide spectrum of options available, including options that may not be labelled 'adaptation' but that have resulted in a significant increase in the capacity to cope with climate change. If policy-makers could see what adaptation looks like in practice, they would get a better feel for adaptation as it becomes more real and tangible, and learn ways to improve new policies. Developers could learn how their actions and projects contribute to the adaptation of a certain region or city to climate change. Scientists could see how their research efforts are finally put into practice and develop methods to evaluate their efficiency.

9.2.2 Practical adaptation examples from the Adaptation Inspiration Book

The *Adaptation Inspiration Book*[10] (Pijnappels and Dietl 2013) is an initiative of the ERANET Programme CIRCLE-2, a European network of 34 institutions from 23 countries committed to funding research, sharing knowledge on climate adaptation and promoting long-term cooperation among national and regional climate change programmes.

In a bold and provocative way, the *Adaptation Inspiration Book* shows what adaptation looks like at the local scale in several European countries and in many different sectors, ranging from health to risk reduction to water management. Visual images and quotes from people involved in implementing the measures relate the story of adaptation and describe the measures that have been applied, such as green roofs, the cultivation of new crops and innovative ways of controlling storm surges. The emphasis on visual storytelling is deliberate: new knowledge is more quickly processed when it is accompanied by visuals rather than just plain text (Eppler & Burkhard 2006, Burkhard 2004). New concepts can be learned more easily when visuals are used, and this holds true for climate change concepts (Sheppard 2012). A comparative analysis of these projects will generate key lessons and guidelines, based on real-life projects and real experiences with adaptation measures. The focus on 'real' implemented measures is what makes this book different from other databases. The book seeks to encourage knowledge transfer between European countries and to strengthen the science-policy interface by showcasing the effects of adaptation – the end result of collaborations between scientists and policy-makers. An inspiring visualisation of the combined efforts of scientists and policy-makers can energise readers, encouraging them to start new projects. The measures in the *Adaptation Inspiration Book* provide inspiration for decision-makers who are searching for a wide range of adaptation options.

In the *Adaptation Inspiration Book*, 22 implemented adaptation measures are presented. The measures included did not necessarily have to be intended or labelled as adaptation to climate change, as long as they have led to a significant increase in the capacity to cope with climate change. The adaptation measures selected include incremental improvements to an existing situation as well as inspiringly creative, innovative or even radically new solutions that improve the quality of the project area or sector and that combat climate change effects. This is what sets the *Adaptation Inspiration Book* apart from other initiatives. Measures can involve the reinvention of old practices, such as the Dutch *terpen* (raised areas in the landscape) or the construction of broad new dikes. Although specific attention has been devoted to finding projects from all over Europe, northern and western Europeans are responsible for the majority of the projects in the *Adaptation Inspiration Book*, reflecting the leading role of these regions in climate change adaptation in Europe.

The *Inspiration Book* provides detailed information on each measure: the sector to which it belongs (e.g. water safety, agriculture, cities), the length of the project, its costs and benefits (including the proportion of the budget allocated specifically to adaptation) and the specific climate effects addressed. The identified measures most often involve solutions for flooding (72%). Other projects focus on water management (19%), biodiversity (19%), city planning (14%) and disaster risk reduction (14%). In 14 of the 22 projects, adaptation to climate change was the main motivation. Flood prevention and erosion are examples of the triggers for projects not driven directly by climate change. Half of the projects had budgets that exceeded €500,000; only one cost less than €10,000. The part of the budget allocated specifically to adaptation ranged from <5% (3 projects) to the total budget (5 projects).

An example of an inspirational adaptation project from the *Adaptation Inspiration Book* is the 'renaturing' of a concrete waterway in the city of Arnsberg, Germany. After two extreme rainfall events in 2007 – 130 litres of rainfall per square metre in only a few hours led to the flooding of four streams, causing serious damage to a built-up area – the city decided to take measures to prevent flooding, now and in the future.

Within just two years, 2.7 kilometres of these streams were successfully 'renatured', i.e. arranged in a more natural and ecologically sound way. The speed of this operation was made possible by a productive cooperation between the municipality and local inhabitants, fuelled by the fear that extreme weather events might cause serious harm again in the near future. As part of the renaturing measures, stream banks inside and outside of the city were enlarged and flattened. Water-obstructing elements, such as stairs close to the stream bed, were removed, reducing the power of the water.

In Arnsberg, the involvement of local inhabitants from the very beginning of the planning process made all the difference. Inhabitants agreed to give up parts of their properties to allow the streams to expand again. In practice, this meant that two to three metres of property and gardens were returned to the streams. Most inhabitants were more than willing to cooperate, as the flood adaptation measures resulted in a higher safety level. A hundred-year flood can now flow through these streams by Arnsberg without causing major damage.

As an important side effect, the implemented climate change adaptation measures have led to increased biological diversity through nature (re)-development, improvement of the city landscape and increased touristic value for the city of Arnsberg. The adaptation measures undertaken for the four smaller streams in combination with larger measures on the river Ruhr contributed to the city meeting the objectives of the EU Water Framework Directive (WFD) on time (2012); it is one of only a few towns in Germany to do so. The WFD commits the Member States of the European Union to achieving good (ecological) status for all water bodies by 2015.

In Spain, flooding occurs as well, but drought is a bigger problem. In the Green Deserts project (part of European Life+), trees are planted on dry, arid land to prevent erosion, mitigate flooding and increase the water-holding capacity of the land during dry periods. This project team included universities, local municipalities and businesses. The reforestation of dry or arid landscapes is a strategy that several Mediterranean countries pursue. Often, reforestation projects have only limited success because of the continuous need for irrigation, extremely low survival rates and high overall costs. Green Deserts makes use of an innovative technique called 'waterboxx', which permits the planting of trees without irrigation. This has inspired people and companies as well as local and regional governments to start planting more trees in dry and degraded areas. The waterboxx collects rain and condensation. The design and storage section of the box prevents evaporation of the water once it is absorbed. With the help of a wick, the box distributes a daily dose of water to the roots of the young tree. The Green Deserts project seeks to convert 63 hectares of bare, dry land into forest by planting around 55,000 trees in five Spanish regions (Valladolid, León, Zamora, Zaragoza and Barcelona). For 22,000 of these trees, the waterboxx will be used. The test sites in the five regions vary in terms of climatic conditions, altitude, land use and soil composition. They are all already feeling the effects of climate change and desertification; no trees, plants or crops can be grown without artificial irrigation, if at all. This makes the Green Deserts project special, as its aim is reforestation of these areas without the use of artificial irrigation, a strategy that no one has attempted before. Local inhabitants are very enthusiastic about the project. Often, the young trees are planted in collaboration with children from local schools (cf. Figure 9.2.1). This creates awareness and interest in the community. The realisation of the Green Deserts project will promote additional reforestation projects in Spain and create small industries (e.g. maintenance, fruit-picking and processing) that will be beneficial for local employment.

FIGURE 9.2.1 Children planting trees in the waterboxes in Viladecans (source: Life+, the Green Deserts Team (Sven Kallen))

9.2.3 Concluding thoughts

Most good-practice reports and databases focus on plans, the outcomes of studies and guidelines, but few provide detailed information on measures that have been actually implemented in Europe. Thus, to local-level policy-makers and professionals, the concept of adaptation remains somewhat elusive. Filling this gap, the *Adaptation Inspiration Book* describes adaptation measures that have been implemented all over Europe, providing the target group (policy-makers and professionals) with a concrete image of adaptation in practice. The book literally visualises what adaptation in Europe looks like, and the comparative analysis derives useful lessons learned. The emphasis on visual storytelling is deliberate. New knowledge is more quickly processed when it is accompanied by visuals than when only plain text is used. The adaptation measures identified include incremental improvements to an existing situation as well as inspiringly creative, innovative or even radically new solutions that improve the quality of the project area or sector, and combat climate change. The book also demonstrates that adaptation may actually be further along than previously estimated, if we take into account measures that increase adaptive capacity despite not being specifically labelled as adaptation.

The practice examples found in the *Adaptation Inspiration Book* are already very much sought after. They have been used in science-practice sessions at the first

European Climate Adaptation Conference in Hamburg (ECCA) in March 2013 and form the basis for cases to be taught in summer schools and training events. It would be interesting to broaden the geographical scope of this book to Asia, the United States and Africa, comparing the state of implementation across the world to see what lessons can be learned.

9.3 Identifying adaptation options: Practical experience from the application of the Adaptation Wizard

Megan Gawith, Roger Street and Kay Johnstone

9.3.1 The Adaptation Wizard

The purpose of adaptation interventions is to alter the chain of events that flows from a climatic event to its environmental, societal and economic consequences in order to minimise negative impacts and exploit positive opportunities. Adaptation to climate change can take many forms, depending on the aims, objectives, motivations, desired outcomes and specific context of the adapting organisation.

This case describes the use of an adaptation support tool called the Adaptation Wizard[11] by two UK organisations to identify and assess options for addressing key climate risks.

The Adaptation Wizard (Environment Agency 2013) provides a simple five-step process to help organisations adapt to a changing climate, as illustrated in Figure 9.3.1. Its question-driven approach provides clear instructions and additional information to help users complete each step. Originally developed by UKCIP with government funding, this popular resource has helped organisations in the UK and abroad to adapt and has inspired the development of other local and national adaptation support tools. The Adaptation Wizard is one of the practical resources available from the UK Government's new adaptation support service, Climate Ready.

The identification and assessment of adaptation options to address key climate risks is the fourth step of the adaptation process described in the Adaptation Wizard. It should only be properly undertaken once the aims, objectives, motivations and intended outcomes of the process have been identified and the organisation's vulnerability to current and future climate has been fully assessed.

Experience has also shown that in identifying and assessing adaptation options, due consideration must be given to the management mechanisms and approaches through which the selected adaptation options are to be implemented, as well as to the regulatory, governance and business environment in which the organisation operates. Failure to do so can result in an adaptation bottleneck in which actions will be difficult to implement or may fail to meet their intended objectives (Brown et al. 2011).

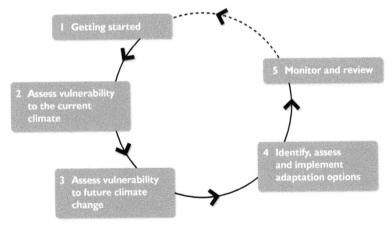

FIGURE 9.3.1 The Adaptation Wizard, showing the options identification stage in the context of the overall adaptation process (source: UKCIP 2010)

9.3.2 Practical experience with identifying adaptation options

Two broad categories of planned adaptation options are widely recognised in the literature (e.g. West & Gawith 2005, UKCIP 2005). *Building adaptive capacity* (BAC) entails establishing the necessary enabling conditions for adaptation actions to take place. It includes gathering the required information, establishing supportive social structures and developing supportive governance structures to build the foundation for delivering adaptation actions. *Delivering adaptation actions* (DAA) refers to more tangible actions undertaken to help reduce vulnerability to climate risks and to exploit opportunities.

Adaptation actions themselves might include the following options:

- Accepting the impacts and bearing the losses that result from those risks (e.g. managed retreat from sea-level rise)
- Off-setting losses by sharing or spreading the risks or losses (e.g. through insurance)
- Avoiding or reducing one's exposure to climate risks (e.g. building new flood defences, or changing locations or activities)
- Exploiting new opportunities (e.g. engaging in a new activity, or changing practices to take advantage of changing climatic conditions).

An alternative approach is to consider options that might be:

- Temporary or transitional (e.g. using large umbrellas to reduce solar heat gains)
- Non-structural (e.g. raising awareness; demanding management, monitoring and data management, skills development and early warning systems; working from home)

- Technical and structural (e.g. refurbishing buildings, enhancing flood defences, improving asset management)
- Strategic or policy-related (e.g. commissioning new buildings with a climate-resilient design as part of planned capital-building programmes; making, implementing and reviewing policies related to strategic intentions).

In practice, the distinction between BAC and DAA is somewhat artificial, as many capacity-building actions are also tangible adaptation actions, and many BAC options are integral prerequisites to DAA. However, distinguishing between the two can usefully stimulate broad thinking regarding options. In reality, a range of different options will be employed to address climate risks. Dependencies between the two types of responses should be recognised and reflected in the suite of adaptation options assessed and implemented.

Identifying and evaluating adaptation options for a UK water company

The authors worked with a UK water company to begin the process of scoping out potential adaptation options to address four key climate change risks that had been identified in an internal risk assessment. Individuals with relevant expertise from across the organisation attended a workshop to collectively identify and assess potential adaptation options based on the Adaptation Wizard methodology.

TASK 1: IDENTIFYING THE PRIORITY RISKS THAT REQUIRE AN ADAPTATION RESPONSE

Of the risks identified in the assessment, two priority risks were selected for analysis. The intention was to focus on these in some detail to develop the approach so that it could be applied internally to address other risks independently as required. The two risks examined were:

1 Waste water and nuisance risk: Higher temperatures are expected to result in increased levels of septicity. This could increase odour levels at sewage treatment facilities and pumping stations and increase the number of odour and nuisance complaints.
2 Water resources risk: Lower summer rainfall may decrease water levels and reduce water quality.

TASK 2: IDENTIFYING A RANGE OF ADAPTATION OPTIONS FOR EACH PRIORITY RISK

Two breakout groups were formed to address each of the identified risks. The groups were populated by employees from across the organisation that had the most relevant expertise with respect to understanding and potentially managing each risk. Each group identified a 'long list' of adaptation options that could conceivably address the priority risks. All contributions were captured; none was

TABLE 9.3.1 Excerpt of adaptation options to address priority risks

Risk	Example of proposed options
Wastewater risk: Higher temperatures increase septicity rates and odour problems	Limit development around treatment plants
	Cover tanks to reduce odour
	Use traditional odour-control methods such as filtration
	Plant trees around treatment plants as a natural barrier
	Explore different treatment methods to reduce odour levels
	Invest in odour modelling and forecasting to better understand and manage events
	Identify trigger levels and circumstances under which odour-control measures should be mobilised
	Monitor (and forecast) temperatures and link to odour control to better anticipate and manage odour events
Water resources risk: Lower summer rainfall may cause reductions in water levels and water quality	Manage catchments more effectively
	Upgrade water treatment equipment and sewage treatment
	Increase proportion of groundwater use
	Employ artificial recharge
	Increase or improve water storage capacity – e.g. reuse abandoned sources; de-silt storage facilities so that they store water, not silt
	Identify simple solutions that can be deployed seasonally (e.g. temporary mobile plants)
	In-river treatment (e.g. oxygenation of rivers)
	Supply two qualities of water to households: potable and non-potable (as done in Spain, for example)

Source: UKCIP 2010

excluded. This list therefore contained some options that were unlikely to be implemented and at first glance may not meet the assessment criteria; however, having all options on the table was found to be helpful at this stage, as it broadened the thinking of the group and triggered the suggestion of more creative options. The two groups were then brought back together to explore their respective ideas and to identify any conflicts or synergies before further addressing these in the assessment. Some of the proposed options are illustrated in Table 9.3.1.

TASK 3: ASSESS THE ABILITY OF EACH ADAPTATION OPTION TO ACHIEVE YOUR STRATEGIC OBJECTIVES

The next task in the process was to agree on the criteria against which the 'long list' of adaptation options could be assessed in terms of their ability to fulfil the organisation's strategic objectives. A common list of criteria described in UKCIP (2010) was presented to the group. These were collectively refined in a plenary discussion. In this way, a common understanding of each criterion was reached, and each was tailored to suit the specific requirements and nature of the organisation.

Further discussion generated definitions of what would constitute a 'high' (good) score and a 'low' score. Each adaptation option was assessed according to these criteria to generate a weighted list of adaptation options, thereby concluding the initial scoping exercise.

In the next step, the relevant specialists within the organisation concerned explored the options that had scored highly in greater detail, seeking to determine what further information and analysis would be required to develop them into viable adaptation options that could be taken forward within the organisation for evaluation and implementation.

Identifying and assessing adaptation options for a UK housing association

A similar exercise was conducted with a housing association in England to explore the robustness of their organisation to future climate risks and to identify what adaptation actions, if any, should be taken.

TASK 1: IDENTIFYING THE PRIORITY RISKS THAT REQUIRE AN ADAPTATION RESPONSE

A series of workshops was held to work through the first three steps of the Adaptation Wizard. During these workshops, key risks were identified, including:

1 Design implications for new homes and existing stock
2 Impacts on wet trades processes
3 Threats to supply chains and the procurement of goods and services.

TASK 2: IDENTIFYING A RANGE OF ADAPTATION OPTIONS FOR EACH PRIORITY RISK

Having identified the priority climate risks, an internal workshop with representatives from a broad range of relevant business areas was held to identify possible adaptation options, following the approach outlined in Step 4 of the Adaptation Wizard. The group considered each risk in turn, working collectively to ensure that everyone's expertise was applied to each risk. As in the example above, a 'long list' of potential adaptations was suggested (cf. Table 9.3.2), some of which could be classified as capacity-building activities (e.g. ongoing monitoring and research) while others represented more tangible adaptation actions.

The value to this group of having previously completed Step 2 of the Adaptation Wizard was underlined in this options identification exercise. Having a sound understanding of the organisation's vulnerability to present-day weather and climate, as well as evidence of the effectiveness of measures used to deal with past weather events, was a useful starting point for generating options and mechanisms for addressing future risks.

TABLE 9.3.2   Excerpt of potential adaptation options to address risks to new homes and existing stock

Risk 1: Design implications for new homes and existing stock	
New homes	Conduct further research into future building materials and their likely performance in a future climate. Will timber or steel frames be better suited to future conditions?
	Seek opportunities to take advantage of financial initiatives associated with photovoltaic (PV) installation to fund win-win climate adaptations
	Review 'sealing' of new properties. Are sealed buildings more or less adaptive to future climate conditions?
	Build in climate adaptation 'headroom' so that adaptive measures can be installed in the future, even though the technology or financing may not yet be available
	Engage and raise the awareness of the design team and encourage them to review design codes and standards
	Monitor developments in design
	Monitor customer complaints to better understand the issues currently experienced in existing housing stock
Existing stock	Explore opportunities for embedding adaptation measures
	Install water conservation measures
	Link into and draw from climate adaptation research projects
	Find out more about retrofit options
	Review the effectiveness of existing stock (and policies?)
	Accept that some properties have a lifespan that they are not expected to exceed (accept the loss)

Source: UKCIP 2010

TASK 3: ASSESS THE ABILITY OF EACH ADAPTATION OPTION TO ACHIEVE YOUR STRATEGIC OBJECTIVES

The organisation involved in this instance possessed an established set of criteria for appraising new business opportunities. It was found that with minor modifications these criteria could be applied effectively to the evaluation process, with each adaptation option being evaluated as a new business opportunity. By using the organisation's existing systems, outputs from the assessment of adaptation options could be more easily interpreted by managers and so were more readily integrated into their decision-making process.

The transferable lesson from this example is that an existing business framework can be readily adapted to evaluate the feasibility and effectiveness of potential adaptation options. In this way, the assessment of adaptation options can be mainstreamed into existing systems and practices, in turn facilitating the consideration and implementation of adaptation options.

9.3.3 Concluding thoughts

The guidance provided by the Adaptation Wizard and its supporting tools and resources proved extremely helpful for organisations seeking to identify and select options to address priority climate risks. Some useful transferable lessons were learned in practice.

Given the context-specific nature of adaptation, it was helpful to tailor the general guidance provided by the Adaptation Wizard to reflect the nature of the risks being addressed, the existing processes within the organisation and the operating environment of the organisation concerned. Tailoring adds greater relevance and meaning to the assessment, creates a sense of ownership regarding the results of the process and places the process and its results in the appropriate context. Furthermore, by placing the organisation's perspective centre-stage and combining this with a good understanding of the drivers, enablers and barriers to change, it is possible to demonstrate how a better appreciation of climate change can help to realise an organisation's objectives.

A valuable starting point for identifying and implementing adaptation options is an assessment of the organisation's vulnerabilities to current weather and climate, as described in Step 2 of the Adaptation Wizard. Addressing the existing adaptation deficit can lead to quick wins and can make a seemingly remote issue more tangible, helping those involved to make a case for further action where appropriate. It also creates an opportunity to focus on positive outcomes and to make use of the organisation's own systems, knowledge and data to identify positive adaptation options.

Finally, the principle of working in partnership with individuals within and beyond one's organisation is absolutely critical to effective adaptation. Partnerships create a collective sense of buy-in and ownership; well-supported actions are more likely to be successfully implemented. Furthermore, an organisation will inevitably face risks the management of which lies beyond their control. In these cases, effective adaptation can only be achieved by influencing others or working in partnership with them.

9.4 Lessons learned for the identification of adaptation options

Inke Schauser

When searching for adaptation options, it is helpful to start with the known and then to examine what remains unknown. In most cases, climate change is exacerbating the existing challenges caused by current climate variability. Thus, when confronted with the need to adapt to a climatic impact, known solutions should be considered in the first step. Addressing existing climate impacts and adaptation deficits can lead to quick wins and will facilitate the process of adaptation (cf. Section 9.3). However, future climate change impacts might require a further search for sustainable solutions or consideration of how known adaptation pathways might be supported or replaced by new ones, if in the future they prove to be no longer sufficient (cf. Section 9.2). Other options coming from other sectors or areas should therefore also be taken into account (cf. Section 9.1).

If there are risks that lie beyond the control of the decision-makers, they should consider, for example, trying to influence or work in partnership with other affected organisations, involving those individuals who will be responsible for delivering measures or implementing adaptation (cf. 9.3).

Since adaptation options are in many cases already known options or integral parts of already implemented measures, adaptation implementation may be further developed than previously thought, especially considering the many options that are not explicitly labelled as adaptation but still decrease vulnerability to climate change, in addition to fulfilling other objectives (cf. 9.2). Thus, when looking for a wide range of suitable options for possible future climates, it is helpful both to collect familiar options (including those not formally labelled as adaptation) and to use brainstorming techniques to create innovative solutions. Working in partnerships is crucial for the identification of effective and feasible adaptation options. In addition, available databases of adaptation options and good-practice examples can be used to identify adaptation possibilities (cf. 9.1). However, many best-practice reports and databases focus primarily on plans, outcomes of studies and guidelines; few provide detailed information on measures already implemented in Europe (cf. 9.2). Thus, obtaining the necessary detailed information on a wide range of options can be a very time-consuming process.

Practical experiences in Sections 9.2 and 9.3 have shown that although general advice and guidance can be extremely useful in informing the adaptation process, tailoring this information to the specific context is almost always necessary and can be very time-consuming. However, tailoring the information adds relevance and meaning to the assessment, creates a sense of ownership in the results of the process and places the process and its results in an appropriate context. In discussing options with local-level policy-makers and professionals, detailed information on concrete implemented measures is important, as otherwise the concept of adaptation remains somewhat elusive.

After options have been collected, Section 9.1 indicates how a stepwise sorting and assessment process including a gap analysis can facilitate the creation of a robust portfolio. Sorting the options into different categories provides an overview of the theoretically available adaptation potential. However, the sorting process and the criteria used must be adjusted to the specific vulnerable system and the aims of the adaptation process.

Thus, we can conclude with a summary of the following lessons learned:

- Adaptation is much broader than often thought, since it includes all measures that help natural or human systems to cope with extreme weather events, climate variability and climate change in addition to fulfilling other objectives.

- Existing databases, collections of good-practice examples and stakeholder participation can be helpful in the process of collecting and tailoring a wide range of adaptation options to concrete climate impacts.
- Working in partnership with individuals inside and often also outside the affected system is crucial for the identification of effective and workable adaptation options.
- Analysing and sorting adaptation options can help actors to create new solutions, combine existing options and find gaps in current or planned adaptation activities.

Notes

1 http://www.klimawandelanpassung.at/ms/klimawandelanpassung/de/kwadatenbank
2 http://www.cipra.org/de/klimaprojekte/cc.alps/wettbewerb/online
3 http://climate-adapt.eea.europa.eu
4 http://www.climwatadapt.eu/inventoryofmeasures
5 http://adam-digital-compendium.pik-potsdam.de/adaptation-catalogue
6 http://cigrasp.pik-potsdam.de
7 http://www.tatenbank.anpassung.net, developed by KomPass – Climate Impacts and Adaptation in Germany, Federal Environment Agency (UBA).
8 http://www.bmu.de/en/topics/climate-energy/climate/
9 http://www.unep.org/provia/
10 http://www.circle-era.eu/np4/552.html
11 http://www.environment-agency.gov.uk/research/137639.aspx

References

Bizikova, L., Neale, T. and Burton, I. (2008) *Canadian Communities' Guidebook for Adaptation to Climate Change*, including an approach to generate mitigation co-benefits in the context of sustainable development, 1st edn, Vancouver, CN: Environment Canada and University of British Columbia.

Brown, A., Gawith, M., Lonsdale, K. and Pringle, P. (2011) *Managing Adaptation: Linking Theory and Practice*, Oxford, UKCIP. http://www.ukcip.org.uk/wordpress/wp-content/PDFs/UKCIP_Managing_adaptation.pdf.

Burkhard, R.A. (2004) *Learning from Architects: The Difference between Knowledge Visualization and Information Visualization*, in proceeding of 8th International Conference on Information Visualization (IV04), London, UK.

Burton, I. (1996) 'The growth of adaptation capacity: practice and policy', in Smith, J., Bhatti, N., Menzhulin, G., Benioff, R., Budyko, M.I., Campos, M., Jallow, B. and Rijsberman, F. (eds.) *Adapting to Climate Change: An International Perspective*, New York, USA: Springer Verlag, pp. 55–67.

Burton, I. (2008) 'Moving forward on adaptation', in Lemmen, D.S., Warren, F.J., Lacroix, J. and Bush, E. (eds.) *From Impacts to Adaptation: Canada in a Changing Climate 2007*, chapter 10, Ottawa, CN: Government of Canada, pp. 427–440.

EC – European Commission (2009) *Adapting to Climate Change: Towards a European Framework for Action, Impact Assessment*, Commission Staff Working Document accompanying the White paper, Brussels.

Environment Agency (2013) *Adaptation Wizard*. http://www.environment-agency.gov.uk/research/137639.aspx.

Eppler, M.J. and Burkhard, R.A. (2006) 'Knowledge visualization', in Schwartz, D.G. (ed.) *Encyclopedia of Knowledge Management*, Hershey: Idea Group, pp. 551–560.

Laaser, C., Leipprand, A., de Roo, C. and Vidaurre, R. (2009) *Report on Good Practice Measures for Climate Change Adaptation in River Basin Management Plans*, ETC Water Report EEA/ADS/06/001 – Water to the European Environment Agency. http://icm.eionet.europa.eu/ETC_Reports/Good_practice_report_final_ETC.pdf.

McGray, H., Hammill, A. and Bradley, R. (2007) *Weathering the Storm*, World Resources Institute Report.

Pijnappels, M.H.J. and Dietl, P. (2013) *Adaptation Inspiration Book – 22 Implemented Cases of Local Climate Change Adaption to European Citizens*, Eranet, Climate Impact Research and Response Coordination for a Larger Europe (CIRCLE 2).

PROVIA (2013) *PROVIA Guidance on Assessing Vulnerability, Impacts and Adaptation to Climate Change*. Consultation document, United Nation Environment Programme, Nairobi, Kenya, 198 pp. Available at http://www.unep.org/provia.

Sheppard, S.R.J. (2012) *Visualizing Climate Change: A Guide to Visual Communication of Climate Change and Developing Local Solutions*, Abingdon, UK: Earthscan/Routledge.

Smit, B., Burton, I., Klein, R.J.T. and Wandel, J. (2000) 'An anatomy of adaptation to climate change and variability', *Climatic Change* 45 (1), pp. 233–251.

UKCIP – Climate Impacts Programme (2005) *Identifying Adaptation Options*, Oxford.

UKCIP – Climate Impacts Programme (2010) UKCIP Adaptation Wizard v3.0. UKCIP, Oxford.

Van Ierland, E.C., de Bruin, K., Dellink, R.B. and Ruijs, A. (2007) *A Qualitative Assessment of Climate Change Adaptation Options and Some Estimates of Adaptation Costs*, Wageningen.

Watkiss, P., Hunt, A. and Horrocks, L. (2009) *Scoping Study for a National Climate Change Risk Assessment and Adaptation Economic Analysis*, Defra Study.

West, C.C. and Gawith, M.J. (2005) *Measuring Progress. Preparing for Climate Change Through the UK Climate Impacts Programme*, UKCIP, Oxford.

10

PRIORITISE ADAPTATION OPTIONS

Rob Swart, Andrea Prutsch, Torsten Grothmann, Inke Schauser and Sabine McCallum

Explanation of the guiding principle

Explorations of adaptation options (cf. Chapter 9) typically identify more adaptation options than can reasonably be implemented, especially in the short term. Deciding what to do first is often as important as deciding what to do at all (Füssel 2007). Thus, specific adaptation actions need to be selected based on criteria. In the majority of cases, different types of adaptation actions should be realised, as there will hardly ever be a clear optimum or one definitive solution. Stakeholder discussions and agreement (cf. Chapter 8) on the main objectives, the criteria and their weighting for the prioritisation of adaptation options are very useful for transparency in the decision-making process.

In practically all cases, economic criteria play a key role: what do the options cost, when should they be implemented and what is their effectiveness? These questions are relevant at different scales, from the economic implications of adaptation at the global, European or national levels to the costs and benefits of specific investment projects or other concrete local adaptation measures. Because adaptation to climate change has entered the policy and scientific agenda in Europe later than mitigation, there is as yet very little experience with the application of methods to determine costs and benefits and the use of such information to evaluate adaptation options. However, over the past few years, a range of projects have begun to build a (still quite limited) evidence basis.

A comprehensive cost-benefit analysis (CBA) also considering indirect costs and benefits, non-monetary values and externalities can be useful for prioritising adaptation options (EEA 2007, van Ierland et al. 2007). Many adaptation actions can be implemented at relatively low costs with significant benefits, e.g. the use of habitat-adapted crops and tree species in agriculture and forestry. However, the information currently available on adaptation costs (both market and non-market costs) and adaptation benefits is limited. Due to uncertainties in climate change impacts, it is also uncertain how much damage will be avoided through adaptation options. However, more information on the costs and benefits of adaptation will probably become available in the near future due to ongoing and emerging research efforts.

For many projects, a multi-criteria analysis (MCA) of adaptation options is a meaningful and feasible method of evaluating adaptation options, especially when insufficient data are available to perform a meaningful CBA, or when values that cannot meaningfully be monetised are considered to be significant. This analysis should include criteria such as importance, effectiveness, urgency, sustainability, efficiency, co-benefits and side effects, reversibility, flexibility, fairness, resilience, robustness, acceptability and other criteria (based on Smith & Lenhart 1996, McCarthy et al. 2001, Willows & Connell 2003, Adger et al. 2005, Mendelsohn 2006, Füssel 2007, Bizikova et al. 2008, De Bruin et al. 2009, Hallegatte 2009, Füssel 2009, Dyszynski 2009). These criteria are very generic; however, the relative importance attached to each one depends on the country, the system and the vulnerability of the key system for which adaptation options are being assessed. Thus, the ranking of the options is based on criteria weighting, which should be done with the involvement of stakeholders (De Bruin et al. 2009) (cf. Chapter 8).

However, while a climate-focused MCA can prove a useful method for prioritising adaptation options to rank them and find the optimal solution, it also has limitations; these include the fact that in the real world, climate change adaptation is often just an additional consideration among other objectives. In addition, the method does not easily allow the evaluation of portfolios of options that can be deployed over time in a flexible manner as knowledge on climate change evolves. Here, other approaches are necessary, such as Adaptation Pathways Analysis.

Chapter overview

In this chapter, in Section 10.1, Watkiss and Hunt first discuss methodologies for assessing the costs and benefits of adaptation, including economically integrated assessment methods, investment and financial flows, computable general equilibrium models, scenario-based and extreme-event-focused impact assessment, risk assessment and econometrics-based methods. All of these present serious challenges regarding the collection of required data, as well as methodological problems. The authors then elaborate on the three methods that are most commonly used to support decision-making under uncertainty: social cost-benefit analysis (CBA), social cost-effectiveness analysis (CEA) and multi-criteria analysis (MCA). Noting the serious challenges involved in applying these methods in different circumstances, they propose a set of methods relatively new to the evaluation of adaptation options: real options analysis, adaptation pathway analysis, portfolio analysis and robust decision-making.

In Section 10.2, one of the 'traditional' methods is discussed: van Ierland, De Bruin and Dellink summarise a multi-criteria analysis that was performed in the Routeplanner Project to support the development of the National Adaptation Strategy in the Netherlands, which focuses mainly on spatial planning and water safety. In this project, five key criteria were selected to evaluate a number of options using a specific long-term climate scenario.

In the final section of this chapter, Haasnoot, Kwadijk and Asselman explore the innovative approach of Adaptation Pathways Analysis. In this method, the analysis starts with current policies and objectives, assessing how long these will suffice and when a change in policies or practices will be required as a result of climate change. These points in time are referred to as 'adaptation tipping points'. This approach, which has also been applied in the context of the Thames Barrier, is illustrated here with an example from the Netherlands concerning the development of a sustainable delta-management plan. The approach focuses on sets of robust and flexible adaptation options that can be adjusted over time; this may be a better match for the way in which project developers work than the classical economic approaches.

10.1 Economic appraisal

Paul Watkiss and Alistair Hunt

10.1.1 Introduction

There is increasing policy interest in the economics of adaptation, particularly as part of the appraisal of adaptation options. This information is potentially relevant at a number of different aggregation levels:

- The European level, where information on costs and benefits can raise awareness of the need for adaptation, as well as providing input for the discussion of options, resources and funds for European-level or trans-boundary adaptation (for an example, cf. Watkiss 2011).
- The national level, as input for national adaptation policy and strategy development, to support the case for adaptation and financing and to identify efficient and effective strategies. Examples include analyses in Sweden (SCCV 2007) and the Netherlands (van Ierland et al. 2007), with major ongoing examples in the UK and Germany.
- The regional or local level, as input for the design, appraisal and prioritisation of adaptation programmes and projects.

Estimates of the costs and benefits of adaptation at all of these levels are emerging (cf. recent reviews by EEA 2010, 2013, Agrawala et al. 2011), although there is still a relatively limited evidence base and a low confidence level in the magnitude of estimates. This is due to a number of methodological issues. It is clear that any estimates of costs and benefits depend on the methods used, the objectives set and the spatial, sectoral and temporal context. The current uncertainties over future climate change risks and the need to consider non-climate factors contribute to the complexity of the challenge.

10.1.2 Methods for assessing adaptation costs and benefits

Over the past few years, a wide range of methodologies have emerged for assessing the costs and benefits of adaptation (cf. Table 10.1.1). These use various metrics, modelling approaches and assumptions, and they focus on different time periods. No one method is right or wrong – different methods may be more or less appropriate according to their aim: providing headline information, scoping out possible options, identifying the costs of increasing climate resilience or undertaking detailed economic assessment of specific plans or projects.

The coverage of the costs and benefits of adaptation varies across risks and sectors. A synthesis of the current state of the art, primarily focused on the European and national levels, is shown in Table 10.1.2. A rating of the relative quality of the sectoral studies considered based on the results of a literature review is also presented.

This information shows that the coverage of adaptation costs is partial. Furthermore, many of these assessments focus on technical adaptation, investigating a small set of available options for a limited number of defined scenarios, and almost all estimates ignore policy costs. Importantly, all adaptation costs are influenced by the framework of the analysis, e.g. cost-benefit or cost-effectiveness analysis, and the objectives set for adaptation.

In moving to practical adaptation policy implementation, a recent key focus in such estimates is on the *uncertainty* over future climate change, a factor that makes it difficult to predict how much adaptation will be needed and when, as well as whether responses will be effective. This has led to an emerging focus on decision-making under conditions of uncertainty, and there are a range of techniques that are potentially useful for the application to adaptation.

10.1.3 Adaptation economics: Decision-making under uncertainty

The need to recognise uncertainty is acknowledged in the mainstream adaptation literature, and is now beginning to be adopted in the consideration of adaptation costs and benefits. The consideration of uncertainty, especially as a part of iterative adaptive management, focuses on (UNFCCC 2009):

- Supporting adaptive capacity as an early adaptation option
- Identifying no- and low-regrets and low-cost measures justified by current climate change or socioeconomic developments that will also provide future resilience

TABLE 10.1.1 Methodologies for assessing the costs and benefits of adaptation (see EEA 2013 for detailed analysis and references)

Approach	Description	Advantage	Limitations
Economic Integrated Assessment Models (IAM)	Global aggregated economic models that assess costs of climate change and costs and benefits of adaptation.	Provides headline values for awareness. Range of economic outputs. Used to provide economic information on global climate policy.	Very aggregated approach with highly theoretical form of adaptation; no technological detail. Insufficient detail for national or sub-national adaptation planning
Investment and Financial Flows (IFF)	Early studies estimate costs of adaptation as % uplift. More recent national studies determine the marginal cost increase required to reduce climate risks.	Highlights scale of short-term investment requirements in sectoral or development plans.	Often insufficient linkage with climate change scenarios and little consideration of uncertainty.
Computable General Equilibrium models (CGE)	Multi-sectoral and macro-economic analysis of economic costs of climate change and emerging analysis of adaptation.	Captures cross-sectoral linkages across economy, including autonomous market adaptation. Can represent global trade effects. Can link to sector studies.	Utilises aggregated representation of impacts and adaptation; no technical detail; no consideration of uncertainty. Omits non-market effects. Not suitable on its own for detailed national or sectoral-based planning.
Impact-assessment: scenario-based	Projects physical impacts and welfare costs from climate model outputs using impact functions, plus costs and benefits of adaptation options.	Sector-specific analysis at the regional, national or sub-national scale. Physical impacts as well as welfare values. Can capture non-market sectors.	Not able to represent cross-sectoral, economy-wide effects. Treats adaptation as a menu of technical options for defined scenarios. Medium- to long-term focus has less relevance for short-term policy.

Impact assessment: extreme weather events	Variation of the above, using historic damage-loss relationships. Adaptation costs from replacement expenditures or analysis of options.	Consideration of future climate variability. Provides information on short-term priorities (with current climate extremes).	It may be inappropriate to apply historical relationships to future socio-economic conditions. Robustness is limited by the current high degree of uncertainty in predicting future extremes.
Risk assessment	Risk-based variations include probabilistic analysis and thresholds.	As above, but risk-based context allows greater consideration of risk and uncertainty.	Risk-based approach introduces an extra dimension of complexity with a probabilistic approach.
Econometrics-based	Econometrics used for relationships between economic production and the climate are applied to future scenarios.	Provide information on multiple factors and can capture autonomous adaptation.	Mostly focused on autonomous or non-specified adaptation. Simplistic relationships for complex parameters. No information on specific attributes.
Adaptation assessments	Economic analysis of adaptive management (iterative adaptation pathways).	Focus on immediate adaptation policy needs, soft and hard adaptation, and decision making under uncertainty.	Resource intensive.

Source: Updated from Watkiss & Hunt 2010

TABLE 10.1.2 Coverage of studies for European adaptation costs and benefits

Sector	Coverage	Cost estimates	Benefit estimates
Coastal zones	High coverage (infrastructure/erosion) for Europe, regions, Member States (MS); medium coverage for cities/local level.	+++	+++
Infrastructure inc. floods	Medium. Estimates exist at the EU level and for several countries (floods), but coverage is lower for other infrastructure.	++	++
Agriculture	High coverage for farm-level adaptation benefits; medium for costs and planned adaptation.	+	++
Energy	Medium–Low. Cooling/heating demand* for EU and MS, plus MS-planned adaptation. Less coverage for energy supply.	++	++
Health	Medium. Adaptation costs for heat alerts and food-borne disease, but lower coverage of other health risks.	++	+
Water	Medium–Low. Some national, river basin and sub-national studies on water supply.	++	+
Transport	Low–Medium. Some national and individual sub-sector analysis.	+	+
Tourism	Low–Medium. Studies of winter tourism and some studies of changing summer tourism flows.*	+	+
Forestry and fisheries	Low. Limited number of quantitative studies.	+	+
Biodiversity/ ecosystems	Low. Limited number of studies on restoration costs.		
Business and industry	Very low. No quantitative studies.		
Adaptive capacity	Low. Selected studies only, and only qualitative descriptions of benefits.		
Indirect and cross-sectoral	Low. Some recent studies confined to consideration of the cross-sectoral impacts of floods.		
Macro-economic	Low. Small number of studies, with sectoral coverage limited (reflecting evidence in the sectors listed above).	+	+

Key
+ Low coverage with a small number of selected case studies or sectoral studies.
++ Some coverage, with a selection of national or sectoral studies.
+++ More comprehensive geographical coverage, with quantified costs (or benefits).
* Can be considered as both an impact of climate change and an adaptation to it.

Source: EEA 2013 (updated from EEA 2010, Watkiss & Hunt 2010)

- Identifying robust rather than optimal options that will perform well across future scenarios
- Identifying those long-term issues requiring early pro-active investigation (though not necessarily definite action), e.g. where long life/lead-times or irreversibility are involved.

This involves a suite of new decision-support approaches that can address different elements of uncertainty; a summary of such tools is presented in Table 10.1.3, alongside more traditional techniques. Note that while a range of tools is described, they are not mutually exclusive. A brief description of the approaches is presented below, with a particular focus on the new approaches.

- *Cost-Benefit Analysis* (CBA) is the method of choice in most governmental economic appraisals or impact assessments. Social cost-benefit analysis evaluates all the relevant costs and benefits to society for all options and then estimates a net present value or a benefit/cost ratio. In this regard, CBA is an absolute measure providing an absolute justification for intervention; however, it is often difficult to evaluate all the costs and especially the benefits of a particular project or policy. The routine CBA applied in economic appraisals does not fully address many of the complex issues of adaptation, especially the issue of uncertainty (UNFCCC 2009). In general, the more unique and less routine the decision-making context is, the less useful CBA will be; the technique is appropriate only for certain adaptation decision-making contexts, although it can be combined with many of the new techniques below.
- *Cost-Effectiveness Analysis* (CEA) is used to compare the costs of alternative methods of producing the same or similar outputs. In this respect, it is a relative measure. It has been widely used to assess the least-cost way of reaching given targets, thresholds or pre-defined levels. CEA can also be used to compare options using a physical rather than monetary metric. Indeed, CEA has become the primary method of analysis for greenhouse gas mitigation, using the metric of the cost per tonne of abating GHG, with the use of marginal abatement cost curves to look at total abatement potential. However, the lack of common metrics makes a similar approach challenging for adaptation. The approach does have potential applications, notably in relation to flood protection (where there are defined levels of risk), but it does not lend itself to the analysis of uncertainty without additional extension of the technique.
- *Multi-Criteria Analysis* (MCA) allows consideration of quantitative and qualitative data together using multiple indicators. It can integrate broad objectives (and related decision criteria) in a quantitative analysis without assigning monetary values to all factors, thus comparing some criteria that are expressed in monetary terms and others that are not. MCA does have considerable potential for adaptation and, through its process of scoring and weighting, for prioritisation, especially within a stakeholder context. However, much of the analysis can be subjective in nature, especially the consideration of uncertainty.

TABLE 10.1.3 Traditional and new decision support tools for evaluating adaptation options (see EEA 2013 for a more detailed discussion, references and caveats)

Decision support tool	Brief description	Usefulness & limitations in climate adaptation context
Social Cost-Benefit Analysis (CBA)	CBA evaluates all relevant costs and benefits to society for all options and estimates the net costs/benefits in monetary terms. CBA seeks to directly compare costs and benefits, allowing comparisons within and across sectors.	Most useful when: • Climate risk probabilities are known • Climate sensitivity is likely to be small compared to total costs/benefits • Good quality data exists for major cost/benefit components.
Social Cost-Effectiveness Analysis (CEA)	CEA compares the relative costs of different options and can assess alternative ways of producing the same or similar outputs, identifying least-cost outcomes using cost-curves. Used extensively in climate change mitigation.	Most useful when: • As with CBA, but also can be used with non-monetary metrics (e.g. health) • There is agreement on the sectoral social objective (e.g. acceptable risks of flooding).
Multi-Criteria Analysis (MCA)	MCA allows consideration of quantitative and qualitative data using multiple indicators to integrate broad objectives (and related decision criteria) in a quantitative analysis. It provides systematic methods for comparing these criteria, some of which may be expressed in monetary terms, some in other units.	Most useful when: • There are broad objectives and qualitative data (including non-monetary metrics) • There is an opportunity/need for stakeholder input in reaching agreement

Traditional appraisal

Real Options Analysis (ROA)	ROA extends the principles of CBA to allow economic analysis of learning, information and future option values, thus providing context to decisions under uncertainty. It can also provide economic analysis of the benefits of flexibility and value information.	Most useful for: • Large irreversible capital-intensive investment, with the potential for learning (especially with long decision/construction lifetimes) • Climate risk probabilities are known or the range is bounded.
Iterative adaptive management/ adaptation pathways	Adaptation pathways and turning points use adaptive management to monitor, research, evaluate and learn to improve or update strategies, usually to defined risk, biophysical or socio-political thresholds.	Most useful when: • Clear risk thresholds • Where potential for learning and updates • Mix of quantitative and qualitative information, and also non-monetary areas (e.g ecosystems).
Portfolio Analysis (PA)	PA allows an explicit trade-off to be made between returns (measured e.g. in terms of net benefit by CBA) and the uncertainty of returns (measured by the variance in alternative combinations or portfolios of adaptation options) under alternative climate change projections.	Most useful when: • A number of adaptation actions are likely to be complementary in reducing climate risks • Climate risk probabilities known or good information.
Robust Decision-Making (RDM)	RDM identifies robust (rather than optimal) decisions under deep uncertainty, by stress testing large numbers of scenarios.	Most useful when: • There is deep uncertainty or high uncertainty in direction of climate change signal • Mix of quantitative and qualitative information/non-monetary areas.

Economic decision making under uncertainty

Source: Updated from Hunt & Watkiss 2011

- *Real Options Analysis* (ROA) provides quantitative economic information on uncertainty and risk in cases in which there is flexibility regarding the timing of investment decisions and some potential for learning. It can be applied when an investor (or a policy-maker) has the option to invest at some point in the future, when there are uncertain returns on that investment and when there is the potential to adjust the timing of this decision based on underlying conditions or new information. ROA provides quantitative economic information to assist the decision of whether to invest or to wait. It can also examine the economic benefits of flexibility. The approach is particularly relevant for the analysis of large capital investments under conditions of uncertainty, especially those that involve irreversibility. While it has many strengths, ROA is most applicable for large infrastructure projects, and it is dependent on probabilistic information on outcomes. It should be noted that Real Option Analysis has been widely cited as a key decision tool for adaptation, although there has been only limited practical application of the tool to date, and many commentators confuse ROA with iterative adaptive management.
- *Iterative adaptive management with decision trees/adaptive pathways* is emerging as a very promising approach for integrating uncertainty into adaptation decision-making. While there are different approaches, the underlying focus is on the management of uncertainty, allowing adaptation to work within a process of monitoring, research, evaluation and learning to improve and update management strategies. Recent applications have also used the term 'adaptation pathways' or 'adaptation turning/tipping points'. Most applications identify possible risk thresholds (or biophysical, human, social or economic thresholds) and accompanying indicators, and assess options (or portfolios of options) that can respond to these threshold levels. These are accompanied by monitoring plans that track key indicators, and through a cycle of evaluation and learning, allow the adjustment of plans over time. The appraisal of options within these pathways can be undertaken using some of the tools above, using qualitative or semi-quantitative decision-support tools, such as MCA, or more formalised economic appraisal with CBA. The results of these iterative assessments are often presented as adaptation pathways or route maps.
- *Portfolio Analysis* (PA) seeks to develop and assess portfolios of options rather than individual options. The formal approach, derived from financial markets, aims to spread investments over a range of asset types to diversify risks, thereby reducing dependence on single assets. The same principles have high relevance for climate change adaptation in assessing the effectiveness of portfolios of options against uncertainty. This process selects a range of options that together will be effective over the range of possible projected future climates, rather than one option that is best suited to one possible future climate. The main strength of the approach is the structured way in which it allows the analysis of several options together in

response to risks in a way that consideration of single adaptation options does not allow. PA can also be undertaken using physical and economic metrics. However, it is a highly quantitative and data-intensive approach, relying on the availability of data on effectiveness (return) and co-variance, and thus probability.

- *Robust Decision-Making (RDM)* is a decision-support tool that is used in situations of deep uncertainty, i.e. in the absence of probabilistic information on scenarios and outcomes. In the most thorough and formal applications of the approach, it involves the combination of both qualitative and quantitative information through a detailed modelling interface, considering climate and non-climate uncertainty. The key aim of RDM is to find strategies that are robust rather than optimal over many future outcomes. Its formal application considers large ensembles of future scenarios reflecting different plausible future conditions (using hundreds to thousands to millions of input combinations). Iterative and interactive techniques are then applied to 'stress test' different strategies in order to identify the potential vulnerabilities or weaknesses of proposed approaches. More recently, applications to climate change have adopted a simplified form, focusing on climate scenario and climate model uncertainty, but still looking for robustness. RDM is a particularly useful tool when future uncertainties are poorly characterised and probabilistic information is limited or unavailable.

10.1.4 Discussion

While conceptually these new techniques are highly informative, they are also all technically complex. In their formal forms, they are all data- and resource-intensive and require a relatively high degree of expert knowledge. They also have particular strengths; as a result, individual techniques are likely to be more or less suited to specific types of adaptation problems. All of these factors indicate that the formal applications of these techniques will be directed at specific contexts in which there are large investment decisions to be made or major risks, rather than more generally applied in adaptation decision-making. However, the general concepts of these new techniques do have wider applications for adaptation appraisal – whether it involves considerations of flexibility and future options, the potential for iterative pathways and decision trees, the combination of portfolios or options, or an emphasis on robustness. Identifying these concepts and providing guidance on the likely applicability of these concepts to different types of adaptation problems is a priority for future investigation.

Acknowledgements

The underlying review and analysis work here and the funding for writing this paper was provided by the EC 7FWP MEDIATION and IMPACT2C projects.

10.2 Prioritisation of adaptation options for the Netherlands: A Multi-Criteria Analysis

Ekko C. van Ierland, Karianne De Bruin and Rob B. Dellink

10.2.1 Background

In the Netherlands, adaptation to climate change is closely related to spatial planning, as the Netherlands is a densely populated country where adjustments to economic development policies have significant spatial consequences. The study presented in this chapter was part of the Routeplanner Project, which focused on the identification and prioritisation of adaptation options that could reduce the vulnerability of the Netherlands. The Routeplanner Project was initiated in 2006 to provide scientific input in the process of formulation of the National Strategy on Adapting Spatial Planning to Climate Change (ARK),[1] with the aim of 'climate-proofing' the Netherlands. The project provided a 'systematic assessment' of potential adaptation options to respond to climate change in conjunction with spatial planning in the Netherlands, including 1) the identification of adaptation options, 2) a qualitative assessment of these options, 3) a quantification of the direct and indirect costs and benefits associated with the adaptation options and 4) the identification of institutional aspects related to the implementation of the adaptation options. A Multi-Criteria Analysis was carried out that enabled us to categorise and rank promising and feasible adaptation options. Due to the lack of data on the incremental costs and benefits of the adaptation options, it was impossible to perform a proper Social Cost-Benefit Analysis. A total of 96 adaptation options have been identified. These options, based on an inventory of options found in the literature or proposed by stakeholders, include a wide variety of policy measures, technological solutions and adjustments in behaviour. The challenge was to create a balanced list of options with very different characteristics, including realistic and more far-fetched options, hard and soft options and options focused on different spatial scales.

The study focuses on the Netherlands; the assessment was carried out using a typical climate change scenario developed by the Royal Netherlands Meteorological Institute for the period up to 2050 (KNMI 2006). Adaptation options were identified for various sectors, namely agriculture, nature, water, energy and transport, housing and infrastructure, health and recreation, and tourism. Experts on spatial planning and adaptation to climate change as well as public and private stakeholders were involved in the identification and ranking of the adaptation options, including representatives from various research institutes, NGOs, universities and ministries. A detailed list of the different adaptation options and experts and stakeholders involved is provided in De Bruin et al. 2009 and van Ierland et al. 2007.

10.2.2 Multi-Criteria Analysis

Decision-support tools are applied to support decision-makers in their search for the optimal decision in adaptation to climate change from an economic perspective. Multi-Criteria Analysis is used to evaluate adaptation options under existing climate change scenarios based on a set of criteria. Multi-Criteria Analysis requires the identification of alternatives, the selection of criteria and assessment of scores and the selection of the weight for each criterion (Janssen and Van Herwijnen 2006). Each criterion is given a weight that reflects the preferences of the decision-makers, and the weighted sum of the different criteria is used to rank the options. Decision-support tools that assess adaptation strategies have been further developed to incorporate uncertainty and local knowledge about the impacts of climate change (e.g. Mathew et al. 2012, Ceccato et al. 2011, De Bruin & Ansink 2011).

The priority ranking of adaptation options for the Netherlands was based on evaluation and feasibility criteria. Through stakeholder analysis and expert judgement, the options have been identified and criteria selected and weighted to derive a priority setting for alternative adaptation options.

The evaluation criteria were: (i) importance, (ii) urgency, (iii) no-regrets characteristics, (iv) ancillary benefits and (v) mitigation effects of the adaptation options. Furthermore, the options have been scored regarding their feasibility through analysis of the technical, societal and institutional complexity associated with the implementation of the proposed measures. In principle, it is possible to integrate the ranking of the adaptation options based on the evaluation and feasibility criteria; however, we consider the feasibility issue to be relatively distinct from the evaluation criteria and therefore prefer a separate listing. A detailed definition of the criteria for scoring the adaptation options can be found in De Bruin et al. (2009).

Table 10.2.1 provides an overview of the adaptation options that have been identified by stakeholders and that have the highest priority. The scores range from 1 to 5, where 5 indicates the highest priority and 1 the lowest priority. The scoring is based on subjective expert judgements; a workshop with external experts was held to discuss and validate the scores. The exercise shows that there are a large number of adaptation options available. Adaptation options related to the integrated management of nature and water and several other options in the nature and water sectors are evaluated as being the most important, followed by key options in the energy and transport and housing and infrastructure sectors. Changing the mode of transport refers to changes in shipping on rivers in scenarios of extremely low water levels. Table 10.2.1 shows equal weights for all criteria (20%). Choosing different criteria weights would barely affect these high-priority options, as they have (very) high values for all criteria. Criteria weighting has significance for more specialised options that are only valuable according to some criteria.

TABLE 10.2.1 Ranking of high-priority options

Nr.	Sector	Adaptation option	Importance (20%)	Urgency (20%)	No-regrets (20%)	Ancillary benefits (20%)	Mitigation effect (20%)	Weighted sum
34	Nature	Integrated nature and water management	5	5	5	5	4	4.8
	Nature	Integrated coastal zone management	5	5	5	5	4	4.8
40	Water	More space for water: a Regional water system b Improving river capacity	5	5	5	5	4	4.8
41	Water	Risk-based allocation policy	5	5	5	5	4	4.8
65	Water	Risk management as basic strategy	5	5	5	5	4	4.8
68	Water	New institutional alliances	5	5	5	4	5	4.8
28	Nature	Design and implementation of ecological networks (National Ecological Network)	4	5	5	5	4	4.6
75	Energy and transport	Construct buildings differently (less need for air-conditioning/heating)	5	4	5	4	5	4.6
84	Energy and transport	Change modes of transport and develop more intelligent infrastructure	5	5	4	4	5	4.6
87	Housing and infrastructure	Make existing and new cities robust (avoid 'heat islands', provide for sufficient cooling capacity)	5	5	4	5	4	4.6

Note: High scores indicate high priority for implementing the option

10.2.3 Conclusion

MCA is an appropriate decision-support tool for assessing and ranking adaptation options. This ranking can be used in further discussions and decisions on an adaptation strategy. The MCA method can also be applied to other countries or regions through the involvement of local stakeholders and experts who are familiar with the specific circumstances in the relevant country or region. In addition, it is useful in communication with stakeholders and in awareness-raising about the challenges of adaptation to climate change. The study showed that in the Netherlands, integrated nature and water management and risk-based policies have the highest priority, followed by policies aiming at 'climate-proof' housing and infrastructure. However, we also observed that many important and significant adaptation options may have low feasibility because they are ranked high in terms of institutional complexity, as reported in De Bruin et al. (2009).

Some critical points need to be taken into account when considering the assessment of adaptation options based on MCA. As Füssel (2009) has indicated, it is important to present "clear definitions of the set of criteria to improve the transferability of the method" to other countries. Furthermore, he notes that when there is lack of information on costs and benefits, "it is important to include indicators on the costs and economic feasibility to provide insight into both the benefits and costs of adaptation options". In addition, the further analysis of institutions and distributions of tasks and responsibilities with regard to policy making and execution of policies is very important for the successful implementation of adaptation options (Smith et al. 2009).

Recommendations:

- Use of Multi-Criteria Analysis as an appropriate decision-support tool to assess and rank adaptation options, especially when there is a lack of data on the costs and benefits of the adaptation options
- Clear definition of the criteria increases the transferability of the approach
- Involvement of local stakeholders and experts helps to bridge the gap between top-down and bottom-up approaches to adaptation.

Further reading

De Bruin, K., R.B. Dellink, A. Ruijs, L. Bolwidt, A. Van Buuren, J. Graveland, R.S., De Groot, P.J. Kuikman, S. Reinhard, R.P. Roetter, V.C. Tassone, A. Verhagen and E. C. Van Ierland, 'Adapting to climate change in the Netherlands: An inventory of climate adaptation options and ranking of alternatives', *Climatic Change* 95 (1–2), 2009, 23–45.

Acknowledgements

We acknowledge the contributions of A. Ruijs, L. Bolwidt, A. Van Buuren, J. Graveland, R.S., De Groot, P.J. Kuikman, S. Reinhard, R.P. Roetter, V.C. Tassone and A. Verhagen to the Routeplanner study on the qualitative assessment of climate adaptation options and some estimates of adaptation costs performed within the framework of the Netherlands Policy Programme ARK as Routeplanner projects 3, 4 and 5.

10.3 Prioritising actions using adaptation tipping points and adaptation pathways

Marjolijn Haasnoot, Jaap Kwadijk and Nathalie Asselman

10.3.1 Starting from policy objectives

Uncertainties arising from climate change, population growth, economic developments and their impacts challenge decision-making on long-term investments. Urbanised river deltas and estuaries are particularly vulnerable due to pressures from changes in river discharges, precipitation, relative sea-level rise, groundwater seepage and population growth. In these densely populated areas, pressure on space is also high. Decision-making under conditions of uncertainty in these deltas is difficult because the impact of these decisions may reach far into the future, the implementation of policies takes time and some policies may be feasible today but not in the future. Meanwhile, the investments required are often large, and many people and economic sectors are affected.

The Dutch government is aware of the vulnerability of the Rhine-Meuse delta and has established the Second Delta Committee to identify actions to prevent future disasters, as the expected future climate change and sea-level rise "can no longer be ignored" (Delta Programme 2011: 5). The Commission's advice resulted in the enactment of the Delta Act and is presently being elaborated in the Delta Programme. The chair of the Delta Programme summarised the main challenge of this Programme as follows: "One of the biggest challenges is dealing with uncertainties in the future climate, but also in population, economy and society. This requires a new way of planning, which we call adaptive delta planning. It seeks to maximise flexibility; keeping options open and avoiding 'lock-in'" (Kuijken 2010). The question is: how can this be done?

Traditionally, planners develop a static *optimal* policy using a single 'best estimate' of the future situation based on central estimates and extrapolations of trends. This wrongly implies that we can predict the future. In past decades, planners have focused on a static *robust* policy that will perform acceptably in most plausible future scenarios. However, if the future turns out differently from these hypothesised futures, the policy might fail. For complex problems with deep uncertainty, such as long-term delta planning under changing conditions, a different approach is needed.

Among planners, a new paradigm has emerged. This paradigm includes the acceptance of uncertainty and focuses on the question: *Given the uncertainties about the future, what actions can we take now to achieve our objectives now and in the future?* This paradigm holds that in light of deep uncertainty, one must design dynamic, flexible plans that will be able to adapt over time (e.g. Walker et al. 2001, Lempert et al. 2003, Albrechts 2004, Haasnoot et al. 2012). The idea common to these studies is committing to short-term actions and establishing a framework for learning and guiding future actions. Typical questions from planners adhering to this paradigm are:

- How can actions be prioritised in a time of uncertainties and limited budgets?
- How can we ensure that our investments will be robust to future conditions and/or flexible to adaptation if needed?
- How can options be kept open in case we need to adapt to new environmental or societal conditions, and how can we capitalise on opportunities?

This section describes a method for developing sustainable delta-management plans under changing conditions. It focuses on two concepts for prioritising actions: Adaptation Tipping Points (Kwadijk et al. 2010) and Adaptation Pathways (Haasnoot et al. 2011, 2012), illustrating them with experiences from Dutch delta management.

10.3.2 Adaptation Tipping Points

The Adaptation Tipping Point (ATP) approach focuses on assessing whether and when adaptation strategies are needed. The term *tipping point* is introduced to indicate the point at which a system change, initiated by an external forcing, no longer requires the external forcing to sustain the new pattern of change (an example from the natural sciences is the irreversible decay of the Greenland ice sheet). Adaptation Tipping Points do not relate to changes in geo-ecological systems, but rather to management strategies, policies or actions. They are points in time at which the magnitude of change (e.g. due to climate change or sea-level rise) is such that the current management strategy can no longer meet its objectives (Kwadijk et al. 2010). By applying this approach to climate change, the following basic questions of decision-makers can be answered: what are the first issues that we will face as a result of climate change, and when can we expect them to occur? In addition to the ATP approach, an alternative adaptive strategy is required. The approach differs from a traditional top-down impact-assessment approach because it emphasises the management objectives or strategies rather than the potential climate impacts.

The ATP approach was developed in the Netherlands in response to the publication of a new generation of climate scenarios developed by the national meteorological institute (KNMI). Just when Dutch water managers had formally agreed to adopt the central scenario of three originally developed scenarios for the design of adaptation strategies, these scenarios were replaced by four new KNMI scenarios triggered by new insights from the IPCC. It should be noted that in the Netherlands, the climate scenarios developed by KNMI draw on the same climate model output database as the IPCC, but are specifically selected to match the Dutch situation. Consequently, there was suddenly a lack of a central scenario, making it difficult to select a scenario for strategy design and rendering the formal agreement insufficient.

An ATP analysis starts from the perspective that a water system provides the natural boundary conditions for living and working, i.e. for all socioeconomic activities. The system needs management in order to maintain the proper conditions and achieve objectives such as safety. Should these conditions change due to disturbances such as climate change, the current policy may fail, e.g. a certain safety standard may no longer be met. At that moment, an ATP of the management strategy is reached. To define ATPs, planners assess the conditions under which an ATP can be reached. In combination with a particular scenario, the moment at which an ATP occurs can be assessed (Kwadijk et al. 2010). This is sometimes called the 'sell-by-year' of a policy. Using a range of scenarios, a span of time in which the ATP may be reached is determined, reflecting uncertainty. Using transient scenarios (time series with changing conditions) to determine the effectiveness of policy actions over the course of time is another method for determining the sell-by-year (Haasnoot et al. 2012). Reaching ATPs might have physical, ecological, technical, economic, societal or political causes. An example of a physical boundary is the possible shift in aquatic habitats due to sea-level rise, limited by natural dunes or artificial barriers such as dikes. Economic ATPs may occur if the investments required for adaptation are larger than the economic benefits. Society may change its values and standards, resulting in different objectives, which may cause an ATP or may shift the timing of an ATP.

10.3.3 Exploring Adaptation Pathways

Adaptation Pathways (APs) describe a sequence of policy or management actions (Haasnoot et al. 2011, 2012). The APs are a logical follow-up of ATPs, since after an ATP is reached, a new policy or management strategy is needed. By exploring pathways, planners acknowledge the dynamics of natural variability and the interaction between the water system and society. For example, over the course of time, events may trigger policy or management responses and may change societal perspectives, including the objectives and evaluation criteria for policy or management strategies. As a result, the future is determined not only by what is known or anticipated at present, but also by what will be experienced and learned as the future unfolds, and by the responses to events (Haasnoot et al. 2012).

APs are developed by exploring all relevant policy and management options after an ATP. Different pathways into the future depend on the realisation of different climate scenarios, different realisations of the same climate scenario over time (i.e. different run-off variability), different external socioeconomic future events or trends, different evolving societal preferences and different policy responses. The final adaptation map presents an overview of possible relevant adaptation pathways. Similar to a Metro map, this map presents different routes that can be taken to arrive at a desired point in the future. Occurrences of ATPs (terminal stations) and the available policy options are also shown (via transfer stations). Some routes are only available in some of the scenarios (dashed lines). With such a map it is possible by means of one or more combinations of socioeconomic and climate scenarios to identify opportunities, no-regrets strategies, lock-ins and the timing of a policy in order to support decision-making in a changing environment. That is, an adaptation map can be used to prepare a plan for actions to be taken

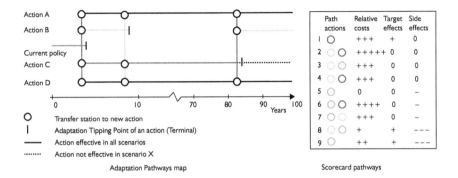

FIGURE 10.3.1 Example of an Adaptation Pathways map (source: Haasnoot et al. 2013a)

immediately, as well as preparations that can be made to enable activation of policy options in the future.

An example of an Adaptation Pathways map (Figure 10.3.1, left) and a scorecard (right) presenting the costs and benefits of the nine possible pathways shown in the map can be found below. In the map, starting from the current situation, targets begin to be missed after four years. Following the medium-grey lines of the current policy, one can see that there are four options. Actions A and D should be able to achieve the targets for the next 100 years in all climate scenarios. If Action B is chosen after the first four years, a tipping point is reached within about five years; a shift to one of the other three actions will then be required to achieve the targets (following the light-grey lines). If Action C is chosen after the first four years, a shift to Action A, B or D will be required in the case of Scenario X (following the solid medium-grey lines). In all other scenarios, the targets will be achieved for the next 100 years (the dashed medium-grey line). The colours in the scorecard refer to the actions: A (very light grey), B (light grey), C (dark grey) and D (very dark grey). While the pathways illustrate potential sequences of possible management actions over time, the scorecard next to the map presents additional information about the options, such as their relative costs, the level at which the main objectives are met and potential positive or negative side effects.

10.3.4 Experience from the Dutch Delta Programme

The Delta Programme adopted the concepts of ATP and AP in the design of an adaptive delta-management plan. In this section, we give an example of possible APs for the flood risk management along the River Rhine in the Netherlands (cf. Figure 10.3.2). The APs for the Waal branch are shown in Figure 10.3.3. The main objective of the Delta Programme is to reach an acceptable flood risk in the years 2050 and 2100 and thus develop flood-prevention strategies for the future that will allow the design discharge to pass through the river branches without flooding adjacent areas. Given the uncertainties in discharge estimates for these years, we

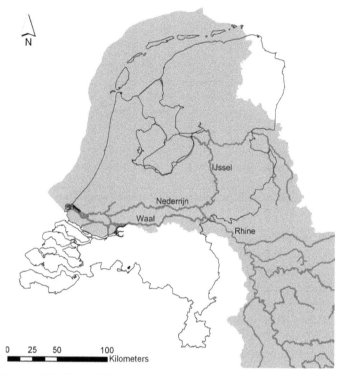

FIGURE 10.3.2 The River Rhine and its main tributaries in the Netherlands (Waal, Nederrijn, IJssel)

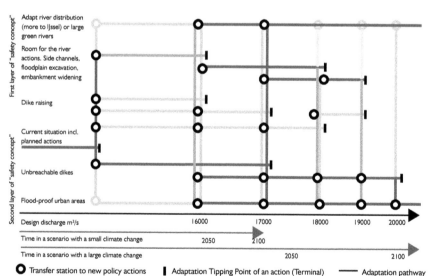

FIGURE 10.3.3 Adaptation Pathways map for the River Rhine (Waal branch) (source: Haasnoot et al. 2013b)

chose to show the possible increase in design discharge on the x-axis instead of time. Depending on the scenarios, the design discharge may increase at a faster or slower rate. Different ATPs on one pathway indicate the stepwise deployment of a particular option. The actions that are planned for the near future consist of some dike strengthening and actions seeking to lower the water level by giving more space to the river. These actions are insufficient to control the flood risk over a longer time span. Ultimately, five policy options were defined. Different shades of grey represent the five options listed in the left-hand column of Figure 10.3.3. The first option consists of actions that will result in lowering of the water level during floods by giving more space to the river (e.g. lowering flood plains). It is expected that these actions will only be able to solve part of the problem. Therefore, if we start with the implementation of these actions, we will eventually have to continue with or add another action. Other actions might include dike strengthening (either through a large effort in one action or in consecutive smaller steps), the development of 'unbreachable dikes' (e.g. De Bruin et al. 2012), the adaptive construction of houses and other buildings and the implementation of very large measures, such as the development of a 'green river' (i.e. a new river area that will only be flooded during extreme events). As it is not very likely that this type of action will be selected at present, we have indicated the first part of this route in the lightest shade of grey.

10.3.5 Concluding remarks

Adaptation Tipping Points (ATP) and Adaptation Pathways (AP) are concepts that support decision-making under conditions of deep uncertainty; these concepts have been applied in the area of safety against flooding and fresh water supply, but have potential for other policy fields as well. With the ATPs of current policies or management strategies, planners can identify where and when problems may first arise. This approach places less emphasis on (updating) the climate projections or socioeconomic scenarios than the classical top-down approach, in which scenarios of future conditions are used as the starting points for a risk assessment. The APs provide insight into the sequence of actions over time, potential lock-ins and path dependencies. This contributes to the creation of an adaptive plan that includes the dynamic interaction between the water system and society. Such a plan can be seen as an improvement in comparison to the static policies developed using a classical scenario approach based on one or two projected years.

To assess the conditions under which an ATP of a policy action will occur, quantitative targets are required. However, in reality, policy-makers sometimes choose to keep these targets vague, making it difficult to determine the timing of an ATP or the efficacy of a pathway. Exploring different quantifications of the targets can show their impact, which may support a discussion of the targets. The occurrence of an ATP also depends on climate and socioeconomic scenarios. For some actions, the time span before an ATP could occur is large, which makes it difficult to decide when to start the implementation of another policy action. Consequently, for actions that are affected by developments with a large magnitude of change relative to the natural variability, the occurrence of an ATP is easier to pinpoint than for actions whose ATPs are primarily determined by (changing) natural variability. For example, the ATP of a coastal barrier that is the result of sea-level rise (a relatively gradual process) is fairly easy to determine. In contrast, the ATP of river dikes depends on the occurrence of extreme river discharges; these have high (monthly, annual) variability, and changes in extreme discharges are difficult to detect.

The strengths of ATP and AP could be integrated with other concepts of adaptive policies to support decision-makers with a clear stepwise approach for the development of an adaptive plan. Signposts and triggers (Dewar et al. 1993, Walker et al. 2001) could be identified to monitor the need to implement a pre-specified action. In addition, anticipatory and contingency actions, such as those described in the approach of adaptive policy making (Walker et al. 2001, Kwakkel et al. 2010), can be implemented in order to get on track and stay on a preferred pathway. Recent efforts have been made in this direction (Haasnoot 2013b).

Practical examples of the ATP and AP approach can be found in the UK Thames Barrier (Jeuken & Reeder 2011, Reeder & Ranger 2011, Wilby & Keenan 2012) and the Dutch Delta Programme (Jeuken & Reeder 2011, Delta Programme 2011). The approach is increasingly being adopted by a variety of sectors and stakeholders, for example in urban spatial planning.

10.4 Lessons learned for the prioritisation of adaptation options

Rob Swart

In this chapter, various methods were explored that can be applied to evaluate different (sets of) adaptation options. Only after the turn of the century has adaptation emerged as an issue of political and scientific interest in Europe and other developed countries; thus, as yet there has been very little practical experience with the analysis of the costs and benefits of adaptation options and few experiments on different approaches to prioritise adaptation options. In addition, it appears that adaptation has a number of characteristics that make the evaluation of options a still greater challenge, including many diverse sectors in which adaptation should be implemented, and an even wider variety of types of options, including many 'soft' options. This means that different adaptation situations are likely to require divergent methods for assessment and prioritisation. Deriving generic recommendations from individual cases determined by the local context should be done with caution. The evidence basis is still very small, and there is still a long way to go before we can determine best practices with any degree of confidence. However, the cases described above are drawn from a greater body of experiences and studies, and therefore a number of lessons learned can be derived that can be considered to be generally applicable to the evaluation of adaptation options:

- Different assessment methods for appraising adaptation options are available; each have their specific strengths and weaknesses and are thus suitable for different situations and different objectives. The applicability of formal techniques such as CBA, CEA, MCA and new techniques depends on the availability of information and agreement on objectives and common criteria.
- If data on costs and benefits are poor and objectives other than monetary costs are important, MCA can be a good method of prioritising options. Clear definitions and agreement on the criteria applied are required to make the approach transferable and transparent.
- New techniques involving flexibility, the potential for iterative pathways, the combination of portfolios of options and robust options offer new possibilities but also new challenges in terms of methodology and required data.
- The concept of Adaptation Tipping Points and the related Adaptation Pathways approach to the design of a set of adaptation options over time can support decision-making under conditions of deep uncertainty, and therefore hold promise for applications beyond the water safety applications in which they have thus far been applied.
- Using more than one method offers opportunities to learn and could help decision-makers avoid decisions based on incomplete information that takes only certain objectives and certain uncertainties into account.
- For all evaluation techniques, involvement of both local stakeholders and experts will help to increase the relevance and legitimacy of the analysis.
- Due to the broad range of potential future climate change impacts and their implicit uncertainties (cf. Chapter 11), multiple-benefit, no-regrets and low-regrets adaptation options should be favoured in most adaptation situations. Multiple-benefit options provide synergies with other goals, such as mitigation or sustainability (e.g. ecosystem-based approaches). The socioeconomic benefits of no-regrets options, such as early warning systems and insurance against floods, exceed their costs in all plausible climate futures. Low-regrets options produce relatively large benefits with only limited associated costs. An example is restrictions on the type and extent of development in flood-prone areas (UKCIP 2005).
- In other adaptation situations, actions with potentially high effectiveness that are likely to become more costly (e.g. large and long-lasting infrastructure projects), more difficult to implement (e.g. spatial planning for nature conservation) or redundant (e.g. raising awareness) when postponed should be undertaken immediately.

Note

1 http://www.climateresearchnetherlands.nl/ARKprogramme

References

Adger, W.N., Arnell, N.W. and Tompkins, E.L. (2005) 'Successful adaptation to climate change across scales', *Global Environmental Change* 15 (2), pp. 77–86.

Agrawala, S., Bosello, F., Carraro, C., de Cian, E. and Lanzi, E. (2011) 'Adapting to climate change: costs, benefits, and modelling approaches', *International Review of Environmental and Resource Economics* 5 (3), pp. 245–284.

Albrechts, L. (2004) 'Strategic (spatial) planning reexamined. Environment and Planning B', *Planning and Design* 31 (5), pp. 743–758.

Bizikova, L., Neale, T. and Burton, I. (2008) *Canadian communities guidebook for adaptation to climate change. Including an approach to generate mitigation co-benefits in the context of sustainable development,* Environment Canada.

Ceccato, L., Giannini, V. and Giupponi, C. (2011) 'Participatory assessment of adaptation strategies to flood risk in the Upper Brahmaputra and Danube river basins', *Environmental Science & Policy* 14 (8), pp. 1163–1174.

De Bruin, K. and Ansink, E. (2011) 'Investment in flood protection measures under climate change uncertainty', *Climate Change Economics* 2 (4), pp. 321–339. DOI 10.1142/S2010007811000334.

De Bruin, K., Klijn, F. and Knoeff, J.G. (2012) *Unbreachable embankments? In pursuit of the most effective stretches for reducing fatality risk,* accepted for publication in the proceedings of the Flood Risk Management 2012 conference, Rotterdam, The Netherlands. DOI: 10.1201/b13715-131.

De Bruin, K., Dellink, R.B., Ruijs, A., Bolwidt, L., van Buuren, A., Graveland, J., de Groot, R.S., Kuikman, P.J., Reinhard, S., Roetter, R.P., Tassone, V.C., Verhagen, A. and van Ierland, E.C. (2009) 'Adapting to climate change in The Netherlands: an inventory of climate adaptation options and ranking of alternatives', *Climatic change* 95 (1–2), pp. 23–45. DOI 10.1007/s10584-009-9576-4.

Delta Programme (2011) *Working on the delta. Investing in a safe and attractive Netherlands, now and in the future.* http://www.deltacommissaris.nl/english/Images/Deltaprogramma_ENG1_tcm310-286802.pdf. Access date 08-06-2012.

Dewar, J.A., Builder, C.H., Hix, W.M. and Levin, M.H. (1993) *Assumption-based planning: a planning tool for very uncertain times,* Tech. Rep. Report MR-114-A, Santa Monica, USA: RAND.

Dyszynski, J. (2009) *The economics of adaptation,* WeAdapt Briefing Note.

EA (2011) *TE2100 Strategic outline programme,* Environment Agency.

EEA (2007) *Climate change – the cost of inaction and the cost of adaptation. Technical report* No. 13/2007, No. 978-92-9167-974-4, European Environmental Agency, Copenhagen.

EEA (2010) *State and outlook report,* European Environment Agency, Copenhagen.

EEA (2013) *Adaptation in Europe. Addressing risks and opportunities from climate change in the context of socio-economic developments,* EEA Report 3/2013, Copenhagen.

Füssel, H.M. (2007) 'Adaptation planning for climate change: concepts, assessment, approaches, and key lessons', *Sustainability Science* 2 (2), pp. 265–275.

Füssel, H.M. (2009) 'Ranking of national-level adaptation option. An editorial comment', *Climatic Change* 95 (1–2), pp. 47–51.

Haasnoot, M., Middelkoop, H., van Beek, E. and van Deursen, W.P.A. (2011) 'A method to develop sustainable water management strategies for an uncertain future', *Sustainable Development* 19 (6), pp. 369–381. DOI 10.1002/sd.438.

Haasnoot, M., Middelkoop, H., Offermans, A., van Beek, E. and van Deursen, W.P.A. (2012) 'Exploring pathways for sustainable water management in river deltas in a changing environment', *Climatic Change* 115 (3–4), pp. 795–819. DOI 10.1007/s10584-012-0444-2.

Haasnoot, M., Kwakkel, J.H., Walker, W.E. and Ter Maat, J. (2013a) 'Dynamic adaptive policy pathways: a method for crafting robust decisions for a deeply uncertain world', *Global Environmental Change* 23 (2), pp. 485–498.

Haasnoot, M. (2013b). *Anticipating change. Sustainable water policy pathways for an uncertain future.* Ph.D thesis University of Twente. DOI 10.3990/1.9789036535595.

Hallegatte, S. (2009) 'Strategies to adapt to an uncertain climate change', *Global Environmental Change* 19 (2), pp. 240–247.

Hunt, A. and Watkiss, P. (2011) *The UK adaptation economic assessment. Methodology report. Report to Defra as part of the UK Climate Change Risk Assessment,* Final Report to Defra, Deliverable 2.2.1., published by Defra.

Janssen, R. and Van Herwijnen, M. (2006) 'A toolbox for multicriteria decision-making', *International Journal of Environmental Technology and Management* 6 (1), pp. 20–39.

Jeuken, A. and Reeder, T. (2011) *Short-term decision making and long-term strategies: how to adapt to uncertain climate change. Examples from the Thames Estuary and the Dutch Rhine-Meuse Delta,* Water Governance 1, The Netherlands.

KNMI (2006) *Climate in the 21st century, four scenarios for the Netherlands,* Royal Netherlands Meteorological Institute, De Bilt, Netherlands.

Kuijken, W. (2010) *The Delta programme in the Netherlands: the Delta Works of the Future.* http://www.deltacommissaris.nl/english/news/presentations/thedeltaprogramme inthenetherlandsthedeltaworksofthefuture.aspx. Access date 08-06-2012.

Kwadijk, J.C.J., Haasnoot, M., Mulder, J.P.M., Hoogvliet, M.M.C., Jeuken, A.B.M., van der Krogt, R.A.A., van Oostrom, N.G.C., Schelfhout, H.A., van Velzen, E.H., van Waveren, H. and de Wit, M.J.M. (2010) 'Using adaptation tipping points to prepare for climate change and sea level rise: a case study in the Netherlands', Wiley Interdisciplinary Reviews, *Climate Change* 1 (5), pp. 729–740. DOI:10.1002/wcc.64.

Kwakkel, J.H., Walker, W.E. and Marchau, V.A.W.J. (2010) 'Adaptive airport strategic planning', *European Journal of Transportation and Infrastructure Research* 10 (3), pp. 227–250.

Lempert, R.J., Popper, S. and Bankes, S. (2003) *Shaping the next one hundred years: new methods for quantitative, long-term policy analysis,* Santa Monica, CA: RAND.

Mathew, S., Trück, S. and Henderson-Sellers, A. (2012) 'Kochi, India case study of climate adaptation to floods: Ranking local government investment options', *Global Environmental Change* 22 (1), pp. 308–319.

McCarthy, J., Canziani, O.F., Leary, N.A., Dokken, D.J. and White, K.S. (ed.) (2001) *Climate change 2001: impacts, adaptation, and vulnerability,* Contribution of Working Group II to the Third Assessment Report of the IPCC.

Mendelsohn, R. (2006) 'The role of markets and governments in helping society adapt to a changing climate', *Climate Change* 78 (2), pp. 203–215.

Reeder, T. and Ranger, N. (2011) *How do you adapt in an uncertain world? Lessons from the Thames Estuary 2100 project,* World Resources Report, Washington DC. http://www.worldresourcesreport.org.

Smith, J.B. and Lenhart, S. (1996) 'Climate change adaptation policy options', *Climate Research* 6 (2), pp. 193–201.

Smith, J.B., Vogel, J.M. and Cromwell, J.E.C. III (2009) 'An architecture for government action on adaptation to climate change: an editorial comment', *Climatic Change* 95 (1), pp. 53–61.

UKCIP (2005) *Identifying adaptation options,* UK Climate Impacts Programme, Oxford.

UNFCCC (2009) *Potential costs and benefits of adaptation options: a review of existing literature.* UNFCCC Technical Paper. http://unfccc.int/resource/docs/2009/tp/02.pdf.

van Ierland, E.C., De Bruin, K., Dellink, R.B. and Ruijs, A. (ed.) (2007) *A qualitative assessment of climate adaptation options and some estimates of adaptation costs,* Reports on the Routeplanner Projects 3, 4 and 5, Netherlands: Wageningen UR, pp. 1–145.

Walker, W.E., Rahman, S.A. and Cave, J. (2001) 'Adaptive policies, policy analysis, and policymaking', *European Journal of Operational Research* 128 (2), pp. 282–289.

Watkiss, P (ed.) (2011) *The ClimateCost Project*, Final Report, vol. 1, Europe, published by the Stockholm Environment Institute, Sweden.

Watkiss, P. and Hunt, A. (2010) *Review of adaptation costs and benefits estimates*, in *Europe for the European Environment State and Outlook Report*, Technical Report prepared for the European Environmental Agency for the European Environment State and Outlook Report 2010, Copenhagen, published online by the European Environment Agency, Copenhagen.

Wilby, R.L. and Keenan, R. (2012) 'Adapting to flood risk under climate change', *Progress in Physical Geography*, pp. 1–31. DOI 10.1177/0309133312438908.

Willows, R. and Connell, R. (2003) *Climate adaptation. Risk, uncertainty and decision-making*, UKCIP Technical Report.

11

WORK WITH UNCERTAINTIES

Torsten Grothmann, Andrea Prutsch,
Inke Schauser, Sabine McCallum
and Rob Swart

Explanation of the guiding principle

Even with the continuous improvement of climate change scenarios and models, uncertainties in projecting climatic changes will always remain. Research, exchange of good practice (cf. Chapter 7) and stakeholder involvement (cf. Chapter 8) can help reduce informational uncertainties (e.g. regarding probable climate change impacts) and normative uncertainties (e.g. regarding acceptable risks in societies) (Lebel et al. 2010, Pahl-Wostl 2009, Pahl-Wostl et al. 2007). Nevertheless, uncertainties will remain.

Although uncertainty exists, a wait-and-see-attitude is not appropriate due to the potential of serious or irreversible harm resulting from climate change impacts (Smith 1997, UNFCCC 1992). Different objectives, principles and approaches are available that take uncertainties into account, including the precautionary principle, adaptive management, robustness and resilience. In addition, many flexible options, no- or low-regrets options that serve other purposes, and options with safety margins (or redundancy) provide attractive opportunities regardless of how climate change eventually materialises (Hallegatte 2009).

The *precautionary principle* demands that the absence of complete scientific certainty should not be used as an excuse to postpone actions when there is the potential of serious or irreversible harm (UNFCCC1992). *Adaptive management* is a flexible, step-by-step process of planning, implementing and revising adaptation activities on the basis of new research results, regular monitoring and evaluation (Pahl-Wostl 2007). This approach recognises that our understanding of climate change and adaptation will change and improve over time. An adaptive management approach reduces uncertainties through continuous monitoring (cf. Chapter 14). It is particularly well suited for sectors with short-term planning horizons (e.g. agriculture). However, difficulties may occur in sectors in which long-term planning is required (e.g. forestry, where trees planted today must be adapted to the climate in 50 to 100 years). In such cases, the principles of maintaining or increasing robustness and resilience are useful. *Robust systems* continue to function under a wide range of potential future (climatic) conditions (e.g. wastewater systems that are effective in times of light and heavy precipitation) (cf. Chapter 5). This approach does not emphasise specific climate projections, instead using a wide range of plausible (climate change) futures to identify robust adaptation actions (Dessai & Hulme 2007, Hallegatte et al. 2012). In contrast, *resilient systems* are flexible and can quickly adapt to changed conditions (e.g. communities with diversified sources of income): "The resilience approach emphasizes non-linear dynamics, thresholds, uncertainty and surprise, how periods of gradual change interplay with periods of rapid change and how such dynamics interact across temporal and spatial scales" (Folke 2006: 253). In particular, the consideration of surprises (events not included in climate projections, such as the heat wave in Europe in 2003) differentiates the resilience approach from the robustness approach.

Chapter overview

In the following chapter, three cases illustrate the practical realisation of the considerations mentioned above regarding how one can work with existing uncertainties and implement adaptation measures despite these uncertainties. The first case (Section 11.1) describes how adaptive management has been successfully implemented for conservation of the Bosherston Lakes SAC. From the perspective of robust decision-making, the second case (11.2) analyses climate change adaptation of water resource systems in England and Wales. The third case (11.3) presents current psychological studies, developing recommendations from them on how to communicate climate change uncertainties.

11.1 An adaptive approach to conservation management in Bosherston Lakes SAC

Clive Walmsley and Tristan Hatton-Ellis

11.1.1 Background: The Bosherston Lakes Special Area of Conservation

Bosherston Lakes is a very shallow (1- to 3-metre deep) lake system located almost at sea level, on the Pembrokeshire coast in southwest Wales, UK. It was created artificially by damming three connected limestone valleys in the 18th and 19th centuries as an ornamental feature for the nearby Stackpole Estate. The lakes are predominantly groundwater-fed with some surface water inputs, especially in the Eastern Arm. Their principal nature conservation value lies in the clear, calcium-rich lake ecosystem dominated by stoneworts (*Chara* spp., cf. Figure 11.1.1). They

FIGURE 11.1.1 Submerged beds of *Chara hispida* in the Central Arm of Bosherston Lakes. This vegetation helps to maintain the very clear water conditions (source: photo copyright Lisa Whitfeld www.celticimages.co.uk)

are designated as a Special Area of Conservation (SAC) under the Natura 2000 Framework established by the Habitats Directive.

The SAC covers 35 ha and has been principally designated for its "hard oligotrophic lakes with benthic vegetation of *Chara* spp." Annex 1 habitat. The lakes are also valued for their landscape and ornamental value, as exemplified by the extensive lily pads (cf. Figure 11.1.2).

Despite ongoing management within the lakes and the wider catchment, they are currently reported as being in "unfavourable condition" due to eutrophication and insufficient *Chara* cover (Burgess et al. 2006). The ecological status of the lakes is closely linked to the complex hydrological and hydrochemical regime, which is influenced by the limestone aquifer, the surrounding catchment and inflowing streams, the sea and the artificial hydraulic control structures.

11.1.2 Reasons for adaptive conservation management in the Bosherston Lakes SAC

Climate change projections indicate that the sea level in south Wales is projected to rise at a rate of 3–4 mm per year, with a best estimate of an 18–26-cm rise by 2050 (UKCP09 projections, Lowe et al. 2009). Analyses of water temperature, salinity and tidal height monitoring data, together with projected climatic change and sea-level rise from national models (UKCIP02 climate change scenarios, Hulme et al. 2002), suggest that over the next 50 years,

FIGURE 11.1.2   Water lilies at Bosherston Lakes. The lilies only grow in the cleaner parts of the lake and are greatly valued by visitors (source: photo copyright Lisa Whitfeld www.celticimages.co.uk)

climate change will predominantly increase pre-existing pressures on the site, such as droughts, sediment input and eutrophication rather than introduce new pressures (Holman et al. 2009). There is a low probability of occasional short-term episodes of saline intrusion in the coastal margins of the lakes due to a combination of sea-level rise, tide and tidal surges (cf. Figure 11.1.3). The risk of inundation from sea-level rise was identified as being highest in the autumn, when lake levels are comparatively low and storm surges more likely. However, there are considerable uncertainties associated with such projected changes, especially in relation to the sea-level rise estimates, which are critical to determining the future status of the site. These uncertainties, combined with the complexity of the interacting drivers of change within the site, necessitate an adaptive approach to conservation.

A climate change vulnerability assessment of the protected area network in Wales that principally considered the generic sensitivity of habitat and species features of sites and each site's connectivity with similar habitats (Wilson et al. 2013) identified Bosherston as the 17th most vulnerable of the 85 non-marine SAC sites in Wales considered in the analysis. As Bosherston is in the top quartile of vulnerable SAC sites, it is important that potential adaptation measures be considered, but financial and resource limitations necessitate their phased implementation; this can be facilitated by an adaptive approach that informs the implementation priorities.

11.1.3 Practical experiences from the adaptive management approach

Some stakeholders involved in the management of the Bosherston Lakes have suggested that the SAC may not be sustainable as a freshwater lake system in the

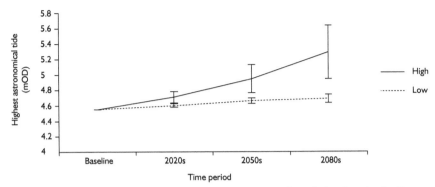

FIGURE 11.1.3 Assessment of the risk of tidal overtopping of the dam in the future, based on the UKCIP02 low-emissions and high-emissions sea-level rise scenarios. The crest of the dam is at 5.6 m AOD. Under the high-emissions scenario, there is a risk of overtopping by the 2080s (source: Holman et al. 2009)

future, principally due to sea-level rise and summer droughts (National Trust 2007). Partly in response to this perception, the Countryside Council for Wales (CCW) initiated research to identify and (where possible) quantify the climate change risks (outlined above) and develop an approach for the site's future management. Contrary to the stakeholders' suggestion, the monitoring data for the site and climate change scenarios did not indicate that immediate or dramatic changes in site management were required (Holman et al. 2009). However, impacts beyond 2050 were much more uncertain and the researchers recommended that the site be managed adaptively.

This adaptive management cycle is illustrated in Figure 11.1.4; it is similar in format to other decision-making frameworks that have been developed for adaptation more generically within the UK (e.g. Willows & Connell 2003). It begins by assessing climate risks and collating new knowledge and research. At Bosherston, this principally involved collating and reviewing the ongoing monitoring data for the site and identifying the most appropriate downscaled climate and sea-level scenario data using the UKCIP02 climate scenario data. The evaluation of these data sources was then reviewed in the context of the biological features of the site and the drivers of change identified within the Site Management Plan. The next step was to decide whether to revise the site conservation objectives. This was relatively straightforward, as there was already work in progress to address the eutrophication of the site, manage the water levels and monitor the key environmental drivers of change (such as water levels and quality). These were also identified as the adaptation options that needed to be continued, while other activities were identified as future priorities either in the medium or long term.

The prioritisation of adaptation options was conducted for three broad time periods based on the likely time-scale for projected impacts on the site features as determined by projected trends in climate and sea-level rise. These are:

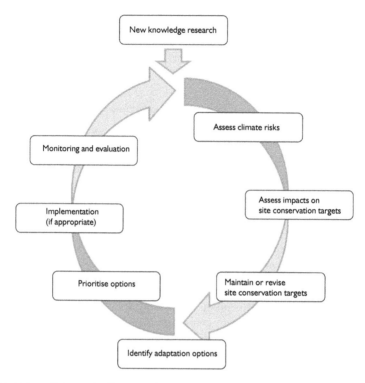

FIGURE 11.1.4 Schematic diagram of the adaptive management cycle adopted for the Bosherston Lakes SAC. The process begins with the assessment of climate risks and ends with monitoring and evaluation; it is periodically restarted by a review of climate risks (source: Holman et al. 2009)

- In the short term (over the next 20 years), further actions to reduce non-climate sources of harm, such as the continuation of extensive catchment and in-lake efforts to improve water quality and reduce eutrophication, were identified as the priority.
- In the medium term (probably 20 to 50 years' time), measures that will increase the resilience of the lake ecosystem will be required, such as the implementation of an alien species management warning system or the raising of spring lake water levels to reduce the risk of summer drought impacts. Improvements to the coastal protection infrastructure, such as maintenance of the coastal dam and the development of a surge management plan to block seawater ingress at times of high risk (for example, by installing tidal flaps) will reduce the likelihood of saline intrusion.
- In the long term (probably beyond the next 50 years), a managed transition from freshwater lake to brackish lagoon may need to be considered in part of the site, but the necessity of this should be reviewed as new evidence related to observed and projected sea-level rise and its likely impacts becomes available.

This research has already informed current management and the initial implementation phase. A key priority has been improving water quality through both on-site measures (such as sediment traps that reduce nutrient inputs) and catchment-wide measures. The surrounding area is now part of a Nitrate Vulnerable Zone within which farmers are required to inject slurry and cannot spread it aerially. This type of action to reduce non-climate sources of harm is a vital component of adaptation. In addition, other activities will become increasingly important, such as the control of invasive alien species. Currently, the water fern *Azolla filiculoides* is being controlled through its removal from silt traps. Water-level management using weirs is expected to be increasingly necessary due to reduced summer rainfall so they have been modified to allow the ingress of elvers. New adaptation measures have also been implemented. Some of the grasslands adjacent to the lakes are also important Annex 1 conservation features in which the timing of management regimes has been altered, e.g. grazing has been reduced to create longer swards that will provide cooler conditions for certain invertebrates in a warming climate.

11.1.4 Conclusions

The widespread use of an adaptive management approach for nature conservation has been recommended (e.g. Mitchell et al. 2007). This work has shown that it is vital for the risk of climate change impacts and the vulnerability of ecosystems to be evaluated at the local scale to avoid poor adaptation decision-making resulting from false assumptions about the importance of climate change. Clarke (2007) has highlighted the substantial uncertainties arising from limited understanding of system behaviour, particularly within freshwater systems, that make this adaptive approach so important.

The phased adaptive management approach adopted at Bosherston Lakes enables CCW and its successor body, Natural Resources Wales (NRW),[1] to manage the uncertainties related to the magnitude of sea-level rise and the nature of impacts upon the ecosystem. This approach is relevant to most other sites, as there is almost universal uncertainty in regard to the significance of climate change exposure, ecosystem component sensitivity and adaptive capacity. The routine monitoring of water quality, water levels and biological SAC features will provide vital input to the adaptive management process. The assessment of site vulnerability using local monitoring data in combination with climate change scenarios for the UK has been especially important in dispelling suggestions that urgent, radical changes to conservation management are required. In collaboration with other UK conservation agencies, CCW has developed guidelines for biodiversity adaptation that include recognition of the need to thoroughly analyse the causes of change and manage non-climatic sources of harm (Hopkins et al. 2007). Their use is currently being promoted for protected areas and at the landscape scale across Wales to support best practice.

In terms of challenges, an adaptive management approach that requires actions to be reviewed and implemented over time runs the risk of recommended actions being delayed or overlooked due to lack of funding or other logistical barriers. There is also a significant ongoing cost for monitoring impacts on the site for the approach to be implemented in a meaningful way. There remain recommended adaptation actions that have not yet been adopted due to resource limitations, so a clear timetable for re-evaluation of the site management plan and continued support for the implementation of adaptation measures is essential. A current review of conservation actions required on all Natura 2000 sites across Wales provides an opportunity for the continued consideration of adaptation actions. Further research and periodic refinement of climate change projections will be provided independently by the climate science community, but it is important to recognise that this is another essential component of the adaptive management approach that was presented in this chapter.

11.2 Robust decision-making? Managing uncertainties in adapting water resource systems to a changing climate in England and Wales

Suraje Dessai and Geoff Darch

11.2.1 General background

Dealing with uncertainty has been identified as a major challenge for climate change adaptation. Uncertainty abounds in terms of what the future will look like (both for the climate and for other important drivers), the efficacy of different adaptation strategies and the acceptability of/preferences for strategies. Deep uncertainty and high decision stakes make climate change adaptation a complex problem.

A wide variety of techniques exist to aid decision-makers faced with such uncertainties. When levels of uncertainty are low to medium, it is feasible to conduct a probabilistic risk analysis that attempts to quantify and combine known uncertainties; this type of analysis often includes an optimisation to assess the best strategy under particular constraints.

When the uncertainty level ranges from medium to high, scenario analysis is often used. Scenario planning produces coherent, internally consistent and plausible descriptions (scenarios) of possible futures from which strategies can be developed, e.g. to avoid particular outcomes or maximise benefits under a range of scenarios.

Hallegatte et al. (2012) have defined "deep uncertainty" as a situation in which analysts do not know or cannot agree on (1) models that describe key forces that will shape the future, (2) probability distributions of key variables and parameters in these models and/or (3) the value of alternative outcomes. A range of different methods have recently been developed (e.g. Dessai & Hulme 2007, Lempert & Groves 2010, Wilby & Dessai 2010) under the broad banner of robust decision-making (RDM) approaches to tackle deep uncertainty in climate change adaptation.

Whereas traditional decision-making processes seek optimality, RDM approaches accept uncertainty and focus on robust strategies, i.e. strategies that are immune to wide ranges of uncertainty. Hallegatte et al. (2012:11) describe the process of a typical RDM approach: "A robust decision process implies the selection of a project or plan which meets its intended goals – e.g., increase access to safe water, reduce floods, upgrade slums, or many others – across a variety of plausible futures. As such, we first look at the vulnerabilities of a plan (or set of possible plans) to a field of possible variables. We then identify a set of plausible futures, incorporating sets of the variables examined, and evaluate the performance of each plan under each future. Finally, we can identify which plans are robust to the futures deemed likely or otherwise important to consider." RDM approaches are often conducted interactively with one or more stakeholders.

A good example of decision-making under uncertainty in climate change adaptation is the process of long-term water resource planning, which seeks to manage the strategic balance between water supply and demand. In England and Wales, every five years private water companies must prepare a Water Resources Management Plan (WRMP), which is scrutinised and negotiated with the Environment Agency (the environmental regulator) and Ofwat (the economic regulator). This is a statutory requirement that requires public consultation and ultimate approval from the Secretary of State. The WRMP sets out the likely changes in supply and demand over the subsequent 25-year period, taking into account uncertainties such as population growth, and climate change and risks such as pollution events. An optimal set of solutions to meet any supply–demand deficit is generated on the basis of lowest cost (including environmental and social factors); a Strategic Environmental Assessment is also produced. Since the late 1990s, assessments have been made of the impact of climate change on both supply and demand, and this period of about 15 years allows some conclusions to be drawn regarding adaptation (cf. Arnell 2011).

11.2.2 Robust Decision-Making in water resource planning?

In this section, the treatment of climate change uncertainties in water resource planning is examined in greater detail to assess whether the principles of RDM are being applied in practice.

The incorporation of climate change scenarios in water resources planning was first required in the plans published in 1999. Water companies were required to produce water resources management plans for the period 2000–2005 with an outlook until 2025 that included climate change in the form of the UKCIP98 climate change scenarios, which had been converted into factors of change in monthly streamflow and annual groundwater recharge. This first exercise had little impact on planning.

The next Periodic Review (PR04; 2005–2010) of WRMPs and water company business plans was more sophisticated. Instead of using a deterministic approach to calculating headroom – the buffer between supply and demand designed to account for specified uncertainties – a probabilistic approach was adopted. The uncertainties of each headroom component were defined as probability distributions and combined using probabilistic simulation techniques (sometimes called Monte Carlo simulation). National climate change scenarios (UKCIP02) were again translated into streamflow and groundwater changes. During this Periodic Review, the impacts of climate change on demand were also considered. However, the Environment Agency stated that the evaluated impact "does not provide sufficient certainty for a major investment" (EA 2003:132). Furthermore, the Agency guidelines note: "The results arising from the application of climate change scenarios must be used sensibly, pragmatically and consistently, accepting that continued research effort will no doubt go on to refine and develop better and more reliable climate predictions. It is important that companies think holistically about climate change when considering options. Overall the Agency encourages a "low-regrets" approach, built around the robustness of options and flexibility of companies plans in adaptation to alternative scenarios" (EA 2003:84). This statement demonstrates that the environmental regulator only partly subscribed to RDM. It argued for the robustness of options and flexibility, but in contrast to RDM principles, it implicitly asserted that more research will eventually reduce uncertainty.

The most recent set of climate scenarios (UKCP09) was not available in time for the 2009 Periodic Review; as a result, flow factors were constructed using outputs from six climate models (used in the IPCC Fourth Assessment Report), but hydrological model uncertainty was also included. The guidance provided by the Environment Agency was more explicit than before, stating that companies should calculate 'mid', wet and dry scenarios from the ensemble (although some companies selected actual climate scenarios to provide this). The 'mid' estimate of climate change impact was incorporated directly in the best estimate of supply (i.e. its impact was assumed to occur), with the range between the wet and dry scenarios included in headroom, the planned buffer between supply and demand (Arnell 2011).

The analyses "suggested a need to invest about £1.5 billion in the period up to 2015 to address the effects of climate change" (Ofwat 2009:82). However, because "business plans were based on the outdated UKCIP02 scenarios" (Ofwat 2009:82) and not UKCP09, Ofwat did not allow companies to be funded for this investment, although they allowed the possibility of such funding based on UKCP09 in situations where companies can establish clearly and robustly that they need to invest by 2015 to address the impact of climate change. Funding was allowed to improve resilience; for example, to protect water treatment facilities from flooding and to increase supply connectivity. While the water companies applied some of the principles of RDM in their planning, the economic regulator did not accept their analyses because they were based on perceived "outdated scenarios".

The methods employed in water resource planning in England and Wales have focused on quantifying the impacts of climate change and their uncertainties, but to date little investment in adaptation has occurred. Adaptation decisions have been subsumed by the optimisation of options to meet the overall supply–demand deficit, which is related to a number of other uncertainties, notably demographic change. Although some planning documents mention robustness, in practice water resource planning in England and Wales has evolved into a curious hybrid between scenario and optimisation analysis. A few scenarios are run to explore climate change uncertainties, and then one scenario is chosen for the optimisation of options. Thus, the principles of RDM are rather half-heartedly followed at present.

Recent research has demonstrated how a full RDM approach could be applied to the adaptation of water systems to an uncertain changing climate in England. Darch et al. (2011) assessed the impact of factoring long-term climate uncertainty into investment planning in Thames Water's London Water Resource Zone. They found that some options are selected under all scenarios – what would be termed 'robust options' – while the selection and sequencing of certain other options is dependent on the climate scenario and the planning horizon used in the analysis. They also note that RDM approaches are particularly relevant when the additional influence of non-climatic risks on the supply–demand balance is taken into account. A potentially larger deficit may occur in the future due to the net effect of sustainability reductions (reducing deployable output), population increases (increasing demand) and demand management measures (reducing demand) (Darch et al. 2011).

Matrosov et al. (2013) compare an economic optimisation approach to an RDM approach with regard to the Thames regional water resource system (including London). The benefits of RDM include detailed plan evaluation according to several metrics, and the simultaneous sampling of a wide variety of conditions with assessments of their impacts on proposed plans; limitations include computational intensiveness, which limits the number of plans and factors of uncertainty that can be considered. The paper concludes that economic optimisation and RDM approaches are complementary, providing different answers and fulfilling different goals. Joint use is suggested as a possible way to improve water planning until a method that unifies the benefits of both can be developed.

Korteling et al. (2013) use Information-Gap decision theory (an RDM approach) to evaluate the performance of a proposed water resource plan beyond the bounded uncertainty range that would traditionally be considered. Info-Gap decision theory provides a method of sampling future design at successively wider ranges of uncertainty, allowing evaluation of the performance of water resource management options under conditions of severe uncertainty. The authors utilise an integrated method based on the Info-Gap approach to quantitatively assess the robustness of various supply-side and demand-side management options over a broad range of plausible future scenarios. Their findings show that beyond the uncertainty range explored with the headroom method, a preference reversal can occur, i.e. some management options that underperform at lower uncertainties actually outperform at higher levels of uncertainty. The study also shows that when 50% or more of the population adopts demand-side management, efficiency-related measures and innovative options such as rainwater collection can perform equally well or better than some supply-side options. In general, the paper illustrates how an Info-Gap-based approach can offer a comprehensive picture of potential supply/demand futures and a rich variety of information to support the adaptive management of water systems under conditions of severe uncertainty.

11.2.3 Concluding thoughts

Adaptation to climate change in water resource planning in England and Wales has taken an incremental approach that has largely focused on understanding and quantifying the impacts of climate change on supply and demand over 25-year periods. The most recent probabilistic UKCP09 climate change projections are being incorporated into the latest WRMPs, which should provide the most detailed estimate of climate change impacts, albeit with a large range of possible outcomes. However, decision-making remains predominantly based on 25-year horizons and the limited testing of solutions against uncertainties, despite the inclusion of some sensitivity and scenario testing in the Water Resources Planning Guideline (cf. EA 2012, Section 8.1). This can be partially explained by the complexity of water resource planning, the cautious nature of the water industry and a lack of appreciation of the depth of uncertainty that climate change (and other drivers) may entail. Although RDM principles are simple in theory, recent papers have shown its practical application in water resource planning in England to be computationally intensive and time-consuming. The principles of RDM can be applied using simpler approaches, but even these would require a change in the philosophy of water resource planning, a change that may well be underway (e.g. UKWIR 2012). Ultimately, RDM approaches might be able to help both water resource planners and regulators to make difficult choices regarding how best to adapt to an uncertain changing climate.

Acknowledgements

Any views expressed are those of the authors rather than the organisations they represent. Suraje Dessai was supported by the ARCC-Water project funded by EPSRC (EP/G061181/1) and the EQUIP project funded by NERC (NE/H003509/1).

11.3   How to communicate climate change uncertainties: Recommendations from psychological research

Torsten Grothmann

11.3.1 Introduction

The communication of uncertainties in scientific projections of climate change and climate change impacts can become a barrier to the planning and implementation of adaptation measures (e.g. Agrawala & van Aalst 2005). Or in other words: "Because people are generally averse to uncertainty, this activity [communicating uncertainties regarding climate change] has the potential to undermine effective action more than stimulate it" (Morton et al. 2011:103).

There are various reasons for the potential undermining effect of uncertainty. Humans often demonstrate a great need for predictability; thus, uncertainty can be uncomfortable. "Predictability helps people feel safe and secure, whereas uncertainty can lead to anxiety" (CRED 2009:24). Furthermore, many people do not know how to deal with uncertainty or how to make decisions under conditions of uncertainty. In other words, there is a general lack of uncertainty competence (Grothmann & Siebenhüner 2012). Due to the emotional discomfort caused by uncertainty, the lack of uncertainty competence and the lack of willingness to change current practices (e.g. abandoning ski tourism), people often react to messages of uncertainty with climate scepticism and denial ("The climate scientists actually have no idea what climate we can expect in the future"), wishful thinking ("It won't be that bad"), fatalism ("I actually have no idea how I should react to such a broad range of possible climatic futures"), denial of responsibility ("The government should deal with this") or postponement of adaptation action ("I will deal with this when more reliable studies are available").

During recent years, several encouraging studies have been published that point towards possibilities to communicate uncertainties regarding future climate change and its impacts without demotivating climate action.[2] In the following sections, I describe some of these studies and develop recommendations from their results regarding how the communication of uncertainty can be improved. These recommendations do not emphasise the scientifically correct and comprehensive presentation of uncertainties in projections of climate change and its impacts; rather, they focus on methods of communicating uncertainties that avoid the demotivation of adaptation action.

As a result, this section is different from most other contributions to this book, as it does not concentrate on the experiences of one specific project or process in climate change adaptation. Instead, it is a review of a number of instructive empirical studies on the communication of uncertainties, and it also reflects my personal experiences from more than 10 years in adaptation research and practice.

11.3.2 The words we use

Non-scientists often interpret certain common words differently than scientists do (CRED 2009). The term 'uncertainty' is often understood as 'not knowing', sometimes even as a sign of incompetence. Therefore, instead of speaking of 'the uncertainty of climate change (impacts)', it seems advisable to speak in terms of the 'range of potential climate change (impacts)', avoiding the word 'uncertainty' due to its potential for misunderstanding. In general, when speaking about uncertainties, words should be chosen carefully and not used loosely, recognising that they can mean different things to different people.

In addition, the terminology used to describe the *extent* of scientific uncertainty is often misunderstood. Budescu et al. (2009) have shown that laypeople systematically misinterpreted the words used to describe the extent of scientific uncertainty in the IPCC's 4th assessment report, especially the terms 'very likely' and 'very unlikely'. For example, the study participants did not understand 'very likely' as '>90% probability' (as intended by the IPCC authors), instead assigning much lower probabilities to this expression. Consequently, IPCC results with a relatively high scientific certainty were systematically misinterpreted as being less certain. Based on their results, Budescu et al. (2009) recommend always combining verbal uncertainty terms (e.g. 'very likely') with quantitative specifications (e.g. '>90% probability'). In a further study (Budescu et al. 2012), the authors were able to show that the realisation of this recommendation (a) increases the level of differentiation between the various uncertainty terms, (b) increases the consistency of the interpretation of these terms and (c) increases the level of consistency with the IPCC guidelines (i.e. the interpretations of the uncertainty terms become scientifically more adequate). These positive effects are independent of the respondents' ideological and environmental views.

However, as shown by studies on the comprehensibility of probability statements regarding the side effects of medications, for example, numerical information expressed in a frequency format (e.g. '90 out of 100') is more easily interpreted and utilised and has a greater influence on judgements than identical information presented as a percentage ('90%') – as used by the IPCC – or as probabilities ('0.9') (Newell & Pitman 2010). Based on these studies, the certainty or uncertainty of scientific projections of climate change should also be expressed in a frequency format (e.g. 'in more than 90 of 100 climate change simulations, there is a temperature increase of at least X degrees by 2050').

11.3.3 Explaining climate change research: Science as a debate and non-scientific sources of uncertainty

The communication of uncertainty does not always have to be a barrier to climate action. Rabinovich and Morton (2012) have shown that messages that communicate high levels of uncertainty were more persuasive for participants who shared an understanding of science as a debate (e.g. a debate between different climate change scientists with their respective climate models) than for those who believed that science is a search for absolute truth and who often interpreted scientific uncertainty as an indicator of low-quality science. In addition, participants who had a concept of science as a debate were more motivated by higher (rather than lower) uncertainty in climate change messages to take climate action.[3] The results suggest that fostering an understanding of science as a debate can help uncertainty statements to be perceived as a necessary component of scientific projections of climate change. Furthermore, such an understanding can help prevent uncertainty statements from becoming a barrier to climate action.

CRED (2009) recommends explaining the reasons for the scientific uncertainties in future climate change and its impacts by referring to the very high complexity of the climate system. However, the complexity of the climate system and the difficult and diverse representations of this complexity in different climate computer models are not the only causes of the scientific uncertainties in projections of future climate change and its impacts. Uncertainties in projections of future climate change and its impacts are also caused to a large extent (sometimes even primarily) by uncertainties in future greenhouse gas emissions, which are the major drivers for climate change and are the main input variables for climate models. These uncertainties in greenhouse gas emissions do not stem from a lack of scientific understanding of the complex climate system, but rather from the lack of political and economic decisions regarding allowances for future greenhouse gas emissions. Ekwurzel et al. (2011) recommend stressing this role of human choice as a source of uncertainties in projections of future climate change – in order to clarify that only part (sometimes only a very small part) of the uncertainties in future climate change projections stem from the lack of scientific understanding of the climate system. Stressing the importance of the lack of political and economic decisions regarding greenhouse gas allowances as a major source of climate change uncertainties can also contribute to an understanding that future scientific progress can reduce uncertainties only to a certain extent and that waiting for definite climate change projections is futile.

My personal experience is that it helps to explicitly stress that computer models for projecting future climate change actually reduce uncertainty, even when these projections are uncertain to some extent. In the past, without climate models, there was total uncertainty of the future climate, and people simply assumed the climate would not change. With climate models, this total uncertainty of future climate can be reduced to ranges of uncertainty.

11.3.4 The positive effect of a positive frame for uncertainty

Morton et al. (2011) show the influence of the so-called 'framing' of climate change projections on motivations for climate mitigation action. Whereas the communication of high uncertainty in climate change projections combined with a negative frame (highlighting possible future losses due to climate change) decreased individual intentions for climate mitigation behaviour in the study participants, the communication of high uncertainty in climate change projections combined with a positive frame (highlighting the possibility of damages or losses *not* materialising) produced stronger intentions to act. These results suggest that uncertainty is not an inevitable barrier to action, providing that communicators frame climate change messages in ways that trigger caution in the face of uncertainty.

Hence, uncertainties regarding climate change and its impacts should be framed as messages about the possibilities of damages or losses not materialising – for example, by communicating mitigation or adaptation options in relation to their chances of preventing damages or losses. Such a positive frame will likely motivate people to initiate climate action much more than a negative frame that only highlights possible future damages or losses due to climate change.

A further result of the study by Morton et al. (2011) is that these effects of uncertainty are mediated through feelings of efficacy ("I can do something to stop climate change. I am not powerless."). The positive frame clearly increased such feelings of efficacy.

If the results by Morton et al. (2011) can be generalised to motivating *adaptation* behaviour – the study conducted by Morton and his colleagues only addressed motivations for climate *mitigation* action – this would have serious consequences for the communication of climate change and its impacts. Currently, the dominating communication paradigm is to communicate climate change losses (the negative frame), which may be useful for raising awareness or putting the issue on the political agenda, but is not conducive to motivating action.

11.3.5 Stressing the certainty of the existence of climate change and its human causes

CRED (2009:26) stresses that "one of the first key tasks for communicators is to put that uncertainty into context by helping audiences understand what is known with a high degree of confidence and what is relatively poorly understood". Indeed, there are often huge differences between the uncertainty ranges of different climate variables. For example, whereas the ranges of projections for future temperature changes are often relatively small, the ranges for precipitation changes are larger.

Spence et al. (2012) stress the possibility of an unintended effect of the increased media reports regarding the extent and impacts of future climate change, in which scientific uncertainties are often mentioned. Due to this increased reporting on scientific uncertainties, there could be an 'uncertainty transfer' from uncertainties with respect to the extent and impacts to the existence and causes of climate change, even though there is a high level of scientific certainty regarding the existence and the anthropogenic causes of climate change. People hear about scientific uncertainties more often in the context of climate change; particularly, those groups that do not know much about climate change and its different aspects (causes, extent, impacts, etc.) could misinterpret this as a sign that all results of climate change research have a high degree of uncertainty. To avoid this 'uncertainty transfer', it seems useful to be very specific regarding which uncertainties one is referring to and to stress that even if the extent and impacts of climate change are very uncertain in part, the existence and anthropogenic causes of climate change are scientifically very certain.

11.3.6 Communication principles and strategies to deal with uncertainties

People often seem to lack the competence to deal with uncertain projections of future climate change (Grothmann & Siebenhüner 2012). To prevent uncertainties in future climate change from becoming a barrier to the planning and implementation of adaptation measures, the communication of uncertain climate change projections should be combined with the communication of methods for dealing with these uncertainties.

CRED (2009) recommends invoking the "precautionary principle" when communicating uncertainties. This principle holds that action should be taken to reduce the risk of harm to the public from potential threats such as climate change, despite the absence of 100% scientific certainty about all aspects of the threat. Especially when there is the possibility of high or irreversible losses, precautionary action should be taken.

Further principles and strategies for decision-making under uncertainty are building robustness (cf. Section 11.2), realising adaptive management (cf. Section 11.1), building resilience or realising 'no-regrets' and 'win-win options' and developing options with safety factors or redundancy (Hallegatte 2009). Such options can be presented as actions that have positive consequences regardless of the materialisation of specific climatic changes. A decision option that is assessed as worthwhile in the present (in that it would yield immediate economic and environmental benefits that exceed its cost) and that continues to be worthwhile irrespective of the nature of the future climate is an example of a 'no-regrets option'. 'Win-win options' are options that reduce the impacts of climate change and have other environmental, social or economic benefits (for an overview of principles and strategies for dealing with climate change uncertainties, cf. Willows & Connell 2003). Potential no-regrets or win-win options can also be cited to demonstrate that realising precautionary adaptation can have *certain* benefits even when the benefits in terms of preventing losses from climate change are *uncertain* due to the uncertainty of future climate change.

In addition, catchy expressions (e.g. 'Better safe than sorry') and analogies to everyday situations in which people take precautions for an uncertain future (e.g. buying insurance, preventive medical check-ups) can be used to convey methods of dealing with uncertainties.

11.3.7 Dialogical communication formats and learning

Accomplishing many of the recommendations for uncertainty communication (e.g. fostering a 'science as debate' understanding, conveying principles and strategies for decision-making under uncertainty) require dialogical communication and training formats that last hours to days (e.g. workshops, counselling interviews). It would be very difficult to communicate these complex issues with short unidirectional communication formats such as flyers, websites or short presentations. In general, various recent publications on the communication of climate change (e.g. Wolf & Moser 2011) cast doubts on the effectiveness of unidirectional formats.

Furthermore, dialogical communication formats can lead to mutual social learning and can reduce uncertainties. Lebel et al. (2010) list six ways in which social learning can contribute to adaptation to climate change; two of these are related to reducing uncertainties. First, social learning can help individuals to reduce and cope with informational uncertainty. *Informational uncertainty* refers to deficits in knowledge. Social learning (e.g. between climate change scientists and practitioners who have practical experience and knowledge about a specific sector or region) can decrease informational uncertainty regarding probable climate change impacts and vulnerabilities. Second, social learning can reduce normative uncertainty. *Normative uncertainty* refers to uncertainty with regard to the values, norms, goals and criteria invoked in selecting adaptation measures and in evaluating potential climate change impacts and vulnerabilities. Thus, normative uncertainty is also linked to perceptions of acceptable risk. "Normative uncertainty can be reduced […] especially through participatory decision processes. For example, strong stakeholder participation in a water-sensitive region can clarify priorities (for instance on tourism) and acceptable risks (for instance agricultural losses)" (Lebel et al. 2010:335). As a result, not only are dialogical communication formats often more convincing methods of encouraging individuals to take climate action despite climate change uncertainties, they can also reduce these uncertainties.

11.3.8 Concluding thoughts

The communication of uncertainties in projections of climate change and its impacts often become a barrier to the planning and implementation of adaptation measures, but this does not have to be the case. The various studies presented in this chapter imply recommendations for methods of communicating uncertainties without demotivating adaptation action.

However, it has yet to be tested whether the realisation of these recommendations will really have the desired effects. It would be preferable for such testing to be done systematically on different actor groups (in policy, administration, business, civil society, citizens) in different countries and sectors. Although such a comprehensive investigation is impracticable, a more informal 'test' of these recommendations could be conducted. This informal test could be carried out by applying the recommendations in the communication of climate change and its uncertainties in various settings and making the results accessible to others on platforms such as the European Climate Adaptation Platform (CLIMATE-ADAPT).

11.4   Lessons learned for decision-making under uncertainty

Torsten Grothmann

The various cases presented in this chapter address very different aspects of working with and making decisions under the uncertainties associated with climate change, its potential impacts and suitable adaptation measures. As a result, it is difficult to identify integrative lessons learned common to all cases. Therefore, in the following section, the main lessons learned are presented separately for each case. First, the lessons learned for adaptive management under uncertainty are summarised, then those for robust decision-making, and finally the lessons of uncertainty communication.

The main lessons learned that can be drawn from the case of *adaptive management* in conservation (Section 11.1) are:

- There is a danger that evidence from impact assessments elsewhere will be inappropriately applied to another locality, such that climate change will be wrongly perceived as the key driver of environmental change in this locality. It is vital that the risks of climate change impacts and the vulnerability of ecosystems are evaluated at the local scale to avoid poor adaptation decision-making because of false assumptions regarding the importance of climate change.
- In order to evaluate climate change risks to sites and ecosystems, the on-site monitoring of drivers of change and their impacts provides a critical source of evidence. Establishing and maintaining such long-term monitoring is vital for the effective implementation of an adaptive management approach.
- Adaptive management that provides a flexible approach to the adoption of adaptation measures on multiple time-scales allows uncertainty (in terms of both climate change scenarios and the impact-response of ecosystems and their components) to be integrated within conservation planning.
- While sometimes perceived as a business-as-usual activity, the reduction of non-climate sources of harm is a vital component of adaptation that takes on greater importance in an uncertain, changing climate.

The case of *robust decision-making* in water resource management (Section 11.2) results in the following lessons learned:

- Water resource planning has made use of increasingly sophisticated climate change projections, particularly to assess impacts on supplies; however, decision-making remains largely reliant on 25-year horizons with only limited testing of solutions against uncertainties, despite the inclusion of some sensitivity and scenario testing in the Water Resources Planning Guideline. This can be explained in part by the complexity of water resource planning, the cautious nature of the water industry and a lack of appreciation of the depth of uncertainty that climate change (and other drivers) may entail.

- Robust Decision-Making (RDM) may be a particularly appropriate method for aiding long-term decision-making in water resource planning in an uncertain climate. RDM implies the selection of a project or plan that will meet its intended goals across a variety of plausible future scenarios. Although RDM principles are simple in theory, recent papers have shown its practical application in water resource planning in England to be computationally intensive and time-consuming.
- The principles of RDM can be applied using simpler approaches, but even these would require a change in the philosophy of water resource planning, a process that may well be underway. Ultimately, RDM approaches might be able to help both water resource planners and regulators to make difficult choices regarding how best to adapt to an uncertain changing climate.

The lessons learned from the contribution on *how to communicate climate change, associated uncertainties and positive solutions* to stakeholders and decision-makers (Section 11.3) are:

- The communication of uncertainties in projections of climate change and its impacts can become a barrier to the planning and implementation of adaptation measures. However, this does not have to be the case. Various recent studies suggest recommendations for methods of communicating uncertainties without demotivating adaptation action.
- For example, fostering an understanding of 'science as a debate' can help uncertainty statements to be perceived as a necessary component of scientific projections of climate change.
- Uncertainties in climate change and its impacts should be framed as messages about the possibilities of damages and losses *not* materialising – for example, by communicating positive messages on mitigation or adaptation options and their chances to prevent losses.
- The communication of uncertainties should be combined with the communication of methods for dealing with these uncertainties (e.g. precautionary principle, adaptive management, robust decision-making) and developing suitable solutions.
- Accomplishing many of the recommendations for uncertainty communication will require dialogue and training formats that allow direct (preferably face-to-face) interactions with stakeholders and decision-makers. Such dialogue formats can also reduce informational uncertainties (e.g. regarding probable climate change impacts) and normative uncertainties (e.g. regarding socially acceptable risks).

Notes

1 On 1 April 2013 the functions of the Countryside Council for Wales, Environment Agency Wales and Forestry Commission Wales were merged to create a new body, Natural Resources Wales.
2 *Climate action* refers to mitigation and/or adaptation measures.
3 Certainly, this result does not imply that people with the same understanding of science would necessarily favour the same kind of response actions. In this study, the participants were asked to rate their willingness to implement a given set of climate change mitigation actions.

References

Agrawala, S. and van Aalst, M. (2005) 'Bridging the gap between climate change and development', in *Bridge Over Troubled Waters: Linking Climate Change and Development*, Agrawala, S. (ed.), Paris: OECD, pp. 133–146.

Arnell, N.W. (2011) 'Incorporating climate change into water resources planning in England and Wales', *JAWRA (Journal of the American Water Resources Association)* 47 (3), pp. 541–549.

Budescu, D.V., Broomell, S.B. and Por, H.H. (2009) 'Improving communication of uncertainty in the reports of the Intergovernmental Panel on Climate Change', *Psychological Science* 20 (3), pp. 299–308.

Budescu, D.V., Por, H.H. and Broomell, S.B. (2012) 'Effective communication of uncertainty in the IPCC reports', *Climatic Change* 113 (2), pp. 181–200. DOI: 10.1007/s10584-011-0330-3.

Burgess, A., Goldsmith, B. and Hatton-Ellis, T. (2006) *Site Condition Assessments of Welsh SAC and SSSI Standing Water Features*, CCW Contract Science Report No. 705, Bangor, Countryside Council for Wales.

Clarke, S. (2007) 'Climate change adaptation strategies for freshwater habitats: changing the emphasis of freshwater conservation', in Kernan, N., Battarbee, R.W. and Binney, H.A. (ed.) *Climate Change and Aquatic Ecosystems in the UK: Science Policy and Management*, Proceedings of a meeting on 16 May 2007. http://www.ecrc.ucl.ac.uk/content/view/414/38/.

CRED – Center for Research on Environmental Decisions (2009) *The Psychology of Climate Change Communication. A Guide for Scientists, Journalists, Educators, Political Aides, and the Interested Public*, New York: Columbia University.

Darch, G., Arkell, B. and Tradewell, J. (2011) *Water Resource Planning Under Climate Uncertainty in London*, Atkins Report (Reference 5103993/73/DG/035) for the Adaptation Sub-Committee and Thames Water, Epsom: Atkins.

Dessai, S. and Hulme, M. (2007) 'Assessing the robustness of adaptation decisions to climate change uncertainties: a case study on water resources management in the East of England', *Global Environmental Change* 17 (1), pp. 59–72.

EA – Environment Agency (2003) *Water Resources Planning Guidelines Version 3.3 (+Supplementary guidance notes)*, Environment Agency, Bristol.

EA – Environment Agency (2012) *Water Resources Planning Guideline: The Technical Methods and Instructions*, Environment Agency, Bristol.

Ekwurzel, B., Frumhoff, P.C. and McCarthy, J.J. (2011) 'Climate uncertainties and their discontents: increasing the impact of assessments on public understanding of climate risks and choices', *Climatic Change* 108 (4), pp. 791–802. DOI: 10.1007/s10584-011-0194-6.

Folke, C. (2006) 'Resilience: the emergence of a perspective for social–ecological systems analyses', *Global Environmental Change* 16 (3), pp. 253–267.

Grothmann, T. and Siebenhüner, B. (2012) 'Reflexive governance and the importance of individual competencies: the case of adaptation to climate change in Germany', in Brousseau, E., Dedeurwaerdere, T. and Siebenhüner, B. (ed.), *Reflexive Governance and Global Public Goods*, Cambridge, MA: MIT Press, pp. 299–314.

Hallegatte, S. (2009) 'Strategies to adapt to an uncertain climate change', *Global Environmental Change* 19, pp. 240–247.

Hallegatte, S., Shah, A., Lempert, R., Brown, C. and Gill, S. (2012) *Investment Decision Making Under Deep Uncertainty: Application to Climate Change*, Policy Research Working Paper 6193, The World Bank Washington, DC.

Holman, I.P., Davidson, T., Burgess, A., Kelly, A., Eaton, J. and Hatton-Ellis, T.W. (2009) *Understanding the Effects of Coming Environmental Change on Bosherston Lakes as a Basis for a Sustainable Conservation Management Strategy*, CCW Contract Science Report No. 858, Bangor, Countryside Council for Wales.

Hopkins, J.J., Allison, H.M., Walmsley, C.A., Gaywood, M. and Thurgate, G. (2007) *Conserving Biodiversity in a Changing Climate: Guidance on Building Capacity to Adapt*, published by Defra.

Hulme, M., Jenkins, G.J., Lu, X., Turnpenny, J.R., Mitchell, T.D., Jones, R.G., Lowe, J., Murphy, J.M., Hassell, D., Boorman, P., McDonald, R. and Hill, S. (2002) *Climate Change Scenarios for the United Kingdom: The UKCIP02 Scientific Report*, Norwich: Tyndall Centre for Climate Change Research, University of East Anglia.

Korteling, B., Dessai, S. and Kapelan, Z. (2013) 'Using information-gap decision theory for water resources planning under severe uncertainty', *Water Resources Management* 27 (4), pp. 1149–1172.

Lebel, L., Grothmann, T. and Siebenhüner, B. (2010) 'The role of social learning in adaptiveness: Insights from water management', *International Environmental Agreements: Politics, Law and Economics* 10 (4), pp. 333–353.

Lempert, R.J. and Groves, D.G. (2010) 'Identifying and evaluating robust adaptive policy responses to climate change for water management agencies in the American west', *Technological Forecasting and Social Change* 77 (6), pp. 960–974.

Lowe, J.A., Howard, T.P., Pardaens, A., Tinker, J., Holt, J., Wakelin, S., Milne, G., Leake, J., Wolf, J., Horsburgh, K., Reeder, T., Jenkins, G., Ridley, J., Dye, S. and Bradley, S. (2009) *UK Climate Projections Science Report: Marine and Coastal Projections*, Met Office Hadley Centre, Exeter, UK.

Matrosov, E., Padula, S. and Harou, J.J. (2013) 'Selecting portfolios of water supply and demand management strategies under uncertainty – contrasting economic optimisation and 'robust decision making' approaches', *Water Resources Management* 27 (4), pp. 1123–1148.

Mitchell, R.J., Morecroft, M.D., Acreman, M., Crick, H.Q.P., Frost, M., Harley, M., Maclean, I.M.D., Mountford, O., Piper, J., Pontier, H., Rehfisch, M.M., Ross, L.C., Smithers, R.J., Stott, A., Walmsley, C.A., Watts, O. and Wilson, E. (2007) *England Biodiversity Strategy – Towards Adaptation to Climate Change*, Final report to Defra for contract CR0327. http://nora.nerc.ac.uk/915/1/Mitchelletalebs-climate-change.pdf.

Morton, T.A., Rabinovich, A., Marshall, D. and Bretschneider, P. (2011) 'The future that may (or may not) come: how framing changes responses to uncertainty in climate change communications', *Global Environmental Change – Human and Policy Dimensions* 21 (1), pp. 103–109. DOI:10.1016/j.gloenvcha.2010.09.013.

National Trust (2007) *Shifting Shores – Living with a Changing Coastline*, Cardiff: National Trust.

Newell, B.R. and Pitman, A.J. (2010) 'The psychology of global warming – improving the fit between the science and the message', *Bulletin of the American Meteorological Society* 91 (8), pp. 1003–1014. DOI: 10.1175/2010bams2957.

Ofwat (2009) *Future Water and Sewerage Charges 2010–15: Final Determinations*, Ofwat, Birmingham.

Pahl-Wostl, C. (2007) 'Transitions towards adaptive management of water facing climate and global change', *Water Resources Management* 21 (1), pp. 49–62.

Pahl-Wostl, C. (2009) 'A conceptual framework for analyzing adaptive capacity and multi-level learning processes in resource governance regimes', *Global Environmental Change* 19 (3), pp. 345–365.

Pahl-Wostl, C., Craps, M., Dewulf, A., Mostert, E., Tabara, D. and Taillieu, T. (2007) 'Social learning in water resources management', *Ecology and Society* 12 (2), p. 5.

Rabinovich, A. and Morton, T.A. (2012) 'Unquestioned answers or unanswered questions: beliefs about science guide responses to uncertainty in climate change risk communication', *Risk Analysis* 32 (6), pp. 992–1002. DOI: 10.1111/j.1539-6924.2012.01771.x.

Smith, J.B. (1997) 'Setting priorities for adapting to climate change', *Global Environment* 7 (3), pp. 251–264.

Spence, A., Poortinga, W. and Pidgeon, N. (2012) 'The Psychological Distance of Climate Change', *Risk Analysis* 32 (6), pp. 957–972. DOI: 10.1111/j.1539-6924.2011.01695.x.

UKWIR – UK Water Industry Research (2012) *Water Resources Planning Tools 2012: Summary Report*, Report Ref. 12/WR/27/6, UK Water Industry Research, London.

UNFCCC (1992) *United Nations Framework Convention on Climate Change*, New York: United Nations.

Wilby, R.L. and Dessai, S. (2010) 'Robust adaptation to climate change', *Weather* 65 (7), pp. 180–185.

Willows, R.I. and Connell, R.K. (ed.) (2003) *Climate Adaptation: Risk, Uncertainty and Decision-making*, UKCIP Technical Report, Oxford.

Wilson, L., McCall, R., Astbury, S., Bhogal, A. and Walmsley, C. (2010) *Climate Vulnerability Assessment of Designated Sites in Wales*, CCW Contract Science Report No. 1017, Bangor, Countryside Council for Wales.

Wolf, J. and Moser, S.C. (2011) 'Individual understandings, perceptions, and engagement with climate change: insights from in-depth studies across the world', *WIREs Clim Change* 2 (4), pp. 547–569.

12

AVOID MALADAPTATION

Rob Swart, Andrea Prutsch, Torsten Grothmann,
Inke Schauser and Sabine McCallum

Explanation of the guiding principle

Maladaptive actions and processes do not succeed in reducing vulnerability to climate change impacts; instead, they increase it (McCarthy et al. 2001) and/ or reduce the capacity to cope with the negative effects of climate change. Maladaptation may deliver short-term benefits (e.g. financial profit) but will lead to harmful consequences in the medium- and long-term perspective (Lim et al. 2004). It may be evident that maladaptation should be avoided, but this can be more difficult than one might imagine. It is important to understand what maladaptation is and how it can be caused. According to the IPCC (2001), maladaptation is adaptation that "does not succeed in reducing vulnerability but increases it instead". Later publications have extended this definition to include the following actions (Nelson et al. 2007, Prutsch et al. 2010, Barnett & O'Neill 2010):

- Actions that increase vulnerability, or through which the capacity to cope with the negative effects of climate change is decreased (e.g. adaptation that is ineffective or may reduce short-term vulnerability but increases vulnerability in the longer term, or increases vulnerability elsewhere, such as hard flood prevention measures in one area that increase risks for downstream systems).
- Actions that increase greenhouse gas emissions and thus conflict with mitigation (e.g. the installation of energy-intensive air conditioners).
- Actions that use resources unsustainably (e.g. using groundwater for irrigation in dry regions, resulting in an unsustainably decreasing level of groundwater).
- Actions that distribute the benefits of adaptation unequally across society or that disproportionally burden the most vulnerable (e.g. prevention of climate change-induced diseases or inconveniences only for affluent people).

- Actions that have high opportunity costs, i.e. the economic, social or environmental costs are high relative to alternatives (e.g. expensive infrastructural investments where behavioural changes with similar effects on vulnerability may be possible – for example, investing in a desalination plant where an increase in the efficiency of water use would also be effective).
- Actions that set paths that limit the choices available to future generations (e.g. large investments in particularly vulnerable agricultural practices or flood defences that make future changes more difficult).

Maladaptation may be avoided by identifying and carefully assessing different options (cf. Chapters 9 and 10) to clarify the potential consequences of adaptation actions in the long term and on actors and systems outside the project or policy area. This is consistent with the concept of 'sustainable adaptation', defined by Eriksen et al. (2011) as "a set of actions that contribute to socially and environmentally sustainable development pathways including social justice and environmental integrity". Thus, linking the concepts of sustainability and adaptation can help to prevent maladaptation (Wilson & McDaniels 2007, Bizikova et al. 2007, Laukkonen et al. 2009). Furthermore, engaging stakeholders (cf. Chapter 8) can facilitate the identification of maladaptation that distributes the benefits of adaptation unequally across society or that disproportionally burdens the most vulnerable.

Chapter overview

This chapter discusses three case studies. The first case study (Section 12.1) on winter tourism by Carmen de Jong reveals that maladaptation is not just a theoretical concept, but a problematic reality when short-term interests collide with longer-term concerns. Technologies and practices that were developed in part to counter the reduced snow availability in the Alps due to climate change appear to have resulted in a number of negative environmental, social and economic consequences in a process that seems difficult to avoid or reverse. In the second case (12.2), Aleksandra Kazmierczak presents an example of how adaptation could be successfully integrated with sustainable development in practice, using the example of green roofing in Basel, Switzerland, which combines adaptation and mitigation objectives. In Section 12.3, the third case study addresses the maladaptation challenges from an even broader perspective. Rob Swart and Marjolein Pijnappels describe an assessment of actual experiences in about a hundred large spatial planning and water management investment projects in the Netherlands in which climate change played a role – here, the focus is on avoiding maladaptation by identifying opportunities and exploring sustainable solutions (Pijnappels & Sedee 2010). This case suggests that adequately incorporating climate change concerns in planning processes and simultaneously seeking out opportunities rather than only avoiding risks can not only prevent maladaptation, but can also improve the 'quality' of projects, making them more sustainable in the long term.

12.1 A white decay of winter tourism in Europe?

Carmen de Jong

12.1.1. Globally widespread maladaptation in the winter sports industry

Climate change has had major impacts on Alpine winter tourism by decreasing snowfall distribution and duration (Marty & Meister 2012). Winter tourism has therefore evolved into an industry that is heavily dependent on technological adaptation to climate change. Artificial snow is produced as a surrogate for natural snow, whether present or absent. Investments in this product are costly, and in the long term they are highly insecure because they are ultimately limited by sub-zero temperatures, water and electricity availability, large water storage facilities and increasing maintenance costs. Since the turn of the millennium, low-lying ski areas (around 1200 m) have been switching from snow- to rain-dominated regimes. Paradoxically, the ski season has developed inversely to this trend, regardless of altitude. Ski resorts are prepared to operate during a fixed period, independent of natural snow and numbers of tourists. Nonetheless, since 2002, warming temperatures have put a cap on artificial snow production across the Alps, even at high altitudes (2400 m), and have sometimes forced the ski season to start and end several weeks earlier.

Moreover, artificial snow and its infrastructure have far-reaching negative impacts on hydrology, meteorology, soils, flora and fauna and the environment (de Jong et al. 2008). Its production ranges from 800 to 3200 m in altitude (including glaciers for summer skiing) independent of aspect (horizontal direction) and slope at the sub-catchment to catchment level. Glacial and peri-glacial topography (especially block glaciers and moraines) are increasingly being destroyed to accommodate ski runs adapted to artificial snow and higher skier density. Water over-abstraction during winter low flow causes water scarcity and conflicts with drinking water demands. Surplus artificial snowmelt considerably increases flood peaks. Water reservoirs for snowmaking frequently destroy wetlands by displacing them. Large-scale water transfers from poorer quality lower-lying areas and stagnating or contaminated water in snowmaking reservoirs can deteriorate high-altitude water quality and cause health problems in humans and animals. Artificial snow and repeated snow grooming compact snow and soil. Intense summer precipitation and artificial snowmelt can trigger surface runoff, erosion and mass movements on impermeable ski runs that have been enlarged and levelled to optimise artificial snow cover (de Jong 2011, Lagriffoul et al. 2010, and the author's investigations). The continual remediation of deteriorated ski runs increases maintenance costs.

Snow canons and snow lances in a typical large ski resort can consume about 3 million kWh per season, which equals a third of the total consumption of the average ski resort or the annual household consumption of 550 families of four (Abegg 2011), not taking into account the energy required for water transportation across thousands of metres of altitude. Artificial snow emits about 3600 tonnes of

CO_2 per season (without taking into account emissions from eroded ski runs and the destruction of wetlands). The fuel consumption of 10 snow grooms entails around 1000 tonnes of CO_2 emissions per season. Approximately 75% of CO_2 emissions in ski resorts are due to car transportation for tourists. Other sources of pollution include the oil from snow grooms and the salt applied to keep roads snow-free during the winter in some Alpine countries. When it enters the hydrological network in spring, this salt can cause health problems for fauna.

Originally, winter tourism was driven by natural resources and local food and culture. Over the last two decades, triggers have shifted towards investments in real estate, shares, lifts, artificial snow infrastructure, broader ski runs, snow parks,[1] snow grooming,[2] snow groom pleasure riding, roads, parking lots, heliports, spa and wellness. The carrying capacity of a ski resort limits winter tourism when services become over-saturated or landscape value decreases through excessive urban development and artificial snow infrastructure.

The stakeholder groups with various roles and interests include the ski industry, tour operators, the lift industry, hotel and restaurant industries, real estate, politics, foresters, farmers, meteorologists and environmentalists. In addition to tourists and local citizens, additional stakeholders involved in the winter sports sector are support and medical services, banks and insurance companies, energy, water and construction companies, and town halls and communities. Box 12.1.1 and Figures 12.1.1 and 12.1.2 present examples of recent winter sports expansions with serious environmental impacts. Figure 12.1.3 shows the course of typical developments over time in three phases (as investments and costs increase and environmental quality deteriorates).

Maladaptation has spread to other mountain resorts outside the Alps. French and Austrian companies have invested in the new Olympic site of Sochi for the Winter Olympics in 2014. The environmental impacts of forest clearing and the construction of new ski runs have already been very serious. In winter, multiple avalanches have been triggered on deforested ski runs. In summer, large-scale debris flows and landslides have been generated in the unconsolidated sediments of ski runs.

12.1.2 Practical experiences with planning and implementation addressing adaptation

Winter tourism is an industrial sector that has undergone little systematic or science-based planning; attempts to balance economic factors with social, environmental and climate change issues are generally lacking. Few coherent strategies have been applied to winter tourism planning across the Alps (Abegg 2011). Moreover, planning and implementation has been copied from one place to another, independent of any climate-proofing adaptation plans. Practically all of these strategies follow the same sequence of infrastructural development: artificial snow, water-holding reservoirs, ski-run enlargement, lifts, roads and housing. Since the general trend is to invest in high-speed lifts that can carry more passengers in less time, there is a

---

BOX 12.1.1 MALADAPTATION IN WINTER SPORTS: EXAMPLES FROM GERMANY AND THE FRENCH ALPS

A typical example of maladaptation concerns the small ski resort of Abondance in Upper Savoy in the French Alps, which lacks artificial snow facilities. The resort suffered from poor snow conditions and had incurred debts from high lift costs. In 2007, the resort decided to discontinue the lifts. Despite this, local commerce did not suffer. However, some dissatisfied local inhabitants elected a new local mayor who had campaigned on the promise of re-opening the lifts. In 2009, an American investor was found to finance the re-opening of the lifts, with plans for housing and lift extensions. After the poor winter of 2010/11 incurred new debts, the investor insisted on investments in artificial snowmaking equipment and a lift connection to the larger, neighbouring Portes du Soleil resort. These plans had to be abandoned because of the high costs of snowmaking equipment and the impossibility of constructing a connection across a protected area. As a result, the investor broke his contract and the community was left in a situation worse than before (sources: newspaper articles, author's research).

Another typical case of maladaptation for a large ski resort heavily dependent on artificial snow production is that of Valmorel, Savoy, France, which opened a new Club Med in December 2011. Because the water demand for tourists and snowmaking exceeds availability by approximately 50%, water must be imported from other catchments. However, these suffer from water scarcity as well; as a result, water for snowmaking is stored in a second river catchment and drinking water is transferred from a third river catchment.

A case outside the Alps is the Feldberg (Black Forest) in Germany, where in 2005 German banks and insurance companies refused to invest in a new lift. They based their decision on scientific studies of economic and climate change limits. However, the Bank of Tyrol and Vorarlberg, supported by the Austrian Exportbank, nevertheless decided to invest in the Ahornbuehlift and to lease it to the local Alpincenter lift operator, who is financing more than 7 million Euros for the three local communities.

---

growing need to accommodate a higher concentration of skiers on ski runs. This necessitates ski-run enlargement as well as more housing developments, creating 'cold beds' that are only temporarily occupied. The extremely high density of tourists over short time intervals triggers many negative effects such as water and electricity overconsumption, under-capacity in wastewater treatment and soil erosion (de Jong 2009). The risk of accidents rises when ski runs are spatially limited by artificial snow or are hard and slippery, especially at the beginning/end of the season or during periods devoid of natural snow. Increases in infrastructure such

FIGURE 12.1.1 Problems with maintenance of artificial snow cover at the end of the season, Val Thorens, 18 April 2011. The season had to be ended several weeks prematurely (source: copyright Kees Wolthoorn)

as lift pylons, snow canons and snow-grooming machines also increase the annual number of lethal accidents (sources: newspaper articles, author's investigations).

Because it is subject to very few controls and regulations at the EU level, winter tourism has been able to expand in an unprecedented manner. Initially growing very fast, winter tourism is now stagnating and even declining in some parts of the Alps (Vanat 2011), despite increased investment in artificial snow. This is because consumer behaviour is dependent on snow as an overall element of the landscape rather than confined to an individual ski run. It took the ski industry nearly five years to recover from the three consecutive poor winters at the end of the 1980s. The poor winter season in 2006/7 also considerably decreased visitor numbers.

In a quest to combat climate change and satisfy presumed skier needs, artificial snow production and ski-run re-modelling has intruded into many economic, social and environmental sectors and domains, raising major concerns. Many developments disrespect the European Water Framework and Drinking Water Directives and the proposed Soil Directive and are at odds with the notion of eco-productivity and the Alpine Space Project's goal of achieving climate-neutral Alps by the year 2050.

Unfortunately, planning strategies in the Alps are based on the assumption that climate change is not yet a reality, at least not during the lifetime of investments over the next 15 to 20 years. Several sites for the Winter Olympics in 2018 were recently developed in the Alps (Garmisch-Partenkirchen in Germany and Annecy in France) despite climate change predictions made by leading climate scientists

---

BOX 12.1.2 SKIING IN DENMARK: AN EXAMPLE OF MALADAPTATION?

A grotesque example of maladaptation exists in the form of an EU-funded ski run on the island of Bornholm in Denmark. A lift, snowmaking equipment and a snow-grooming machine were installed with the help of EU funding on the 40-m Bornholm Ski Hill to enable skiing for a handful of local inhabitants in an area that has never had skiing and is not climatologically or topographically suitable for skiing. EU regional structural funds were acquired under the category of sustainable diversification of agricultural activities. Investments were made without taking into account the climate change limits of snowmaking or skier numbers.

---

indicating that the adequate production of artificial snow would be highly unlikely. New technological and economic developments sometimes result in winter sports facilities in areas that are not climatologically or topographically suitable for skiing (see Box 12.1.2).

There is a tendency in some Alpine countries to extensively develop or expand new low-lying sites with snowmaking facilities despite their limitations in terms of temperature, altitude, water quantity or quality. In Upper Savoy, France, Les Brasses (900–1600 m) was newly equipped with artificial snow infrastructure in 2010 and Mieussy (1400–1980 m) in 2011. In Bavaria, snow infrastructure is also being developed in pre-Alpine areas (Brauneck and Sudelfeld) between 850 and 1550 m. Apart from artificial snowmaking, other strategies implemented include snow farming,[3] development at higher altitudes, ski-run enlargement and faster and bigger lifts. In rare cases, ski resorts and runs are closed and permission to abstract water for snowmaking refused.

The guiding principles for winter tourism fall mainly under the political umbrella, in cooperation with economic actors, real estate and lift operators and the banking sector. The main impetus for winter tourism development is engrained in political election and re-election campaigns. Most of these plans are highly sectoral and are based on identical sequential technological developments and intensive marketing rather than tailor-made, geographically and culturally unique solutions.

In general, for the most part, winter tourism in the European mountains is caught in a deadlock of maladaptation; it is adapted neither to climate change and local conditions, nor to lessons learned from earlier (often negative) experiences. Mitigation through energy conservation or at least efficient energy consumption and the use of low-carbon sources is still uncommon. Although winters are shrinking and summers are extending into spring and autumn, the artificial limits of the highly vulnerable winter season are being pushed even further, at high costs. The climate change impacts of artificial snow production costs are rarely considered. Planning strategies are decoupled from reality, as they do not integrate interdisciplinary scientific knowledge on climate change, water availability or economic barriers, preventing adjustment to the future intensification of negative effects.

FIGURE 12.1.2 Attempts to adapt to climate change by means of artificial snow production. Conditions were too warm to produce snow, and the opening of the ski season had to be postponed by two weeks in La Plagne (France) at 2100 m altitude, 23 November 2011 (source: copyright Carmen de Jong)

12.1.3 Good-practice examples

In this manual, most cases discuss good-practice examples. For winter sports development in response to changing economic, social and climatic conditions, such examples are unfortunately very scarce. There have been some efforts to improve the sustainable development of the tourism industry in national and international projects and via environmental labels or certificates (such as ISO 14001, Alpine Pearls and Green Snowflake), but these have generally been small-scale actions primarily directed towards waste recycling and some mitigation of greenhouse gas emissions. In addition, these actions have been encouraged rather than mandated, and they do not address the unsustainable management of natural resources described in this section, which are in part motivated by maladaptation practices such as climate-induced artificial snowmaking and landscape alterations.

One example of good practice is the Dobratsch ski area in Carinthia, Austria. Plans to invest in artificial snow infrastructure were abandoned in 2002. The Audit Court decided that the drinking water supply for the town of Villach would be endangered by the overuse of water from the protected Dobratsch springs (Könighofer 2007). No public investment was made in the artificial snow infrastructure, the ski area was closed and the lifts dismantled. The decision was

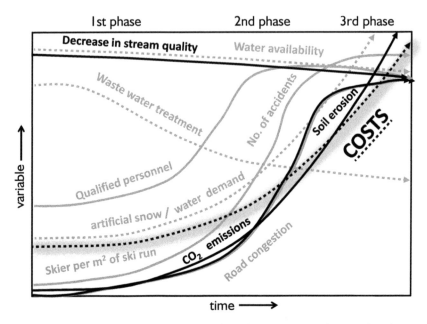

FIGURE 12.1.3 The carrying capacity of a ski resort after three main phases of development. Many newly developed resorts have skipped the first phase (source: Carmen de Jong)

preceded by a heated conflict between politicians, citizens and representatives of the nature park. To support their opinions, local politicians regularly transmitted false information about decisions to the media. Other good-practice examples include small ski resorts, e.g. a resort in Austria (Planneralm) that runs without lifts or artificial snow, and a resort in France (Saint Pierre Chartreuse) that only operates with natural snow. Good-practice examples can also be found in those ski resorts that have the possibility of four seasons of tourism, combining winter sports with agro-tourism, bird-watching, archaeological educational and thermal spa tourism. Wherever winter tourism can be combined with summer tourism and with other sectors such as agriculture, forestry or fishery, it benefits from higher sustainability.

12.1.4 Future challenges

A number of things can be done to turn the tide and avoid or reverse the maladaptation in the winter sports industry described above. Sound and objective information from reliable sources should provide the basis for investment decisions. This is no easy task, since especially at the regional and local levels almost all good-practice initiatives seeking to avoid the mechanisation of winter development are overwhelmed by powerful political and industrial lobbies. It is important to carefully select internationally recognised peers and independent scientists when asking for advice, avoiding advisors who lack a proper background in the field or those with political ties or links to the ski industry. Before developing and designing new winter sport facilities, a number of issues must be addressed in order to avoid maladaptation.

Cost-benefit analyses should be made for communities investing in artificial snow infrastructure via strategic plans with a time horizon of at least 20 years that consider climate change impacts on the present and future limits of snow production (in terms of temperature and water availability), as well as environmental and social costs and benefits that cannot easily be monetised. The per-capita debt of ski-resort communities could be compared to those without resorts. Loss of eco-productivity in zones within and around intensively used winter sports resorts due to erosion, losses in biodiversity and water, soil and air pollution should be addressed. An important further need is data availability on snow precipitation, stream discharge and local stakeholder opinions (Weiss et al 1998). Problem-consciousness and awareness-raising amongst stakeholders should be reinforced. Systematically updated multi-disciplinary analyses should facilitate the identification of emerging problems. Greater cooperation in cross-sectoral and inter-sectoral problem-solving is required. Decision-making should be informed by science and by the reports of organisations such as the OECD. Water availability should be balanced against demand, avoiding inter-basin transfers and inverse gradients. To enhance the knowledge base and communicate information, scientists should be supported in the analysis of good- and bad-practice examples, the identification of stakeholder knowledge and awareness-raising. An EU winter tourism directive could encourage sustainable winter tourism and avoid the further spread of maladaptation.

12.2 Combining climate change mitigation and adaptation: Green roofs in Basel, Switzerland

Aleksandra Kazmierczak

12.2.1 Background

Maladaptation is a term that describes actions that, while focusing on adaptation to climate change, are associated with negative impacts in other spheres; for example, increases in greenhouse gas emissions, unsustainable use of resources or excessive costs. To avoid maladaptation, no-regrets measures ought to be implemented; among these, options that deliver a number of other benefits alongside adaptation should be prioritised. One such solution is urban greening, which, in addition to providing a pleasant environment for urban residents and supporting biodiversity, has been found to lower temperatures in cities and reduce the risk of flash flooding (Gill et al. 2007). In densely built-up areas where providing extensive parks and planting trees may be impossible, vegetated roofs are a feasible greening option. These roofs not only mitigate the urban heat island effect but also act as insulators. By minimising heat gains in buildings, they can lower the indoor temperatures by as much as 5°C (Kumar & Kaushik 2005) and consequently reduce the need for cooling and the associated energy use (Castleton et al. 2010), thus contributing to climate change mitigation. Modelling studies conducted in Manchester, UK, show that greening all the suitable roofs in densely built-up areas could reduce runoff by 17–20% (Gill et al. 2007). Green roofs can also provide 'stepping stones' for migratory species under changing climate conditions (LCCP 2006).

This section presents a case study from Basel, Switzerland, where green roofs have been installed since the 1970s to help reduce storm water runoff, provide summer cooling and thermal insulation in winter and benefit nature conservation (Brenneisen 2005). Located in north-western Switzerland on the Rhine River, the city of Basel (population 187,000) is one of 26 Swiss cantons, with its own constitution, legislature and government (Lawlor et al. 2006). The city is characterised by a mild climate, but it is likely to be affected by future changes such as the increased frequency and severity of extreme precipitation events and rising temperatures (projected +2.5°C in summer by the 2050s) (OcCC 2007). Thus, the role of green roofs in managing runoff, reducing the heat gain in buildings through insulation and mitigating the urban heat island effect can be considered as increasing in importance.

Building on experiences and pilot projects from the 1970s and 1980s, the authorities in Basel have undertaken systematic action to increase the coverage of green roofs, using a combination of financial incentives and building regulations. The drivers were the savings in energy associated with the reduced need for heating, a subsequent interest in biodiversity conservation and the intention to retain high earners in the city by providing an attractive environment (LCCP 2006). Although adaptation has not been the explicit focus of the green-roof schemes, the initiative described here illustrates that urban greening can be used as a no-regrets, cost-effective option, whereby multiple objectives, including adaptation, can be fulfilled.

12.2.2 Greening the roofs: Incentives, research and regulations

In the early 1990s, the city of Basel enacted a law mandating that 5% of all customers' energy bills be put into an Energy-Saving Fund to be used for energy-saving campaigns (Ip 2008). The national Department of the Environment and Energy supported this law after a poll of the general population indicated support for an electricity tax to fund energy-saving measures. The city of Basel then explored a range of energy-saving ideas, including green roofs (Lawlor et al. 2006).

Between 1996 and 1997, the Department of Building and Transport of the city of Basel developed an incentive programme providing subsidies for the installation of green roofs. The programme's development included consultations with city departments and a number of external stakeholders (the national Department of the Environment, Forest and Landscapes, the local business association, non-governmental environmental organisations and others) and it was advertised in newspapers and on posters (Brenneisen 2010). The Energy-Saving Fund provided CHF 1 million in subsidies.

The initial costs of roof greening were estimated at CHF 100/m^2 (Losarcos & Romero 2010). Beneficiaries of the fund received CHF 20 per m^2 of green roof for new housing developments as well as for retrofitting green roofs on an existing building (LCCP 2006). In total, 135 residents and businesses applied for a green-roof subsidy, resulting in the greening of 85,000 m^2 of roofs. The consequent energy savings for Basel were estimated at 4 GWh/year (Ip 2008).

The success of the initial incentive programme spurred interest in the nature conservation benefits of green roofs (Lawlor et al. 2006). A study led by Dr Stephan Brenneisen at the Zürich University of Applied Sciences in Wädenswil (ZHAW) concluded that the increased depth and quality of soil and the use of native plants play a role in supporting endangered invertebrate species (Lawlor et al. 2006). This gave rise to changes in the city of Basel's Building and Construction Law. These alteration efforts were led by the Department of Building and Transport, in close collaboration with ZHAW and a trade association of green-roof contractors (Brenneisen 2010). The amendment, passed in 2002, declares that all new and renovated flat roofs must be greened and also stipulates their design to maximise biodiversity. The green-roof regulation in Basel highlights the following aspects (Ip 2008):

- The growing medium should be native regional soil (the regulation recommends consulting a horticulturalist)
- The growing medium should be at least 10 cm deep
- Mounds 30 cm high and 3 m wide should be provided as a habitat for invertebrates
- The vegetation should be a mix of native plant species characteristic to Basel
- Green roofs on flat roofs over 1,000 m^2 must involve consultation with the city's green-roof expert during design and construction.

The regulation's emphasis on the appropriate depth of the soil or other growing medium has implications for climate change adaptation, as the heat gains (and losses) and the amount of rain water that can be stored are proportionate to the thickness of the soil layer on the roof.

The green-roof subsidy campaign was repeated in 2005–6 (funded by a further CHF 1 million from the Energy-Saving Fund), incorporating the Building and Construction Law design specifications into its guidelines (Lawlor et al. 2006). The energy savings from this second campaign equalled 3.1 GWh/year (Ip 2008).

12.2.3 Concluding thoughts

The initiatives promoting green roofs in Basel can be considered to have been very successful; Basel gained a worldwide recognition as the city with the largest green-roof area per capita. In 2006, 1,711 extensive and 218 intensive green roofs were recorded across the city. Approximately 23% of Basel's flat roof area is now green (Ip 2008, see also Figure 12.2.1). This is forecast to increase to 30% within the next 10 years (Lawlor et al. 2006).

Installing green roofs is now considered to be a routine practice for developers, who make no objections their inclusion (Ip 2008). Basel's green-roof regulations met with no significant resistance due to the stakeholder involvement in the development of the regulations and the success of the incentive programmes. In addition, local businesses have profited from sales of green-roof materials and supplies (Lawlor et al. 2006). Further work on green roofs in Basel includes awareness-raising among architects, planners, builders, gardeners and representatives of other professions on the benefits of green roofs (Baumann 2010). Another positive spin-off of the green-roof campaign is the substantial ongoing research into the biodiversity of green roofs.

Although green roofs in Basel were initially considered primarily in the context of reducing energy consumption for heating, as the climate changes, the insulating qualities of green roofs will help to keep the buildings cool in summer and reduce the need for artificial cooling. No information is available as to whether the energy savings have achieved a return on the initial investment for the individual properties in Basel; however, according to Carter & Keeler (2008), green roofs tend to be 10–14% more expensive than traditional roofs over their lifespan due to their initial costs (the cost of maintaining extensive green roofs is similar to that for traditional roofs). Thus, a 20% reduction in green-roof construction cost (such as that achieved through the subsidy scheme) is sufficient to equalise the costs of green and traditional roofs for investors, which suggests that green roofs are not only a sustainable but also a financially feasible option.

The semi-independent character of the canton allowed changes in Basel's building regulations at the city level; therefore, this case may not be replicable for cities functioning in more restrictive governance frameworks. Nonetheless, it is worth noting that the broad acceptance of green roofs in Basel has been ensured by a comprehensive suite of mechanisms ranging from incentives to statutory regulations, combined with an extensive information campaign and the engagement of local residents and businesses. The leadership of the project by a committed team from the local authorities and the support of the green-roof concept by an academic expert, as well as consultation with other entities, have facilitated the necessary changes in regulations and made the subsidy programmes a success (Losarcos & Romero 2010).

An important lesson is that in the urban context, climate change adaptation can be compatible with mitigation; the case of Basel illustrates that adaptation can be driven by actions aiming at energy savings and climate change mitigation. Such opportunities that utilise the existing and ongoing urban and infrastructure developments driven by other goals for the purposes of adaptation should be sought out and maximised in order to avoid maladaptation. Green roofs are an excellent example of this.

FIGURE 12.2.1 Combining adaptation and mitigation: the green roof of the Basel Exhibition Hall, where around 16,000 m² are extensively greened with moss-sedum cover and herbs, with a photovoltaic installation around the edges. Woody debris has been added to create an artistic display and at the same time enhance biodiversity (source: photograph by Susannah Gill)

Acknowledgements

The information for this case study was initially collected within the Interreg IVC project Green and Blue Space Adaptation for Urban Areas and EcoTowns (2008–2011). Thanks go to Dr Stephan Brenneisen and Natalie Baumann, who contributed their first-hand knowledge of the green-roof programme in Basel to this case study.

12.3   Climate change opportunities and sustainability

Rob Swart and Marjolein Pijnappels

12.3.1 'Climate as Opportunity': 100 investment projects in the Netherlands

This case analyses the lessons learned from a large number of infrastructural investment projects that have been developed and implemented over the last few years in the Netherlands that have considered changing climatic conditions. It focuses on the impact of adaptation to climate change on spatial planning and water management projects at the local, regional and national scales. The lessons from these adaptation projects were analysed using the following array of questions:

- What are the characteristic features of the projects developed and carried out on adaptation to climate change, and what measures have been taken?
- What are the benefits and drawbacks of the climate-proof development of projects for non-climate policies and economic sectors?
- Does adaptation to climate change have clear additional benefits for the development of a region, and if so, what are they?
- What can be said about the possible additional costs of adaptation?
- Why is one project more sustainable and directed towards innovative, integrative solutions offering new opportunities, while another project (starting from a more or less comparable situation) opts for a classical, conservative solution?
- Is there a need to integrate climate change adaptation in (national) spatial planning and water policies?

The study included projects in which climate change was the primary driver (56%) as well as projects that did not originally start from a climate (adaptation) perspective, but that did have a concrete effect in improving the adaptive capacity or climate resilience of an area with respect to climate change (44%). About a third of the projects are currently in progress, and a small number have been completed. This implies that the findings can only be tentative.

One of the most interesting results of the study is the identification of an implicit logical 'order' of the various characteristics of the planning, development and realisation of multifunctional climate-proofing projects; this order explains why a comprehensive, multifunctional approach can result in a project of higher quality. This logical order is useful because it facilitates the understanding of why such adaptation projects have greater added value and exert major influence on spatial planning practices, avoiding the various maladaptation pitfalls described in the Introduction. It explains the difference in development of different types of projects, from defensive single-purpose projects to innovative, integrative projects. On the basis of an analysis of more than 100 projects, this logical order can be visualised (see Figure 12.3.1).

A change in the characteristics of a project's development appears to give rise to the next change (Pijnappels & Sedee 2010). Although no explicit causal link is implied, the eight project characteristics below can be logically organised as indicated in Figure 12.3.1, moving from left to right. The figure also shows how in each step the characteristics of the projects change, from the bottom half of the figure (synthesised as 'single-purpose projects') to the top ('multipurpose projects').

1   *A longer timeframe.* While previously a relatively short-term approach was used in many local and regional land and water management projects, focusing

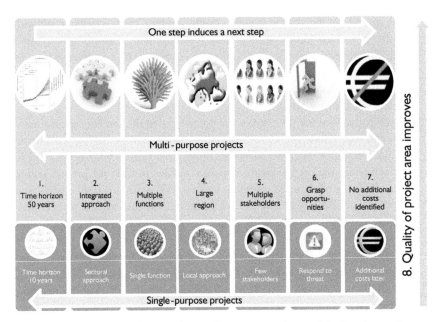

FIGURE 12.3.1 Taking climate change into account in spatial planning and water management projects results in quality enhancement for the region implementing the project (source: Sedee et al. 2013)

on the coming 10–15 years, climate change has now created a time frame in such projects of 40–100 years in the future. When only the short term is taken into account, there is a significant risk that measures will not consider the future impacts of climate change; a longer timeframe prevents the increase of future risks and thus maladaptation. Planning for longer timeframes will stretch the scope of the project from the here and now and the near future to the longer-term future. This prevents reductions in short-term vulnerability that will increase overall vulnerability. The longer timeframe also forces project developers to take into account future generations and the future uncertainties of climate change, thus making it more attractive to opt for a flexible solution rather than a fixed one.

2   *An integrative and sustainable approach.* In the projects, various aspects were included that target the sustainable development of the area involved, broader than climate change alone, often according to the 'people, planet, profit' approach. This more integrative approach will help prevent the unequal distribution of the benefits and costs of adaptation. As more stakeholders and sectors are involved, a greater variety of costs and benefits are taken into account, ensuring the equitable distribution of benefits and costs. This step avoids the type of maladaptation that unequally distributes future benefits.

3 *New functions for the project area.* The combination of climate adaptation with nature protection, housing, economic development, transport, recreation and water management (the integrative approach) can generate new opportunities for combining multiple functions in new designs. Again, the risk of maladaptation is reduced, since multipurpose solutions minimise specific adaptation costs.

4 *A broader spatial context.* The area to be considered in the project design should be enlarged. Examples are sea-water intrusion hampering agriculture in coastal areas not currently affected, flood risks caused by precipitation in a much wider region, decisions by policy-makers outside the project region that affect the amount of water in the project area, and the construction of 'climate buffers' (water retention areas) outside the original project area to limit the need for dike reinforcements. In addition, gaining support for a project is sometimes easier in a larger area that allows more possibilities for co-financing and greater political influence. The broader spatial context will reduce maladaptation risks by ensuring that the costs and benefits of the adaptation measure are distributed more equally and that the negative aspects are not so easily foisted off on a neighbouring region or city.

5 *Participation of multiple stakeholders.* The resulting new opportunities encourage the participation of multiple stakeholders and increase enthusiasm, momentum and innovation. A larger number of diverse stakeholders can result in a larger number of creative and innovative ideas for solutions, avoiding (mal-)adaptation measures that unequally burden specific groups.

6 *Opportunities for entrepreneurs.* Adaptation can provide opportunities for the agricultural, recreation, construction and transport sectors and for nature development. In various projects in the case study, the construction and transport sectors in particular have captured benefits, as have the nature protection and tourism sectors.

7 *Increased cost-effectiveness of projects.* Adaptation does not necessarily lead to cost increases. With only a few exceptions (more sand required for an offshore industrial area taking into account sea-level rise), the additional costs or benefits of adaptation in monetary terms could not explicitly be identified, but according to interviewed stakeholders, they seemed to be rather low. Maladaptation through high opportunity costs is avoided. Confirming this tentative but potentially very important finding and its transferability to other contexts will require additional research.

8 *Enhancing the 'quality' of the project area.* The assessment of the 'quality' of a location is determined by the interests and needs of all stakeholders and users, and thus the definition of the 'quality of a location' differs by location. In the context of the case study, the term refers to the preservation of regional characteristics and cultural heritage and the development of new elements such as variation in the landscape, increased biodiversity or an innovative combination of social, ecological and/or economic functions. If projects enhance these qualities, the risk of maladaptation is minimised.

12.3.2 Concluding thoughts

Based on an analysis of project documents, project websites and interviews with project leaders, the 'Climate as Opportunity' study resulted in the following recommendations that will not only help to prevent maladaptation but will also enhance the overall quality of adaptation projects (Pijnappels & Sedee 2010):

1   *Early incorporation.* Include adaptation in projects, in planning processes and in region-oriented policies (e.g. the policy agenda for the region) from the outset. For example, climate-proofing needs to be taken into account in the Environmental Impact Assessment (EIA), the Strategic Environmental Assessment (SEA) and other relevant evaluations in order to enforce and institutionalise early incorporation. Effective communication of the idea that the incorporation of climate adaptation will add value to the quality of projects will increase public acceptance and help regions to avoid later costs.

2   *Early political commitment.* Politically anchor decisions as quickly as possible, given the intergovernmental cooperation and exchanges among personnel and decision-makers required during the project process. The involvement and responsibility of several authorities in a project will require sustained administrative and intergovernmental discussions. Mutual trust between managers is important here, but such trust must be developed, and the usual frequent changes in decision-makers will not help this process. In addition, over the course of a project, certain decisions will be called into question if they are not properly anchored (e.g. because of the divergent political viewpoint of a new decision-maker). Political power plays may threaten to override maladaptation concerns. The establishment of a management committee with an independent chairman could be considered in order to overcome impasses and ensure a fair and equitable process.

3   *Knowledge co-creation and multi-actor collaboration.* Invest in knowledge through the co-creation of solutions by knowledge institutes, civil servants, the private sector and other societal actors, with special attention devoted to local knowledge. In order to avoid maladaptation, it is essential that knowledge is available regarding the possible effects and impacts of climate change, as well as the potential added value of the incorporation of adaptation for the identification and design of options. The knowledge provided should be as specific as possible for the project. An investment in knowledge will have many benefits at low costs, whether in the form of avoided (current or future) climate change costs or not.

4   *Integrated, multifunctional and forward-looking solutions.* Develop integrative projects and look for possible synergies between different sectors, such as water management, agriculture, nature protection, environment, energy and spatial planning. There will be benefits in terms of spatial quality from a search for functional integration, and maladaptation by shifting vulnerabilities to the future or to other sectors or regions will be avoided. Projects are too often

directed at only one sector because of a narrow perception of the problem, because such an approach seems to be the quickest route or because the administration is still organised sector by sector.

The complexity of climate adaptation concerns implies that maladaptation is always a serious risk when its potential occurrence is not systematically explored. This study suggests steps that can be taken to minimise this risk. It shows that the introduction of climate change adaptation into a spatial planning or water management project can inspire the development of a region or town, lead to quality improvements and accelerate the implementation of projects. At the same time, it reduces the chance of maladaptation because the risk of shifting vulnerability to the future, to other areas or to other sectors or societal groups is explicitly considered. It also provides new opportunities by the addition of new spatial functions to the project area and by linking different sectors, offering new cross-sectoral opportunities. These are all important reasons to integrate climate-proofing into spatial planning and water management, and *vice versa*. Because a majority of the projects had not been finalised at the time of the study, the results can only be tentative. Further research expanding into other countries would be required to determine whether the reported positive effects of integrating climate change adaptation into investment projects and the avoidance of maladaptation are unique to the Dutch context, or whether they can be replicated in other countries as well.

12.4   Lessons learned for the avoidance of maladaptation

Rob Swart

In this chapter, we have addressed the very important principle of avoiding maladaptation. Adaptation actions that may seem effective at first glance can in fact lead to maladaptation. Therefore it is important, as part of the process of identifying and evaluating possible adaptation options, to carefully consider the situations in which maladaptation may occur:

* Avoid actions that increase vulnerability or through which the capacity to cope with the negative effects of climate change is decreased.
* Avoid actions that increase greenhouse gas emissions and thus conflict with mitigation.
* Avoid actions that use resources unsustainably.
* Avoid actions that distribute the benefits of adaptation unequally across society or that disproportionally burden the most vulnerable (e.g. prevention of climate change-induced diseases or inconveniences only for affluent people).
* Avoid actions that have high opportunity costs, i.e. the economic, social or environmental costs are high relative to alternatives.

- Avoid actions that set paths that limit the choices available to future generations.

The contribution by de Jong (Section 12.1) describes a set of connected actions related to winter tourism that are in part derived from the need to respond to a changing climate with decreased natural snow reliability. These actions are at odds with all of the six recommendations above: they increase vulnerability to climate change; increase energy consumption and related GHG emissions; use energy, land and water resources unsustainably; unequally benefit societal actors; have high opportunity costs; and limit choices for future generations. Nevertheless, mostly for reasons of short-term political and financial gains, the measures in these examples have been implemented on a large scale in various countries, demonstrating how difficult it may be in practice to avoid maladaptation. On the positive side, the case presented by Kazmierczak (12.2) elaborates one example of how projects that focus on multiple goals (adaptation, mitigation, etc.) can reduce the risk of maladaptation – in this case, green roofs in Basel, Switzerland. The contribution by Swart and Pijnappels (12.3) complements the green-roof study from the previous section, evaluating a large number of projects with a common methodology rather than one in-depth case to derive lessons learned.

More generally, the following lessons emerge from the three cases:

- When planning adaptation measures, carefully evaluate how the measures may lead to all types of maladaptation (increased vulnerability elsewhere or later, unequal impacts on vulnerable actors, unsustainable resource use, high opportunity costs, limitation of future choices).
- Explicitly explore the implications of integrated, multifunctional, forward-looking and flexible solutions that would entail benefits even if climate change turned out differently from what is currently projected now, or that can be adjusted later (no- or low-regrets; win-win options; adaptive, robust and/or resilient solutions).
- Ensure the early engagement of key decision-makers as well as other stakeholders to achieve political support, to generate a broad variety of possible adaptation solutions for sustainable development and to identify potentially maladaptive solutions *before* they are implemented by engaging experts and stakeholders who will provide a broad range of knowledge and perspectives (co-creation of knowledge).

Notes

1 Outdoor recreation areas with remodelled terrain allowing skiers and snowboarders to perform tricks on snow, similar to skate parks.
2 Snow manipulation using heavy vehicles on ski runs to move, flatten and compact natural and artificial snow.
3 The use of obstacles, equipment and knowledge about the management of snow in order to strategically manipulate snow cover.

References

Abegg, B. (2011) (ed.) *Tourism in Climate Change, Compact, Nr. 1*, Background Report, CIPRA International.

Barnett, J. and O'Neill, S. (2010) 'Maladaptation', *Global Environmental Change* 20 (2), pp. 211–213.

Baumann, N. (2010) *Basel as a GRaBS case study, Personal communication.*

Bizikova, L., Robinson, J. and Cohen, S. (2007) 'Linking climate change and sustainability at the local level', *Climate Policy* 7 (4), pp. 271–277.

Brenneisen, S. (2005) *Green Roofs and Biodiversity – International Context,* Conference paper given at 'Delivering Sustainable Buildings' Conference, Birmingham.

Brenneisen, S. (2010) *Basel as a GRaBS case tudy,* Personal communication.

Carter, T. and Keeler, A. (2008) 'Life-cycle cost-benefit analysis of extensive vegetated roof systems', *Journal of Environmental Management* 87 (3), pp. 350–363.

Castleton, H., Stovin, V., Beck, S. and Davison, J. (2010) 'Green roofs: building energy savings and the potential for retrofit', *Energy and Buildings* 42 (10), pp. 1582–1591.

De Jong, C. (2009) 'Savoy – balancing water demand and water supply under increasing climate change pressures (France)', in *Regional Climate Change and Adaptation. The Alps Facing the Challenge of Changing Water Resources*, Report nr. 8, ISSN 1725–9177, Copenhagen: European Environment Agency (EEA), pp. 81–84.

De Jong, C. (2011) 'Artificial production of snow', in Singh, V.P. and Singh, P. Haritashya (ed.) *Encyclopedia of Snow, Ice and Glaciers,* London: Springer, pp. 61–66.

De Jong, C., Masure, P. and Barth, T. (2008) 'Challenges of alpine catchment management under changing climatic and anthropogenic pressures', iEMSs 2008: *International Congress on Environmental Modelling and Software*, Integrating Sciences and Information Technology for Environmental Assessment and Decision Making, 4th Biennial Meeting of iEMSs, Sànchez-Marr, M., Béjar, J., Comas, J., Rizzoli, A. and Guariso, G. (ed.) International Environmental Modelling and Software Society (iEMSs), pp. 694–702.

Eriksen, S., Aldunce, P., Bahinipati, C.S., D'Almeida, M., Molefe, J.I., Nhemachena, C., O'Brien, K., Olorunfemi, F., Park, J., Sygna, L. and Ulsrud, K. (2011) 'When not every response to climate change is a good one: identifying principles for sustainable adaptation', *Climate and Development* 3 (1), pp. 7–20.

Gill, S., Handley, J., Ennos, A. and Pauleit, S. (2007) 'Adapting cities for climate change: the role of the green infrastructure', *Built Environment* 33 (1), pp. 115–133.

Ip, C. (2008) *Basel. Policy Initiatives, Green Roofs, Tree of Knowledge.* http://www.greenroofs.org/grtok/policy_browse.php?id=63&what=view.

IPCC (2001) 'Climate change 2001: impacts, adaptation, and vulnerability', in McCarthy, D., Canziani, O.F., Leary, N.A., Dokken, D.J. and White, K.S. (eds.) *Contribution of Working Group II to the Third Assessment Report of the Intergovernmental Panel on Climate Change*, Cambridge: Cambridge University Press.

Könighofer, S. (2007) *Conflicts In and Around Dobratsch,* Institute of Geography and Regional Research, Univ. of Klagenfurt.

Kumar, R. and Kaushik, S. (2005) 'Performance evaluation of green roof and shading for thermal protection of buildings', *Building and Environment* 40 (11), pp. 1505–1511.

Lagriffoul, A., Boudenne, J.L., Absi, R., Balle, J.J., Berjeaud, J.M., Chevalier, S., Creppy, E.E., Gilli, E., Gadonna, J.P., Gadonna-Widehem, P., Morris, C.E. and Zini, S. (2010) 'Bacterial-based additives for the production of artificial snow: what are the risks to human health?' *Science of the Total Environment* 408 (7), pp. 1659–1666.

Laukkonen, J., Blanco, P.K., Lenhart, J., Keiner, M., Cavric, B. and Kinuthia-Njenga, C. (2009) 'Combining climate change adaptation and mitigation measures at the local level', *Habitat International* 33 (3), pp. 287–292.

Lawlor, G., Currie, B., Doshi, H. and Wiedetz, I. (2006) *Green Roofs. A Resource Manual for Municipal Policy Makers,* Canada Mortgage and Housing Corporation. http://www.cmhc-schl.gc.ca/odpub/pdf/65255.pdf?lang=en.

Lim, B., Spanger-Siegfried, E., Burton, I., Malone, E. and Huq, S. (ed.) (2004) *Adaptation Policy Frameworks for Climate Change. Developing Strategies, Policies and Measures,* UNDP, Cambridge: Cambridge University Press, p. 255.

London Climate Change Partnership (LCCP) (2006) *Adapting to Climate Change: Lessons for London,* Greater London Authority, London.

Losarcos, L. and Romero, L. (2010) *Green Infrastructure In-depth Case Analysis. Theme 5: Urban Green Infrastructure,* Institute for European Environmental Policy. http://www.ieep.eu/assets/903/GI_Case_Analysis_5_-_Urban_Green_Infrastructure.pdf.

Marty, C. and Meister, R. (2012) 'Long-term snow and weather observations at Weissfluhjoch and its relation to other high-altitude observatories in the Alps', *Theoretical and Applied Climatology,* DOI 10.1007/s00704-012-0584-3.

McCarthy, J., Canziani, O.F., Leary, N.A., Dokken, D.J. and White, K.S. (ed.) (2001) *Climate Change 2001: Impacts, Adaptation, and Vulnerability,* Contribution of Working Group II to the Third Assessment Report of the IPCCN, Cambridge: Cambridge University Press.

Nelson, D.R., Adger, W.N. and Brown, K. (2007) 'Adaptation to environmental change: contributions of a resilience framework', *Annual Review of Environment and Resources* 32, pp. 395–419.

OcCC (2007) *Das Klima ändert – was nun? Der neue UN-Klimabericht (IPCC 2007) und die wichtigsten Ergebnisse aus Sicht der Schweiz,* Bern: OcCC.

Pijnappels, M. and Sedee, A.G.J. (2010) *Klimaat als Kans. Adaptatie aan klimaatverandering in de ruimtelijke ordening,* Utrecht, Netherlands: Programmabureau Kennis voor Klimaat.

Prutsch, A., Grothmann, T., Schauser, I., Otto, S. and McCallum, S. (2010) *Guiding Principles for Adaptation to Climate Change in Europe,* European Topic Centre on Air and Climate Change (ETC/ACC) Technical Paper.

Sedee, A.G.J., Swart, R., Pijnappels, M., de Pater, F., Goosen, H. and Vellinga, P. (2013) 'Climate-proofing spatial planning and water management projects – An analysis of 100 local and regional projects in The Netherlands', *Journal of Environmental Policy & Planning* (forthcoming).

Vanat, L. (2011) *International Report on Mountain Tourism.* Overview of the key industry figures for ski resorts, p. 77.

Weiss, O., Norden, G., Hilscher, P. and Vanreusel, B. (1998) 'Ski tourism and environmental problems; ecological awareness among different groups', *International Review for the Sociology of Sport,* 33 (4) pp. 367–379.

Wilson, C. and McDaniels, T. (2007) 'Structured decision-making to link climate change and sustainable development', *Climate Policy* 7 (4), pp. 353–370.

13

MODIFY EXISTING AND DEVELOP NEW POLICIES, STRUCTURES AND PROCESSES

Andrea Prutsch, Sabine McCallum,
Torsten Grothmann, Inke Schauser
and Rob Swart

Explanation of the guiding principle

Adaptation should not be decoupled from existing policies and political instruments (e.g. legislation, funding systems), organisational structures (e.g. networks) or processes (e.g. in decision-making). Thus, adaptation-relevant instruments, structures and processes currently in place should be reviewed and, where necessary, modified to ensure that they take into account the current and potential future impacts of climate change (SEC 2007). This is often called the 'mainstreaming' of adaptation.

Integrating adaptation through the review and modification of existing instruments, structures and processes should not be restricted to the environmental sector or to the public authorities; it is also relevant for economic sectors and private organisations (Smithers et al. 2008, Kivimaa & Mickwitz 2009). Furthermore, it also makes sense for NGOs and even for households to review, for example, their preparedness plans for natural disasters in consideration of potential climate change impacts. In general, the selection of instruments, structures and processes should be based on their specific suitability to address the major risks identified (cf. Chapters 5 and 10).

Integrating adaptation with existing instruments, structures and processes needs to be recognised as a long-term goal that cannot be resolved in one go (Mickwitz et al. 2009). In particular, it is essential to create learning processes (e.g. openness to new knowledge, implementation of monitoring systems) in order to enable improvements in integration, develop good practices and avoid maladaptation (cf. Chapters 7, 8, 12 and 14).

In comparison to the relatively long period of study and experimentation for various mitigation policies and instruments, the development and implementation of adaptation policies is as yet in its very early stages. Practical experiences in integrating adaptation with existing instruments, structures and processes have shown that, among other aspects, difficulties in integration are related to the specialisation of administrations, strongly sectoral approaches and limitations in terms of knowledge and resources (Mickwitz & Kivimaa 2007, de Bruin et al. 2009). For mitigation to be effective, the emphasis can be on a relatively small number of sectors (such as energy and transport); in contrast, adaptation needs to be integrated into a much wider variety of sector policies reflecting climate change concerns.

However, there are cases in which the modification of existing policies, structures and processes alone is insufficient to handle the adaptation requirements; in these cases, new adaptation-specific policies and instruments must be developed. These seem to be necessary in particular when new cross-sectoral or inter-regional collaborations are involved and/or when 'transformative adaptation' is required. *Transformative adaptation* refers to adaptation that entails fundamental cultural, political, economic, infrastructural or technological changes. A growing number of publications argue the case for transformative adaptation (e.g. Kates et al. 2012, O'Brien 2012, Park et al. 2012), particularly in a potential future world in which the global temperature target of 2°C is exceeded. For example, a recent study on the impacts of a possible global temperature increase of 4°C expects the impacts on water availability, ecosystems, agriculture and human health to lead to the large-scale displacement of populations and adverse consequences for human security and economic and trade systems (PIK 2012). Should these severe impacts of climate change become a reality, many more transformative adaptation measures would be required.

Chapter overview

In the following chapter, we present two cases that focus on practical experiences with mainstreaming adaptation (Sections 13.1 and 13.2). Section 13.1 presents a step-wise approach developed for structured and comprehensible mainstreaming efforts that can be applied at various governance levels. This framework has been applied at the EU level in mid-2012. Lessons learned and further recommendations are presented that seek to provide support for future mainstreaming exercises. The case described in Section 13.2 focuses on the mainstreaming of adaptation in a specific sector, answering the question of how spatial planning in the Alpine countries can be 'climate-proofed'. It proposes a checklist to assess the climate change fitness of spatial planning that can be transferred to other regions across Europe.

The third case focuses on the development of a separate policy for adaptation. Section 13.3 elaborates on activities at the local and regional levels in Finland, with a focus on the initiatives taken in the Helsinki Metropolitan Area from 2009 onwards. It summarises the process of developing an adaptation strategy for this area and draws lessons learned on the factors enabling adaptation and the development of a stand-alone policy document in response to climate change.

13.1   Climate-proofing: A step-wise approach for mainstreaming adaptation applied to EU policies

Sabine McCallum and Thomas Dworak

13.1.1 Introduction

'Mainstreaming', 'climate-proofing' and the 'integration' of adaptation are increasingly important concepts in policy making as various policy areas/sectors are affected by climate change. The objective of mainstreaming climate change adaptation is to ensure that the relevant policies take due account of the climatic changes with which they are concerned, and thus help to increase societal and ecosystem resilience, with an emphasis on the optimal use of potential benefits and new opportunities. Therefore, at the European level, the EU White Paper on adaptation (EC 2009a) strongly recommends 'climate-proofing' of key EU policy areas. This objective was renewed by the EU 'Strategy for smart, sustainable and inclusive growth' (Europe 2020) through its five headline targets and seven flagship initiatives; these highlight the importance of mainstreaming, noting in particular that "we must strengthen our economies' resilience to climate risks, and our capacity for disaster prevention and response" (EC 2010). The issue is also taken up in the 7th Environmental Action Programme (EC 2012a). In addition, the EU Strategy on adaptation to climate change (EC 2013a) clearly proposes climate-proofing of EU action by promoting adaptation in key vulnerable sectors. Thereby, the focus for the period up to 2020 is on the Common Agricultural Policy (CAP), the Cohesion Policy and the Common Fisheries Policy (CFP), infrastructure and policies for insurance and other financial products.

Mainstreaming of relevant EU policies also requires Member States to contribute to achieving EU goals and thus seeks to ensure coherence in the consideration of climate change adaptation across all levels of governance through transposition.

In the context of a service contract supporting the development of the EU Adaptation Strategy, the *status quo* of mainstreaming efforts in key EU sectors has been analysed with the goal of monitoring progress in actions recommended in the earlier White Paper on adaptation and identifying priority policy areas for further mainstreaming efforts under the framework of the EU Adaptation Strategy. A step-wise approach has been developed for a structured and comprehensible analysis that can be applied at various governance levels.

13.1.2 A step-wise approach for climate-proofing policies

For the analysis of key policy areas perceived as being affected by climate change, the following step-wise approach was taken at the EU level; this approach could prove useful for climate-proofing policies at national or sub-national levels, also beyond Europe.

- *Step 1*: Mapping the current status of adaptation efforts in legislation and guidance documents/guidelines
- *Step 2*: Defining criteria for prioritisation with regard to the development of concrete recommendations for mainstreaming
- *Step 3*: Identifying priority policy areas for mainstreaming.

Step 1: Mapping the current status of adaptation efforts in legislation and guidance documents/guidelines

The aims of Step 1 are to screen a set of policies that could be relevant in triggering adaptation to climate change and to map the current levels of effort with respect to the inclusion of adaption in existing and proposed legislation as well as guidance documents/guidelines.

In evaluating the current levels of effort, the following criteria facilitate the screening of existing instruments:

- Is climate change adaptation an explicit objective of the legislation/guidance document/guideline? (yes/no)
- Is there a general reference to the consideration of climate change when implementing the legislation/guidance document/guidelines? (yes/no)
- Are specific mechanisms foreseen to stimulate adaptation when implementing the legislation/guidance document/guideline? (yes/no)
- Are concrete adaptation measures or other measures that facilitate meeting adaptation objectives included/proposed? (yes/no)
- Are monitoring tools for adaptation included/proposed? (yes/no).

The results can then be clustered in the following groups: (i) legislation already including climate change provisions, (ii) legislation that is relevant for mainstreaming but does not yet consider climate change, (iii) guidance documents/guidelines relevant for mainstreaming and (iv) legislation and guidelines for which climate concerns are not relevant. Table 13.1.1 proposes a format for the presentation of this information in a comprehensive manner (completed using an example from the agricultural sector).

Step 2: Defining criteria for prioritisation with regard to the development of concrete recommendations for mainstreaming

Based on the mapping exercise under Step 1, it is necessary to set priorities indicating where mainstreaming should take place first. The following list of criteria (not weighted) can be used to prioritise areas for adaptive intervention:

- Urgency of action with respect to already existing threats, taking into account the geographical distribution of impacts
- Early preparatory action (to avoid or limit the costs of future damage)

TABLE 13.1.1 EU legislation in which mainstreaming is taking/has already taken place

Sector / Policy area	Agriculture
EU Legislation[1] [number and name]	COM(2011) 627 final2,[2] Proposal for a Regulation of the European Parliament and of the Council establishing rules for direct payments to farmers under support schemes within the framework of the common agricultural policy
Main issue addressed by the legislation and why it is relevant for CCA	Rural development, in particular for the agricultural sector. The policy is the main driver for agricultural production in Europe and has a high potential to stimulate adaptation.
Status [proposal/in force]	Proposal
Revision foreseen by [date]	2020
Bodies responsible for the revision and bodies to be involved	DG AGRI
Articles relevant for mainstreaming CCA	Art 14 (1) & (2), Art 30–33, Art 34-35, Art 36 (6), Art 38, 42–44, Art 55
Level of climate change integration	Adaptation is a main policy objective of the proposal. There are mechanisms to stimulate adaptation at all levels and concrete adaptation measures proposed (e.g. Art 30–33). For monitoring, see Articles 74–86 and SF 'umbrella reg'.
Other policy areas affected	Forestry, Water, Biodiversity, Soil, Animal and Plant Health
Comments	The effectiveness of the policy will depend on farmers' uptake of the funding.

1   This refers to Directives, Regulations and Commission Decisions.
2   http://ec.europa.eu/agriculture/cap-post-2013/legal-proposals/com627/627_en.pdf

Source: authors' research

- Adaptive capacity of the sector (including society/businesses/infrastructure likely to be affected and their capability to adapt), including enhancement of learning and fostering of adaptation action
- Potential market failure
- Added value of intervention at the respective governance level (e.g. for the EU level: cross-border impacts (water, nature), interdependency of Member States (marine issues, migration), EU competitive position in world economy (agriculture)
- 'Window of opportunity': Legislation that is subject to revision in the near future (potential political motivations)
- The extent to which climate change adaptation is already covered within the policy

- Common interests between policy-makers, decision-makers and stakeholders
- High mainstreaming impact to be expected (e.g. setting up new funding mechanisms that will allow financing for adaptation in various sectors).

Step 3: Identifying priority policy areas for mainstreaming

A first prioritisation along the criteria suggested in Step 2 can be accomplished based on the results of Step 1, resulting in a list of policy areas for which mainstreaming efforts should be prioritised.

When priorities have been set, concrete recommendations for mainstreaming can be formulated for each policy area/sector that define:

- Main arguments supporting the need for mainstreaming
- Objective(s) to be achieved by mainstreaming of climate change adaptation into a legislation/guidance document/guideline
- Actions to be taken
- Responsibilities (who needs to take action in collaboration with whom)
- Ongoing activities to take into consideration
- Studies and service contracts required (support).

13.1.3 Applying the methodology to EU policies: Results from an analysis of mainstreaming at the EU level

Mapping the *status quo* of mainstreaming efforts at the EU level in mid-2012 showed that the integration of climate change considerations in some key EU sectors has been achieved in line with what was proposed in the EU White Paper on adaptation (EC 2009b) and beyond. Current efforts can be differentiated into legal issues, funding issues and 'soft' measures. The term *soft measures* refers to the development of guidelines, voluntary agreements and internal Commission procedures.

The consideration of adaptation is already legally binding in the context of some policy areas (e.g. marine issues (EC 2008, EC 2011a), forestry (EC 2003), disaster risk reduction (EC 2011b) and transport (EC 2011c), and on a voluntary basis in the case of inland water (EC 2007, EC 2012b).

Further mainstreaming efforts by the European Commission are reflected in legal proposals under negotiation. In this context, the most important policy proposal is the next Multi-annual Financial Framework (MFF) 2014–2020. The Commission suggests that climate-related expenditures shall represent at least 20% of the overall EU budget and will be tracked according to a specific methodology. Mainstreaming of climate change will further be considered in the preparation of the Member States Partnership Contracts and Programmes for the five funds under the Common Strategic Framework (CSF) covering the period 2014–2020. On the sectoral level, mainstreaming took place in (i) the proposals for the Common Agricultural Policy after 2013, (ii) establishing the Connecting Europe Facility in the field of energy (EC 2011d), (iii) guidelines for the development of the trans-European transport network (EC 2011e) and (iv) a dedicated legislative instrument on Invasive Alien Species. Furthermore, a new forest strategy, a proposal for a new plant health law, a Framework Directive for Maritime Spatial Planning and Integrated Coastal Zone Management and the revision of the EIA Directive are all currently underway and foresee addressing climate change adaptation. Guidelines for climate change adaptation have been developed for the implementation of the Water Framework Directive, for the development of regional adaptation strategies, for disaster risk reduction (various), energy and considerations in Environmental Impact Assessment (EIA) and Strategic Environment Assessment (SEA). Within the context of the EU Adaptation Strategy (EC 2013a), guidelines on integrating climate change adaptation into the programmes and investments of Cohesion Policy (EC 2013b), principles and recommendations for integrating climate change adaptation considerations under the 2014–2020 rural development programmes (EC 2013c) and guidelines on developing adaptation strategies (EC 2013d) have also been developed. In addition, so-called 'non-paper' guidelines for project managers ('making vulnerable investments climate resilient') have been made available as documents related to the EU Adaptation Strategy.

Based on the *status-quo* analysis of EU mainstreaming activities (cf. Step 1 of the approach suggested above), the following conclusions can be drawn:

- In Europe, only a limited number of legislative acts consider climate change, with a varying level of detail. Considerations range from mere mentions of climate change as an issue to take into account in the future to concrete provisions for adaptation. However, the assessment showed that several policy areas at risk of climate change are still not addressing climate change at all, and thus mainstreaming of adaptation into a much wider set of policies and legal actions is required.
- Mainstreaming is likely to be possible only within the set review cycles of the respective policies that open up a window of opportunity for integrating the issue of climate change. While some policies are currently under review and integration of climate change is under way, others are scheduled to enter into the review process only shortly before 2020, and some legislation has no built-in review process for legislative amendments at all, which substantially influences planning for mainstreaming.
- Another issue connected to mainstreaming involves the conflicting objectives of different policy areas that could potentially counteract adaptation efforts. For example, the Renewable Energy Directive required more water use (irrigation, hydropower) even though natural water availability is predicted to decrease in many southern European countries due to climate change.

Taking into account the current status of mainstreaming at the EU level and applying criteria for prioritisation (as proposed in Step 2 of the methodology), several EU policy areas and actions have been identified for which mainstreaming efforts should be prioritised, in particular with regard to existing and projected climate change impacts and the low rate of current policy integration with regard to addressing climate change risks, together with high expected mainstreaming impact. For each of the policy areas/sectors, concrete recommendations were formulated, describing how the integration of climate change adaptation could be achieved (cf. Step 3 of the approach). However, it should be noted that even among the identified priority policy areas, there will be a need to decide on the focus of interventions within the timeframe and the available budget and personal resources for implementation of the EU Adaptation Strategy.

13.1.4 Key elements to ensure successful mainstreaming across governance levels

Adaptation cuts across different jurisdictional levels, from the EU to the national to the sub-national and local levels of policy making. In conclusion, there are a few key issues that should be taken into account to support mainstreaming efforts at all levels.

Stakeholder involvement: Raising awareness and building knowledge

Mainstreaming at any decision-making level can only be successful when the relevant stakeholders are:

*   well informed,
*   aware of potential risks,
*   engaged in the joint development of 'CC-proofing' options, and
*   convinced of the need to take action.

A systematic analysis of mainstreaming efforts and needs as described in the methodology above can serve as a basis for discussion with relevant sectoral policy-makers to ensure the uptake of necessary modifications to related policies. Raising awareness of potential risks that the respective policy area/sector might face in the future will facilitate mainstreaming actions in terms of developing concrete policy options to be integrated in new policies or amendments of existing policies.

Mechanisms for coordination and capacity-building

With a view to ensuring the coherence of mainstreaming efforts at various levels, well-defined mechanisms for coordination among all relevant stakeholders will be needed. Given the cross-cutting nature of climate change impacts – that is, the impacts cut across economic sectors, geographic and administrative boundaries and time-scales – it is essential that adaptation mainstreaming is embedded within a broader process by making use of existing networks, building on current activities and engaging with the private sector. Active information sharing and procedural guidance for coordination will enhance the integration of adaptation across all levels of decision-making. Joint efforts for capacity-building are also needed to increase awareness at the highest decision-making level possible and thus ensure that climate change adaptation remains a continuous process that extends beyond the integration of the issue in policies and guidelines towards implementation.

In the EU, the European Commission can take a leading role in ensuring a coherent framework and coordinating with Member States and other relevant stakeholders via proposed actions in the EU Adaptation Strategy.

Entry points and the use of existing instruments and processes

The mapping exercise described above is also intended to identify the best-suited entry points for mainstreaming, emphasising potentials with the highest mainstreaming impacts. In some areas, legal action might be necessary, whereas for other issues 'soft measures' such as the elaboration of guidance documents will be more effective in ensuring the adequate consideration of climate change.

Furthermore, several existing instruments and processes can be utilised to integrate climate change considerations into all relevant policy areas/sectors (e.g. impact assessments, funding instruments, incentives, cooperation networks).

For EU policies, the EU Impact Assessment Procedure could potentially be used as an important means of guaranteeing the mainstreaming of climate change adaptation over the long term. However, not all policy initiatives and legislative proposals are subject to an Impact Assessment (IA); the precise scope of the procedure's application is decided on an annual basis and is published as a roadmap, together with the Commission's Annual Legislative and Work Programme (CLWP). It would be advisable for the underlying screening criteria for deciding on an IA's scope of application to already take into account aspects of climate change, both in terms of mitigation and adaptation. For the latter, the overall suitability of the policy in terms of strengthening climate change resilience should be considered.

In terms of mainstreaming adaptation in EU policies, the Impact Assessment Procedure offers two potential entry points:

1   *Scope of application*: Consider climate change as a criterion when deciding whether a policy initiative must undergo an IA.
2   *Carrying out an Impact Assessment*: Elaborate more explicit recommendations for the IA guidelines (EC 2009b) for considering the impacts of the assessed options on adaptation potentials.

Using the existing Impact Assessment Procedure as an instrument to take climate change adaptation into consideration (both to decide on the scope of application and to carry out an IA) would ensure comprehensive 'CC-proofing' of all upcoming EU policy initiatives with a harmonised approach over the long term.

Similar instruments for plans and programmes as well as the project level include Strategic Environment Assessment (SEA) and Environmental Impact Assessment (EIA) (McCallum et al. 2013):

- SEA can be an effective tool for climate change adaptation by introducing climate change considerations into development planning. The Intergovernmental Panel on Climate Change (IPCC) has concluded that the consideration of climate change impacts at the planning stage is key to boosting adaptive capacity: "One way of increasing adaptive capacity is by introducing the consideration of climate change impacts in development planning, for example, by including adaptation measures in land use planning and infrastructure design" (IPCC 2007a). SEA provides a framework for assessing and managing a broad range of environmental risks that may contribute to the integration (or 'mainstreaming') of climate change considerations into plans and programmes (PPs). The integration of climate change into strategic planning through the application of an SEA should lead to better-informed, evidence-based PPs that are more sustainable in the context of a changing climate and more capable of delivering progress on human development (OECD-DAC ENVIRONET 2008). Indeed, SEAs can help to ensure that plans and programmes take full account of climate issues within a clear, systematic process.

- Environmental Impact Assessment (EIA) is a procedural and systematic tool that is in principle well suited to incorporating considerations of climate change impacts and adaptation within existing modalities for project design, approval and implementation. The International Association for Impact Assessment (IAIA) defines EIA as "the process of identifying, predicting, evaluating and mitigating the biophysical, social and other relevant effects of development proposals prior to major decisions being taken and commitments made" (IAIA 1999). EIAs can therefore ensure that future developments are themselves resilient and that their environmental impacts do not exacerbate the effects of climate change on human or natural systems. One of the main reasons to view EIA as a tool for facilitating the successful 'climate-proofing' of projects or for avoiding maladaptation to climate change is that EIA represents a well-consolidated and publicly accepted process in the European Union and worldwide. The key aspect for consideration in the context of an EIA is the determination of how and when climate change adaptation will be triggered within an EIA process. Experience suggests that the earlier these considerations are made, the easier it will be to incorporate the results into the project development process at the lowest possible financial cost (OECD 2011).

In general, suitable instruments at the national level will vary across Europe and across sectors. However, mainstreaming of national legislation will likely follow the implementation requirements of EU policies (for EU Member States), undergoing further refinement in various affected sectors. The existence or development of National Adaptation Strategies will provide the framework for mainstreaming, following a systematic approach of coordinating adaptation needs horizontally across sectors.

Once the integration of climate change has taken place at the EU and/or national level, subsequent activities and concrete adaptation measures can be expected to be implemented at the sub-national, regional and local levels.

Meeting the implementation challenge

Mainstreaming entails making more efficient and effective use of available financial and human resources rather than designing, implementing and managing new mechanisms separately from ongoing activities. When discussing and developing appropriate adaptation interventions, attention should also be devoted to budgeting and financing by leveraging available funding sources and modalities. This is to ensure that the implementation of any policy intervention considered will be feasible and that a sufficient budget can be allocated. Financing adaptation will likely require joint efforts by the public and private sectors, e.g. by establishing public–private partnerships for necessary investments.

Because the development of a policy framework does not ensure its implementation and sustainability, adaptation mainstreaming also requires investments in climate change monitoring and evaluation (see also Chapter 14). Robust monitoring and evaluation should be an essential part of mainstreaming, both to ensure that the prospective benefits of interventions are being realised and to improve the design of future interventions. Monitoring and evaluation should be complemented by quantitative indicators; for example, measuring the number of projects that have been developed. Quantitative indicators are needed to assess the change brought about by a policy. Such differentiation allows clarification of the relative contribution of each activity towards the long-term objective. In some cases, surveys, focus-group discussions or other means of direct consultation with beneficiaries will be needed in order to assess the level of change (Lamhauge et al. 2012).

13.2 Assessing the climate change fitness of policies: The case of spatial planning in the Alpine space

Marco Pütz and Sylvia Kruse

13.2.1 Introduction

Spatial planning is an interesting policy field to examine with respect to climate change for two reasons. First, climate change and its impacts strongly affect land use and land-use development. This might change the context of the field of spatial planning and shape its priorities. Second, spatial planning is usually assumed to have a high potential to address climate change because of its cross-sectoral, multi-level character and because spatial planners are accustomed to the vertical and horizontal coordination of issues (Bulkeley 2006, Davoudi et al. 2009, Roggema et al. 2012). However, the extent to which spatial planning or other policies can be implemented and made effective depends largely on what kind of interventions, tools and resources are available. This section addresses this concern by proposing a framework to assess the climate change fitness of spatial planning. This framework can support the mainstreaming of adaptation into spatial planning and other policy fields.

This section is based on the Alpine Space project CLISP – 'Climate Change Adaptation by Spatial Planning in the Alpine Space' (2008–2011). The CLISP project seeks to prevent increases in climate change-induced spatial conflicts and to reduce vulnerabilities, damages and costs by providing 'climate-proof' spatial planning solutions for future sustainable territorial development in the Alps. A main objective of the project is tackling the questions of whether spatial planning in the Alpine countries is fit for the challenges posed by climate change and how spatial planning and spatial development in the Alpine countries can be climate-proofed.

13.2.2 Assessing the climate change fitness of spatial planning in the Alpine space

Aim, focus and scope

The field of spatial planning needs tools and guidance in order to be able to address climate change issues and to assess the climate change fitness of spatial planning policies and instruments. *Climate change fitness* refers to the capacity of spatial planning systems to adapt spatial development processes and existing spatial structures to climate change impacts, i.e. in order to moderate potential damages, to take advantage of opportunities or to cope with the consequences. Spatial planning instruments and processes are considered 'fit' for climate change when they support and deliver adaptation, strengthen preparedness and the ability to react to climate change impacts, increase the resilience of communities, raise the flexibility of spatial planning systems to respond to climatic changes and the associated uncertainties, and reconcile short-term planning horizons with long-term climate change.

The guidance developed in the CLISP project[1] is designed for application by planners at national, regional and local levels who must evaluate whether or not their spatial planning policies and instruments are fit for climate change adaptation (Pütz et al. 2011). The core elements of the guidance have been successfully field-tested and applied in the CLISP model regions during the project and can be applied to other regions as well. The guidance is available in English, German, Italian and Slovenian (www.clisp.eu). The concept underlying the guidance refers to the UKCIP Adaptation Wizard (UKCIP 2010), transforming the principles and contents of the Wizard to the Alpine

Space and focusing on spatial planning's response to climate change. After having completed the assessment, planners will know whether their spatial planning policies and instruments are fit for climate change. Planners will have elaborated a climate change fitness assessment report, including the strengths and weaknesses of spatial planning, and enhancement options for climate-proof planning. The results of the assessment and the experience of having gone through the procedure will provide a valuable basis for the development of climate adaptation strategies and measures. Planners will also benefit from the establishment of a team of experts for climate adaptation issues and from engagement with stakeholders to build commitment and motivation for delivering climate adaptation in their region. The guidance for planners provides a practical step-by-step climate change fitness assessment. It recommends useful tools and resources to implement the assessment and refers to findings and experiences from the CLISP project.

Climate Change Fitness Checklist

The 'CLISP Climate Change Fitness Checklist' (Pütz et al. 2011) is at the core of the climate change fitness assessment and supports the mainstreaming of climate adaptation in spatial planning. The assessment will make it easier to identify the necessary responses to climate change in terms of spatial planning and to develop enhancement options to improve the adaptation performance of spatial planning and of climate adaptation policy in general. The checklist is especially helpful for reviewing the adaptation performance of a spatial plan. It also facilitates identification of the strengths and weaknesses of spatial planning policies and instruments. By compiling and appraising the lessons learned from the assessment, by identifying enhancement options for climate-proof planning and by defining their priorities and trade-offs, planners will be able to make progress with climate adaptation.

The checklist is centred around five assumptions. Each of these is exemplified by different rationales. The idea of the checklist is to verify whether these assumptions hold with respect to the spatial plan under assessment. The checklist offers a large number of examples that could potentially prove whether the assumption makes sense and thus whether the spatial plan of the planner is fit for climate change.

- My spatial planning policy or instrument is fit for climate adaptation if *regional adaptation challenges are addressed*. The rationales behind this assumption are that climate adaptation must be informed and evidence-based, responding to current climate sensitivities and future climatic changes, climate change impacts and vulnerabilities. Climate adaptation action must also consider the regional context and be regionally specific. Regional adaptation challenges are addressed by providing open green and blue spaces, and by dealing with warming effects in urban areas, water resources, tourism, natural hazards, energy, transport and other technical infrastructure.

- My spatial planning policy or instrument is fit for climate adaptation if *decision-making processes are well connected* and coordinated across different levels (vertically) and across policy fields or sectors (horizontally). Climate adaptation is a cross-cutting task that requires the involvement of stakeholders and planning domains from all sectors in order to be effective. The characteristics of well-connected decision-making processes include: (i) a strong expert network is set up across all relevant sectors and institutions, (ii) climate change adaptation is accepted by every stakeholder as an everyday planning issue and (iii) both a risk communication concept and a risk governance process are in place.
- My spatial planning policy or instrument is fit for climate adaptation if the *shared benefits of linking adaptation to mitigation and regional development* are achieved. Climate adaptation must be strategically aligned with other strategies to be most effective. The characteristics of the shared benefits of linking adaptation to mitigation and development include: coordination and cooperation mechanisms with other strategies that are in place; synergies and potential conflicts are identified and addressed; adaptation options have been audited for potential negative effects on sustainability, the environment, social groups and other sectors; adaptation options have been checked for maladaptation risks; and priorities for climate adaptation are set and coordinated with other relevant strategies.
- My spatial planning policy or instrument is fit for climate adaptation if *adaptive capacity is high and/or increasing*. Climate adaptation is an ongoing and iterative process and must bring about transformation. The characteristics of high or increasing adaptive capacity include: (i) political will for adaptation exists and is strong; (ii) policy-makers and stakeholders are aware of the need for action; (iii) sufficient resources are available; (iv) implementation is ongoing; (v) incentives and national/regional climate change adaptation programmes are in place; (vi) uncertainties are dealt with in a pro-active and precautionary approach; (vii) no-/low-regrets measures have been identified and are being implemented; (viii) planning instruments and procedures are flexible enough to cope with climatic changes and to respond to new or enhanced knowledge; (ix) short-term actions consider long-term climatic processes; and (x) adaptive planning and management procedures, including monitoring and evaluation, are being applied and linked in spatial plans with regular revision cycles.
- My spatial planning policy or instrument is fit for climate adaptation if a *sound system of monitoring for regional climate change impacts or risks is in place* (with particular emphasis on spatial planning). Climate adaptation must understand the regional adaptation challenges. A sound monitoring system addresses spatially relevant climate change impacts and includes new indicators, such as the size of heat islands or the damage potential for each zoning area.

The actual assessment is a qualitative exploration of the spatial planning fitness. The assessment is intended to support the mainstreaming of climate adaptation into spatial planning by initiating and moderating the dialogue between planners and other sectors.

13.2.3 Conclusion

This guidance provides planners with a generally applicable framework, tools and resources to complete a climate change fitness assessment for spatial planning policies and instruments. As part of the process of mainstreaming adaptation to climate change, the guidance can be incorporated into regular spatial planning processes and administrative routines. It can be used by planning authorities as part of the approval procedure for plans or projects. It can also be integrated into sustainability appraisals, strategic environmental assessments or environmental impact assessments. Alternatively, the guidance can be incorporated into policy making on climate change adaptation, e.g. by conducting assessments as part of the drafting of national or regional adaptation strategies. The assessment results in the evaluation, identification and prioritisation of adaptation options for spatial planning. However, these priority adaptation options must be implemented. Personnel, time and available information and knowledge are the critical factors for successful completion and effective implementation. As spatial planning is a cross-cutting field that affects many different policy areas, implementing adaptation options will in many cases require coordination with other fields of activity. Moreover, the assessment of the climate change fitness of spatial planning is expected to provide substantial suggestions regarding the enhancement of climate adaptation strategies in other policy fields.

Climate change adaptation is not accomplished when the climate change fitness assessment is completed or when the climate adaptation action plan is implemented. Climate adaptation is an iterative process that should be embedded within an adaptive management approach. This calls for ongoing monitoring of the implementation process and of changes in climatic stimuli, impacts and available knowledge. The climate change fitness of spatial planning policies and instruments must be continually assessed against new information on climate-related and other relevant trends. Planners must verify whether or not adjustments to planning policies and instruments are needed. Therefore, a system for monitoring spatial development, climatic trends, the effectiveness of adaptation and the climate change fitness of spatial planning must be established at the regional level. This will enable planners to assess whether or not observed changes in the climate or trends in spatial development require the reassessment of spatial planning's climate change fitness. As part of an adaptive planning approach, the results of monitoring should be linked to the revision of spatial plans in regular planning cycles. Monitoring thus also encourages a learning process surrounding adaptation.

13.3 The world's second northernmost capital region adapts to climate change: The Helsinki metropolitan adaptation strategy

Lasse Peltonen, Leena Kopperoinen and Susanna Kankaanpää

13.3.1 Nordic countries face special adaptation challenges

Climate change is advancing more rapidly in the northern latitudes than the global average. This means that Nordic countries face special adaptation challenges. For example, northern ecosystems are very vulnerable and are not expected to adapt well to changing conditions (IPCC 2007; Marttila et al. 2005). To address these challenges, Finland was among the very first countries in Europe to adopt a National Adaptation Strategy in 2005. The strategy covered multiple adaptation challenges ranging from urban planning and infrastructure to reindeer husbandry (Marttila et al. 2005).

Following this early start, progress in adaptation has proceeded at varying speeds across sectors and levels of governance in Finland. At the local and regional levels, mitigation has dominated the agenda, and until recently adaptation had not been considered as a distinct policy and planning problem (Haanpää, Tuusa & Peltonen 2009, Juhola 2010, Juhola, Haanpää & Peltonen 2012). One of the more important recent initiatives in Finland is the Helsinki Metropolitan Area adaptation strategy, which addresses the adaptation concerns of the world's second northernmost capital region, located on the northern coast of the Finnish Gulf of the Baltic Sea. The cities in the Helsinki Metropolitan Area – Helsinki, Espoo, Vantaa and Kauniainen – have a total population of some 1.1 million inhabitants. The strategy was jointly prepared with the respective regional authorities, rescue services, the Ministry of the Environment and the Association of Finnish Local and Regional Authorities between 2009 and 2012.

13.3.2 Planning and implementation of the Helsinki metropolitan adaptation strategy

Background of the strategy

The Helsinki Metropolitan Area Climate Change Adaptation Strategy (HSY 2012) was prepared in close cooperation between the four constituent cities, which are committed to realising municipal plans and actions based on the strategy. The process was initiated and coordinated by the Helsinki Region Environmental Services Authority (HSY), an inter-municipal bridging organisation for the Helsinki Metropolitan Area, established in 2010 as a merger between the former Helsinki Metropolitan Area Council (YTV) providing waste management and regional information services and the waterworks of the partner municipalities. Furthermore, municipal, regional and state-level organisations were also involved in the strategy process (cf. Figure 13.3.1). In all, there were 30 participating organisations, with 71 individual participants.

The roots of the adaptation strategy can be traced back to 2007, when the board of the former Helsinki Metropolitan Area Council signalled the need for adaptation in the Helsinki region. At the time, the Council had just finalised the metropolitan climate change (mitigation) strategy; this freed space on the agenda for adaptation, which was gaining attention at the national level and in the media. The actual strategy process started in 2009 with a series of background studies focusing on regional climate and sea-level rise scenarios, the projected impacts of climate change in the region such as increases in floods and threats to infrastructure (energy, roads, etc.) and a survey of existing programmes, plans and legislation related to adaptation. The studies were compiled in a background report (HSY 2010).

Main goals and measures

The vision of the strategy is to enable climate-proof development. The strategy aims to 1) assess the impacts of climate change in the area, 2) prepare for the impacts of climate change and for extreme weather events and 3) reduce the vulnerabilities of the area to climate variability and change. The target is to secure the well-being of the citizens and the functioning of the cities under the changing climate conditions. The strategy focuses on adaptation in key sectors and suggests concrete measures, but does not go into details on individual adaptation measures. However, an LCLIP (local climate impacts profile) exercise (UKCIP 2009) revealed differences in how public and private actors are challenged by adaptation measures.

Measures are defined for the following sectors: 1) Land use, 2) Transport and technical networks, 3) Construction and climate-proof neighbourhoods, 4) Water and waste management, 5) Rescue services and safety, 6) Social and health services and 7) Cooperation in producing and disseminating information. Topics such as agriculture and forestry, while important for Finland, were excluded from the metropolitan-level strategy. The strategy outlines a set of adaptation measures for the near-term future (2012–2020), focusing on those that require inter-municipal coordination (e.g. regional water management and infrastructure issues).

Challenges of coordination

A key challenge for the metropolitan strategy was to engage relevant actors and to overcome the problems of inter-municipal coordination. Thus, a steering (management) group consisting of 33 individuals from 12 key institutions was established and a series of workshops were organised for the sectoral actors, including organisations ranging from the Finnish Meteorological Institute to municipal dental care services. Along with the Helsinki Region Environmental Services Authority (HSY), the group included representatives from the four municipalities, rescue services, the regional transport authority (HSL), the Finnish Association of Local and Regional Authorities and the Ministries of the Interior

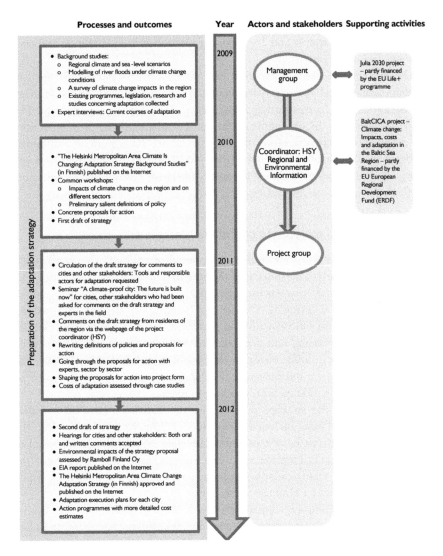

FIGURE 13.3.1 The process of preparing the Helsinki Metropolitan Area Climate Change Adaptation Strategy (source: Leena Kopperoinen)

and the Environment. The steering group took an active role in the preparation of the strategy and in the involvement of their own sectors in the process.

Although adaptation to climate change extends far beyond environmental sectors, the involvement of actors from other areas was challenging at times. Some sectors, such as municipal social services and the health sector – by far the biggest employer in the municipalities – had little previous exposure to climate issues and limited interest in participating. Land-use planning proved to be a difficult case for other reasons; this is an important municipal responsibility that fosters local political autonomy, and external intervention is not welcome. In the end, the land-use measures proposed focused on 1) developing an information base for adaptive land-use planning and 2) developing a joint regional plan for storm-water planning and retention.

The linkage between climate risks and other weather-related risks such as flooding, storms and extreme precipitation proved important in the strategy process, and thus the active participation of the rescue services was important. The rescue services were highly motivated to address adaptation in their operations and easily integrated adaptation concerns in a risk management framework, linking it to their legally mandated duties of contingency planning. Moreover, their involvement strengthened the legitimacy of the process and the commitment of other stakeholders to the process.

Reflexivity through monitoring and impact assessment

Adding to the reflexivity of the strategy, an environmental impact assessment was conducted with regard to the impacts of the measures proposed in the strategy. The synergies and conflicts between the adaptation and mitigation goals constituted a key theme in the assessment. On the one hand, many adaptation measures, such as transport and infrastructure preparedness, were found to contribute to the sustainable functioning and dependability of these systems, thus supporting mitigation targets (e.g. public transport dependability). On the other hand, some measures, such as green infrastructure development, must be carefully balanced with conflicting mitigation measures seeking to increase urban density.[2]

Already during the strategy process, the municipalities initiated practical adaptive actions. These included storm-water management plans and the inclusion of sea-level rise in local planning guidelines (raising the ground level of all new developments in Helsinki, etc.). The municipalities thus took note of the inclusion of climate change adaptation in the national land-use objectives, which set the legal principles for land use and construction.

As new information on climate change becomes available in the future, monitoring progress and implementing adaptive management have been identified as central functions of the adaptation strategy. Monitoring will take place through a set of reporting mechanisms and indicators, which will be jointly established by the key actors in the next step. With regard to adaptive management and the evolving science base, the strategy proposes that an adaptation panel of experts and stakeholders be established to oversee the long-term adaptability of the strategic goals and measures, along the lines of New York's adaptation strategy.[3]

13.3.3 Concluding thoughts

The strategy process succeeded in adopting a strategic perspective on adaptation in the Helsinki Metropolitan Area by engaging key actors and overcoming administrative borders. The process provides lessons on the factors enabling adaptation. In conclusion, we examine these factors as elements of local and regional adaptive capacity, grouped into aspects of awareness, ability and action (cf. Juhola, Peltonen & Niemi 2012).

Awareness among the participating actors was already high at the outset of the process. The level of awareness and the associated momentum for adaptation can be attributed to multiple factors that had been coming together since 2005, including 1) intensive national media coverage of climate change (Lyytimäki 2011); 2) the launching of the national adaptation strategy in 2005 (Marttila et al. 2005) and related sectoral implementation plans; 3) the record-breaking winter storm 'Gudrun' in 2005 (Haanpää et al. 2006), which acted as a "focusing event" (Birkland 1997) for regional adaptation processes; and 4) earlier exposure to adaptation issues in Espoo and Helsinki through national and European research and development projects focusing on adaptation. In addition, long-standing work on climate change mitigation conducted by the HSY had prepared the groundwork for awareness of climate change and adaptation needs. Consequently, all the actors involved were aware of climate change and recognised the related impacts and risks. There was little confusion regarding terminology between adaptation and mitigation perspectives on climate change. Knowledge production was also conducted by participating expert and research organisations, adding to the awareness component of the general adaptive capacity.

Ability was facilitated by institutions and their interplay. First, the regional environmental services authority HSY operated as a bridging institution with a mandate for inter-municipal coordination and a legacy of regional climate policy work. The key role of the HSY as an inter-municipal intermediary can be understood as a partner with the ability to positively disrupt the organisational obstacles created by the municipalities and, perhaps to an even greater extent, their sectoral administrative 'silos' (cf. Dunsire 1996). Second, even though there was a lack of direct state support, national legislation pushed the municipalities to consider adaptation as a serious policy concern. This speaks for the need for multi-level governance to complement local action (cf. European Environment Agency 2012). Third, close regional cooperation and compatible goals building on decades of earlier cooperation provided a strong foundation for collaboration in climate change mitigation and adaptation issues as well. Good regional relationships and the capability for regional collaboration can be seen as crucial for success in the creation of adaptation measures. Fourth, the involvement of sectoral authorities in the process as part of their daily work accustomed them to thinking about the need for climate change adaptation in different spheres of authority.

Action was initiated already years before the actual strategy process started. Here, one cannot overlook the role of leadership and individual 'issue entrepreneurs'. In addition, the role of the HSY as an intermediary organisation was critical for taking action and engaging other actors. The active role of the rescue services added momentum and a sense of urgency to the process. Because both intellectual and monetary resources are key for action, one should note the importance of the contributions of various experts in the process and its background studies. In terms of budgeting, the HSY has its own budget and did not need to raise funds prior to the strategy process. Finally, ongoing EU-funded projects added to the resource base and the expertise of the strategy, underlining the importance of the HSY's capacity to take part in such projects and thus to benefit from access to international best practices.

In addition to awareness, ability and action, *situational factors* also played a role. The officials responsible for the strategy process at the HSY expected that adaptation would not be a top priority for many regional actors. Consequently, they were taken by surprise by the many initiatives that emerged during the three-year process. These included EU initiatives and legislation (Flood and Water framework directives), national legislative developments and many ongoing research and regional development projects. This created the need to keep track of the ongoing developments, making the strategy process an 'observatory' for ongoing adaptation measures. This awareness added to the focus of the strategy.

In terms of implementation, the strategy is still only in its initial phase. The challenges that remain to be addressed include: 1) engaging civil society actors and the general public in the adaptation efforts; 2) addressing broader concerns related to vulnerability and adaptive capacity rather than focusing on the impacts of climate change; 3) addressing the costs and benefits of adaptation beyond the individual studies conducted; and 4) clarifying linkages and trade-offs with climate change mitigation. There is a common understanding that both adaptation and mitigation must be addressed, but practical measures for their integration have yet to be explored and implemented.

13.4 Lessons learned for mainstreaming adaptation

Sabine McCallum and Andrea Prutsch

In this chapter on modifying existing policies, structures and processes, we have gained insights via two cases that focus on the mainstreaming of adaptation into existing policies, structures and processes and another case that concentrates on developing new policies for adaptation.

Regarding the lessons learned in mainstreaming, the experiences of the first case presented by McCallum and Dworak (Section 13.1) show that systematically listing entry points for mainstreaming and identifying priority policies for intervention can help to raise awareness of the need to integrate climate change considerations in key policy areas and can trigger an ongoing dialogue among all affected stakeholders. Preparing and agreeing on a roadmap for mainstreaming priorities in legislation or guidelines (where amending legislation is not feasible or not the most effective solution) could additionally allow greater transparency for all subsequent governance levels when this roadmap is made available and discussed with the relevant stakeholders who will be responsible for implementation. An improved dialogue among decision-makers at all affected governance levels can also prevent potential overlaps in mainstreaming efforts (e.g. EU legislation versus National Adaptation Strategies at the Member State level) and ensure coherence.

However, mainstreaming should not simply focus on introducing adaptation to climate change into legislative actions or developing guidelines. There is also a need to tackle specific bottlenecks such as inconsistencies in policy. Because climate change cuts across a broad range of policy areas, the methodology presented by McCallum and Dworak can also be used to prevent adaptation failures and ensure that policies are coherent with their objectives and provisions. It can serve as a basis for the identification of potential conflicting policy objectives and also highlight synergies that can be achieved by aligning mainstreaming efforts in several policy areas and throughout all governance levels. Thus, a roadmap for mainstreaming provides a framework in which future actions can be designed and implemented coherently even beyond the issue of climate change.

The second mainstreaming case, presented by Pütz and Kruse (13.2), focuses on mainstreaming adaptation into spatial planning instruments, as they play a key role in the well-coordinated and cost-effective adaptation to the impacts of climate change. The practical step-by-step climate change fitness assessment offers guidance for planners and supports the mainstreaming of adaptation. However, the case study stressed that climate change adaptation is not accomplished when the climate change fitness assessment is finished: ongoing monitoring of the implementation process and of changes in climatic stimuli, impacts and available knowledge is imperative. It must be verified whether or not adjustments to planning policies and instruments are needed.

However, there are cases in which the modification of existing instruments alone is insufficient to handle adaptation needs, and new instruments for implementing adaptation must be developed. These might be regulatory, economic, informational, partnering or hybrid (e.g. comprehensive adaptation strategies/plans) instruments.

The case reported by Peltonen et al. (13.3) shares the experiences of the Helsinki Metropolitan Area in the creation of new adaptation policies. It clearly shows that such metropolitan adaptation initiatives face challenges of inter-municipal and cross-sectoral coordination. Regions with long-standing inter-organisational cooperation in other fields, such as public transport planning and water and waste management, are more capable of tackling these challenges of adaptation. In addition, bridging organisations can be key in helping to address inter-municipal and cross-sectoral challenges.

Furthermore, the lessons learned from the Helsinki Metropolitan Area emphasise that linking climate change adaptation to emergency preparedness, safety and public health concerns makes the adaptation challenge more concrete and gives it a sense of urgency, also for stakeholders who might view climate change adaptation as primarily an environmental problem. In line with the other cases in this chapter, this study found that every adaptation process should consider *ex-ante* assessments of proposed adaptation measures and that the monitoring of effectiveness following implementation allows learning and continuous improvement.

In summary, the following main lessons learned can be highlighted:

- Adaptation is much broader than is often thought, as it clearly extends beyond the environmental scope. Consider this when mainstreaming adaptation by also taking social and economic sectors into account.
- Apply a systematic approach in climate-proofing policies, structures and processes to ensure coherence in mainstreaming efforts, avoid overlaps and utilise synergies.
- Establish mechanisms for coordination and capacity-building that involve all relevant stakeholders and decision-makers in order to improve the mutual understanding of adaptation needs, share knowledge and experiences and jointly develop feasible solutions.
- Spatial planning may play a key role in the development of well-coordinated and cost-effective measures to adapt to the impacts of climate change.
- Metropolitan adaptation initiatives face challenges of inter-municipal and cross-sectoral coordination. Bridging organisations are crucial in addressing these challenges.
- Regions with long-standing inter-organisational cooperation in other fields, such as public transport planning and water and waste management, can be more capable of tackling climate change adaptation. It is important to communicate adaptation through more familiar risks, such as flooding and heat waves, and by linking to concerns of safety and security.
- Consider how the special characteristics of adaptation might require new institutional arrangements. Do not hesitate to establish new institutions or collaborations if required.

Notes

1 http://www.wsl.ch/fe/wisoz/dienstleistungen/clisp_guidance/index_EN.
2 The assessment report is publicly available in Finnish at http://www.hsy.fi/seututieto/ Documents/Ilmasto/Sop.Strategia_vaikutusten%20arviointi_raportti_22_02_12. pdf.
3 See http://www.nyas.org/publications/annals/Detail.aspx?cid=ab9d0f9f-1cb1-4f21-b0c8-7607daa5dfcc.

References

Birkland, T.A. (1997) *After Disaster: Agenda Setting, Public Policy, and Focusing Events*, Washington, DC: Georgetown University Press.
Bulkeley, H. (2006) 'A changing climate for spatial planning?', *Planning Theory and Practice* 7 (2), pp. 203–214.
Davoudi, S., Crawford, J. and Mehmood, A. (ed.) (2009) *Planning for Climate Change: Strategies for Mitigation and Adaptation for Spatial Planners*, London: Earthscan.
De Bruin, K., Dellink, R.B., Ruijs, A., Bolwidt, L., van Buuren, A., Graveland, J., de Groot, R.S., Kuikman, P.J., Reinhard, S., Roetter, R.P., Tassone, V.C., Verhagen, A. and van Ierland, E.C. (2009) 'Adapting to climate change in The Netherlands: an inventory of climate adaptation options and ranking of alternatives', *Climatic Change* 95 (1–2), pp. 23–45.

Dunsire, A. (1996) 'Tipping the balance: autopoiesis and governance', *Administration Society 28* (3), pp. 299–334.

EC – European Commission (2003) *Regulation (EC) 2152/2003 Concerning Monitoring of Forests and Environmental Interactions in the Community (Forest Focus)*, Brussels. http://eur-lex.europa.eu/LexUriServ/LexUriServ.do?uri=OJ:L:2003:324:0001:0008:EN:PDF.

EC – European Commission (2007) *Directive 2007/60/EC of the European Parliament and of the Council on the Assessment and Management of Flood Risks*, Brussels. http://eur-lex.europa.eu/LexUriServ/LexUriServ.do?uri=OJ:L:2007:288:0027:0034:en:pdf.

EC – European Commission (2008) *Directive 2008/56/EC of the European Parliament and of the Council Establishing a Framework for Community Action in the Field of Marine Environmental Policy (Marine Strategy Framework Directive)*, Brussels. http://eur-lex.europa.eu/LexUriServ/LexUriServ.do?uri=OJ:L:2008:164:0019:0040:EN:PDF.

EC – European Commission (2009a) *Adapting to Climate Change: Towards a European Framework for Action*, White Paper, 147, Brussels. http://eur-lex.europa.eu/LexUriServ/LexUriServ.do?uri=COM:2009:0147:FIN:EN:PDF.

EC – European Commission (2009b) *Impact Assessment Guidelines*, SEC (2009) 92, Brussels. http://ec.europa.eu/governance/impact/commission_guidelines/docs/iag_2009_en.pdf.

EC – European Commission (2010) *EUROPE 2020 – A Strategy for Smart, Sustainable and Inclusive Growth*, Communication from the Commission, Brussels. http://ec.europa.eu/research/era/docs/en/investing-in-research-european-commission-europe-2020-2010.pdf.

EC – European Commission (2011a) *EU Regulation No 1255/2011 Establishing a Programme to Support the Further Development of an Integrated Maritime Policy*, Brussels.

EC – European Commission (EC 2011b) *Proposal for a Decision of the European Parliament and of the Council on a Union Civil Protection Mechanism*, (COM 2011) 934 final, Brussels. http://ec.europa.eu/echo/files/about/COM_2011_proposal-decision-CPMechanism_en.pdf.

EC – European Commission (EC 2011c) *Proposal for a regulation of the European Parliament and of the Council Establishing the Connecting Europe Facility*, COM (2011) 665 final, Brussels. http://ec.europa.eu/commission_2010-2014/president/news/speeches-statements/pdf/20111019_2_en.pdf.

EC – European Commission (EC 2011d) *Proposal for a Regulation of the European Parliament and of the Council on Union Guidelines for the Development of the Trans-European Transport Network*, COM (2011) 650 final, Brussels. http://eur-lex.europa.eu/LexUriServ/LexUriServ.do?uri=COM:2011:0650:FIN:EN:PDF.

EC – European Commission (2012a) *Proposal for a Decision of the European Parliament and of the Council on a General Union Environment Action Programme to 2020 "Living well, within the limits of our planet"*, COM (2012) 710 final, Brussels. http://ec.europa.eu/environment/newprg/pdf/7EAP_Proposal/en.pdf.

EC – European Commission (EC 2012b) *The Blueprint to Safeguard European Waters – Communication from the Commission*, COM (2012) 673 final, Brussels. http://eur-lex.europa.eu/LexUriServ/LexUriServ.do?uri=COM:2012:0673:FIN:EN:PDF.

EC – European Commission (2013a) *An EU Strategy on Adaptation to Climate Change*, COM (2013) 216 final, *Communication from the Commission*, Brussels. http://ec.europa.eu/clima/policies/adaptation/what/docs/com_2013_216_en.pdf.

EC – European Commission (2013b) *Technical Guidance on Integrating Climate Change Adaptation in Programmes and Investments of Cohesion Policy*, SWD (2013) 135 final, Brussels. http://ec.europa.eu/clima/policies/adaptation/what/docs/swd_2013_135_en.pdf.

EC – European Commission (2013c) *Principles and Recommendations for Integrating Climate Change Adaptation Considerations under the 2014–2020 Rural Development Programmes*, SWD

(2013) 139 final, Brussels. http://ec.europa.eu/clima/policies/adaptation/what/docs/swd_2013_139_en.pdf.

EC – European Commission (2013d) *Guidelines on Developing Adaptation Strategies*, SWD (2013) 134, Brussels. http://ec.europa.eu/clima/policies/adaptation/what/docs/swd_2013_134_en.pdf.

European Environment Agency (2012) *Urban Adaptation to Climate Change in Europe. Challenges and Opportunities for Cities Together with Supportive National and European Policies, EEA Report 2/2012*, Copenhagen. http://www.eea.europa.eu/publications/urbanadaptation-to-climate-change.

Haanpää, S., Tuusa, R. and Peltonen, L. (2009) *Ilmastonmuutoksen alueelliset sopeutumisstrategiat – READNET-hankkeen loppuraportti*. YTK:n julkaisuja C75, Espoo: Teknillinen korkeakoulu, Yhdyskuntasuunnittelun tutkimus- ja koulutuskeskus. [Regional adaptation strategies to climate change. Final report of the READNET-project, in Finnish.]

Haanpää, S., Lehtonen, S., Peltonen, L. and Talockaite, E. (2006) *Impacts of Winter Storm Gudrun of 7th–9th January 2005 and Measures Taken in Baltic Sea Region*, ASTRA project (EU INTERREG IIIB).

HSY (2010) *Pääkaupunkiseudun ilmasto muuttuu*, Sopeutumisstrategian taustaselvityksiä, Helsinki: HSY, Helsinki Region Environmental Services Authority. http://www.hsy.fi/tietoahsy/Documents/Julkaisut/3_2010_paakaupunkiseudun_ilmasto_muuttuu.pdf.

HSY (2012) *Pääkaupunkiseudun ilmastonmuutokseen sopeutumisen strategia*, Helsinki: HSY Helsinki Region Environmental Services Authority. http://www.hsy.fi/tietoahsy/Documents/Julkaisut/10_2012_paakaupunkiseudun_ilmastonmuutokseen_sopeutumisen_strategia.pdf http://www.hsy.fi/tietoahsy/Documents/Julkaisut/11_2012_Helsinki_Metropolitan_Area_Climate_Change_Adaptation_Strategy.pdf [Helsinki Metropolitan Area Climate Change Adaptation Strategy].

International Association for Impact Assessment (IAIA) (1999) *Principles of Environmental Impact Assessment. Best Practice*, IAIA, UK.

IPCC (2007) *Climate Change 2007: Impacts, Adaptation and Vulnerability. Contribution of Working Group II to the Fourth Assessment Report of the Intergovernmental Panel on Climate Change*, Cambridge: Cambridge University Press.

Juhola, S. (2010) 'Mainstreaming climate change adaptation: the case of multi-level governance in Finland', in Keskitalo, E.C.H. (ed.) *Developing Adaptation Policy and Practice in Europe: Multi-level Governance of Climate Change*, Dordrecht: Springer, pp. 149–187.

Juhola, S., Haanpää, S. and Peltonen, L. (2012) 'Regional challenges of climate change adaptation in Finland: examining the ability to adapt in the absence of national level steering', *Local Environment* 17 (6–7), pp. 629–639.

Juhola, S., Peltonen, L. and Niemi, P. (2012) 'The ability of Nordic countries to adapt to climate change: assessing adaptive capacity at the regional level', *Local Environment* 17 (6–7), pp. 717–734.

Kates, R.W., Travis, W.R. and Wilbanks, T.J. (2012) 'Transformational adaptation when incremental adaptions to climate change are insufficient', *Proceedings of the National Academy of Science*, 109 (19), pp. 7156–7161.

Kivimaa, P. and Mickwitz, P. (2009) *Making the Climate Count. Climate Policy Integration and Coherence in Finland*, The Finnish Environment 3/2009.

Lamhauge, N., Lanzi, E. and Agrawala, S. (2012) *Monitoring and Evaluation for Adaptation: Lessons from Development Co-operation Agencies*, OECD Environment Working Papers, No. 38, Paris: OECD Publishing.

Lyytimäki, J. (2011) 'Mainstreaming climate policy: the role of media coverage in Finland', *Mitigation and Adaptation Strategies for Global Change* 16 (6), pp. 649–661.

Marttila, V., Granholm, H., Laanikari, J., Yrjölä, T., Aalto, A., Heikinheimo, P., Honkatukia, J., Järvinen, H., Liski, J., Merivirta, R. and Paunio, M. (2005) *Ilmastonmuutoksen kansallinen sopeutumisstrategia*. [National climate change adaptation strategy] Maa-ja metsätalousministeriön julkaisuja 1/2005, Vammala: Vammalan Kirjapaino http://www.mmm.fi/attachments/mmm/julkaisut/julkaisusarja/5g45OUXOp/MMMjulkaisu2005_1a.pdf.

McCallum, S., Dworak, T., Prutsch, A., Kent, N., Mysiak, J., Bosello, F., Klostermann, J., Dlugolecki, A., Williams, E., König, M., Leitner, M., Miller, K., Harley, M., Smithers, R., Berglund, M., Glas, N., Romanovska, L., van de Sandt, K., Bachschmidt, R., Völler, S., Horrocks, L. (2013) *Support to the Development of the EU Strategy for Adaptation to Climate Change: Background Report to the Impact Assessment, Part I – Problem Definition, Policy Context and Assessment of Policy Options*, Environment Agency Austria, Vienna. http://ec.europa.eu/clima/policies/adaptation/what/docs/background_report_part1_en.pdf

Mickwitz, P. and Kivimaa, P. (2007) 'Evaluating policy integration – the case of policies for environmentally friendlier technological innovations', *Evaluation* 13 (1), pp. 67–85.

Mickwitz, P., Aix, F., Beck, S., Carss, D., Ferrand, N., Görg, C., Jensen, A., Kivimaa, P., Kuhlicke, C., Kuindersma, W., Máñez, M., Melanen, M., Monni, S., Branth Pedersen, A., Reinert, H. and v. Bommel, S. (2009), *Climate Policy Integration, Coherence and Governance*, PEER Report No. 2, Partnership for European Environmental Research, Helsinki.

O'Brien, K. (2012) 'Global environmental change II: from adaptation to deliberate transformation', *Progress in Human Geography* 36 (5), pp. 667–676.

OECD – DAC ENVIRONET (2008) *Strategic Environmental Assessment and Adaptation to Climate Change*. Available at: http://www.oecd.org/dataoecd/0/43/42025733.pdf.

Park, S.E., Marshall, N.A., Jakku, E., Dowd, A.M., Howden, S.M., Mendham, E., and Fleming, A. (2012) 'Informing adaptation responses to climate change through theories of transformation', *Global Environmental Change* 22 (1), pp. 115–126.

PIK – Potsdam Institute for Climate Impact Research (2012) *Turn Down the Heat – Why a 4°C Warmer World Must be Avoided*, A Report for the World Bank by the Potsdam Institute for Climate Impact Research and Climate Analytics, Washington, DC: The World Bank.

Pütz, M., Kruse, S. and Butterling, M. (2011) *Assessing the Climate Change Fitness of Spatial Planning: A Guidance for Planners*, ETC Alpine Space Project CLISP.

Roggema, R., Kabat, P. and van den Dobbelsteen, A. (2012) 'Towards a spatial planning framework for climate adaptation', *Smart and Sustainable Built Environment* 1 (1), pp. 29–58.

SEC – Scientific Expert Group Report on Climate change and sustainable development (2007) *Confronting Climate Change: Avoiding the Unmanageable and Managing the Unavoidable*, Report prepared for the United Nations Commission on Sustainable Development, Washington, DC.

Smithers, R.J., Cowan, C., Harley, M., Hopkins, J.J., Pontier, H. and Watts, O. (2008) *England Biodiversity Strategy – Climate Change Adaptation Principles, Conserving Biodiversity in a Changing Climate*, London: DEFRA.

UKCIP (2009) *A Local Climate Impacts Profile: How to Do an LCLIP*, United Kingdom Climate Impacts Programme, Oxford.

UKCIP (2010) *The UKCIP Adaptation Wizard V 3.0.*, United Kingdom Climate Impacts Programme, Oxford.

14

MONITOR AND EVALUATE SYSTEMATICALLY

Andrea Prutsch, Sabine McCallum,
Torsten Grothmann, Inke Schauser
and Rob Swart

Explanation of the guiding principle

Adaptation is an ongoing process for which systematic monitoring and periodic evaluation are core components. The main aim of monitoring and evaluation (M&E) is to keep the adaptation actions focused on their objectives and to ensure that adaptation responds to changes in the evidence base. In the absence of methods for monitoring and evaluation, opportunities for learning from experiences are minimised (Preston et al. 2011).

A monitoring system can help actors to keep track of the process of implementation and can be seen as an 'early warning system' that supports learning during the adaptation process, e.g. with regard to climate change impacts and vulnerabilities (cf. Chapter 5), to improve adaptation actions (cf. Chapter 10), to avoid maladaptation (cf. Chapter 12) or to identify new adaptation opportunities (cf. Chapter 9). In the evaluation phase, the information gathered from monitoring is systematically analysed to indicate the progress being made towards realising the prioritised adaptation options (cf. Chapter 10). Monitoring and evaluation should address the impacts of climate change (e.g. flood damage before and after adaptation), the performance of the adaptation activities and the direct and indirect costs and benefits of adaptation actions (e.g. the costs and benefits of dike construction). They also support the readjustment of decisions where needed.

Monitoring and evaluation schemes should be developed together with adaptation options before implementation. The more clearly the objectives and the targets of adaptation actions are defined, the easier they can be to monitor and evaluate. In addition, new scientific knowledge should regularly be reviewed and included in the monitoring and evaluation process. This is an important component of the adaptive management approach (cf. Chapter 11) that enables decision-makers to continuously adapt decisions and adaptation actions based on lessons learned and the best knowledge available. To guarantee objectivity, if resources allow, an independent organisation can undertake monitoring and evaluation, publishing status reports on a regular basis that can be shared with stakeholders and the public.

Chapter overview

Experiences involving a systematic approach to monitoring and evaluating are still rather uncommon. The first study in this chapter (Section 14.1) presents challenges that are inherent to monitoring and evaluation. It highlights a number of practical cases as well as emerging lessons from the steadily growing Community of Practice on M&E. In so doing, it identifies some of the key questions that evaluators need to ask in order to develop effective and systematic approaches to monitoring and evaluating climate adaptation interventions.

The use of indicators is one way that can help to describe the progress and shortcomings of adaptation actions. Some work has already been done to develop a set of process-based and outcome-based adaptation indicators for certain situations. The purpose of the second case in this chapter (14.2) is to review some fundamental concepts related to the development of adaptation indicators for monitoring the adaptation process and its outcomes. It examines a number of initiatives undertaken across Europe and draws out lessons learned as well as recommendations to support the future development and use of adaptation indicators.

The final study on M&E (14.3) presents the practical M&E approach that has been applied in France, describing the experiences gained thus far. The aim of the French monitoring scheme is to account for the degree of implementation of the proposed adaptation measures in the French adaptation action plan. The study shows that the establishment of a pragmatic indicator-based monitoring system is characterised by trial-and-error and that placing learning at its core is essential.

14.1 Asking the right questions: Monitoring and evaluating adaptation

Patrick Pringle

14.1.1 Introduction

As international donors, governments and communities invest increasing amounts of money and effort in adaptation, it has become essential to ensure the efficacy, efficiency and equity of adaptation policies and measures (Lamhauge et al. 2012). However, we are still at an early stage in terms of understanding how best to adapt to future climate change, how vulnerability can be most effectively reduced and resilience enhanced, and what the characteristics of a well-adapting society might be. Learning what works well (or not) under what circumstances and for what reasons is critical (Pringle 2011), but monitoring and evaluating adaptation policy and practice are not easy tasks (Anderson 2011). The long-term nature of climate change, its inherent complexity and the uncertainty that exists in interpreting future climate scenarios, impacts and societal responses make it difficult to answer the two key questions that underpin monitoring and evaluation efforts: 'Are we doing the right things?' and 'Are we doing things right?'

14.1.2 What makes adaptation Monitoring and Evaluation (M&E) difficult and different?

By understanding why adaptation M&E is challenging, policy-makers and M&E professionals will be better placed to design more effective approaches. The UKCIP AdaptME toolkit draws on a range of resources and experiences[1] to identify a number of 'tricky issues' inherent to climate change adaptation:

Uncertainty

There are a range of uncertainties that make it difficult to determine whether a proposed adaptation action will be effective in the longer term. These include uncertainties regarding future emissions scenarios, our understanding of climate futures described by climate models, and how society will respond to climate change in the future.

Long time-scales

In human time-scales, climate change will occur over a long period, such that we may not know whether our adaptation actions are effective for decades to come. Adaptation measures will also require different implementation time-scales. For example, the forestry sector must make decisions regarding tree species, the consequences of which will last for decades.

The counterfactual

It is difficult to know what might have happened in the absence of adaptation actions precisely because of the long time-scales and the uncertainty associated with climate change. It is particularly difficult to evaluate effectiveness in terms of the adverse impacts avoided: how do you measure the benefits of preventing a flood that, by definition, did not happen?

Attribution

Making a causal link between an adaptation action and its attributable outcomes is often complicated by long time-scales and uncertainties. This is further complicated because successful adaptation will be determined by the achievement of social, economic or environmental outcomes that are themselves subject to a range of other stressors and influences. This means that it can be difficult to understand how much difference a project or programme of activity really has made.

Defining and measuring success

Adaptation is hard to measure because of its ambiguous nature and because it refers to a process rather than an outcome, which can make the selection of appropriate and measurable indicators of adaptation difficult. Thought must be given to measuring adaptive capacity as well as adaptation actions, and there are important issues of equity and justice involved in determining whose perspective is used to define and measure success.

14.1.3 Placing learning at the core of the M&E process

As individuals, communities and nations, we are still at an early stage of our adaptation journey, thus learning what works (and what does not) and why is essential. As Spearman and McGray (2011) stress, learning must be a core function of M&E if we are to capture successful efforts at adaptation, avoid maladaptation and derive lessons regarding what works. In other words, we must start using "M&E as a potential tool for enabling learning instead of simply measuring results" (Villanueva 2011). Learning must not be confined to a specific intervention, project or programme. Figure 14.1.1 illustrates how learning and experience can 'spin out' of project-level M&E efforts. This can enable improvements to be made to these projects and can also benefit other projects, programmes and organisations. This 'shared space' between individual interventions and organisational boundaries can play a vital role in stimulating innovative and creative responses to climate change.

14.1.4 M&E at different scales

Adaptation initiatives are usually developed in response to actual or projected impacts at the local or regional levels; for example, extreme heat events in a particular city may stimulate action to increase green space through changes in urban planning. However, the decisions and policies that influence local adaptation are made at a variety of spatial scales through a multitude of governance structures that can vary from global, multinational funding mechanisms to national governments, local authorities and informal community groups. Consequently, monitoring and evaluation of adaptation is required across a range of spatial scales.

Wherever possible, there should be vertical integration of M&E outcomes so that national-level decisions are informed by local experiences and nations are

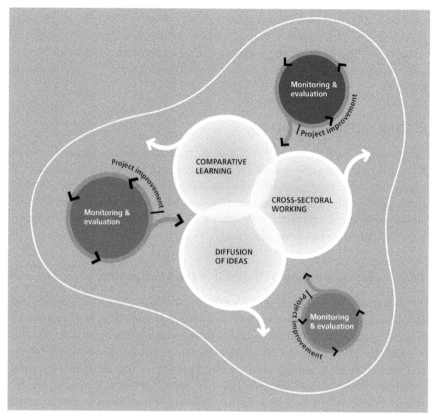

FIGURE 14.1.1 Internal and shared learning through monitoring and evaluation (source: Pringle 2011)

able to share their adaptation progress internationally. Similarly, international mechanisms should reflect local and national experiences, amending their policies and objectives in response to the lessons learned.

14.1.5 Guiding principles for adaptation M&E

As we learn more about the challenges of effective adaptation M&E, a number of guiding principles regarding adaptation M&E are emerging. The following list presents useful starting points:

- *"Think outside the project box"* (Spearman & McGray 2011). Many of the factors that make adaptation M&E difficult lie outside the project cycle. In addition, adaptation will be more effective and efficient when ideas and learning are shared and transdisciplinary approaches are developed. Think beyond the individual project and understand its wider context. These issues are reflected in the approach to monitoring and evaluation now being encouraged by

BOX 14.1.1   MEASURING PROGRESS AT THE NATIONAL LEVEL: EMERGING LESSONS
FROM THE UK

The Adaptation Sub-Committee (ASC) to the UK's Committee on Climate
Change (CCC) was established under the Climate Change Act of 2008. The
ASC has developed an adaptation assessment toolkit that has two main
components:

1. *Monitoring changes in climate risks using indicators.* These fall into three
broad categories:
i) indicators of risk (to assess changes in society's exposure and vulnerability
to weather events),
ii) indicators of adaptation action (to measure risk reduction rather than just
the action itself), and
iii) indicators of climate impact (to track realised impacts on the economy,
society and the environment).

2. *Evaluating preparedness for future climate.* This involves the analysis of
decision-making to assess whether the amount of adaptation occurring is
sufficient to address climate risks, now and in the future. Early adaptation
efforts to address priority climate risks should i) promote the uptake of
low-regrets adaptation options that will deliver benefits in whatever future
climate that unfolds and ii) ensure that decisions with long-lasting or
systemic consequences take future climate change into account.

The ASC's approach recognises that the use of indicators requires
appropriate and robust datasets that provide a reliable time series in order to
track change and differentiate variability from longer-term trends. Such data
must also be spatially disaggregated, enabling the identification of hot-spots
of risk, and provide information relevant at both local and national levels
(ASC 2012). The ASC has applied its toolkit to assess progress in adaptation
for two of the major risks: flooding and water scarcity, as described in a
recent report (ASC 2012). The report provides a useful example of how
national indicators (where possible, disaggregated to the regional or sub-
regional level) can be used alongside local case studies to present a picture
of changing risk, demonstrating how progress in adaptation is responding to
these risks and net impacts.

the German Agency for International Cooperation (GIZ) in the context of
development projects, which will also be highly relevant to the European
context.

• *Challenge assumptions.* Explore and understand the assumptions that were
made in developing and delivering the adaptation intervention; test the logic
underlying the adaptation. A 'theory of change' or adaptation logic model
requires certain assumptions to be made regarding how inputs can generate

BOX 14.1.2   EXAMPLES OF ENGAGING AND INVOLVING STAKEHOLDERS

A survey-based approach to tracking progress

In 2010, the Australian Department of Climate Change (DCCEE) and the Commonwealth Scientific and Industrial Research Organisation (CSIRO) (cf. Gardner et al. 2010) undertook a longitudinal survey of public- and private-sector organisations that were expected to play a significant part in Australia's climate change adaptation response. The purpose of the survey was to benchmark the current level of adaptation activities in the sampled organisations and to enable the future tracking of changes in adaptation activities and the attribution of observed changes to planned adaptation efforts. Such an approach places the views of stakeholders (those expected to 'do' adaptation) at the heart of the M&E process.

Learning from global experiences: Participatory monitoring and evaluation

Participatory approaches have been a central theme in the delivery of effective development projects for many years. Although the adaptation context may be very different in developing countries, those undertaking the monitoring and evaluation of climate adaptation projects in Europe can gain valuable insights from approaches developed elsewhere. CARE International's Participatory Monitoring, Evaluation, Reflection and Learning (PMERL) Manual presents a participatory methodology for monitoring and evaluating on the basis of Community-Based Adaptation (CBA) indicators. This approach enables local stakeholders to articulate their needs and provides M&E with valuable insights as to how adaptation efforts are perceived and experienced at the local level. If implemented effectively, it can provide those who are most vulnerable to climate change with a voice and a means of shaping future adaptation interventions. Although the methods outlined in the PMERL Manual may be focused on developing countries, many of the messages regarding engagement and empowerment within the M&E process are universal and highly relevant to the European context.

activities that will result in the desired outputs, outcomes and impacts. Using such models can help to reveal these assumptions, enabling M&E processes to be developed to test them. Useful examples can be found in AdaptME (Pringle 2011) and WRI/GIZ's recent guidance (Spearman & McGray 2011).

- *Don't reinvent the wheel: make use of existing systems.* Many indicators of adaptation performance may already be measured, and existing M&E systems (e.g. for sustainability or biodiversity) can be adjusted to better account for adaptation.

BOX 14.1.3   EVALUATION OF THE IMPLEMENTATION OF FINLAND'S NATIONAL
STRATEGY FOR ADAPTATION TO CLIMATE CHANGE

Finland is one of the few European countries to have already undertaken
an evaluation of its National Adaptation Strategy. The main objective of the
evaluation of the Adaptation Strategy was to determine the progress made
in various sectors since the strategy's implementation in 2005. The approach
employed was to compare the adaptation measures identified in the Strategy
to the measures actually launched in different sectors. This was achieved by
the use of surveys to determine whether and how the measures presented in
the strategy have been launched in different sectors. A preliminary indicator
of the level of adaptation attained was developed using five steps, each
of which incorporates a statement describing adaptation progress. This
approach enabled evaluators to compare progress across 15 sectors and a
number of cross-cutting themes.

- *Acknowledge trade-offs.* M&E approaches should be proportionate to the
  investment. Recognise and reflect on the trade-offs that have been made in
  the design of your M&E approach and consider whether they can be justified.
  For example, policy in the UK now reflects an acceptance that it will not be
  possible to defend *all* coastal communities using 'hard' protection measures.
  In this case, trade-offs have been made between the cost to the government,
  the practicality of protecting all existing homes from coastal erosion and
  the desire of citizens to not relocate, resulting in some communities being
  protected and others not.
- *Consider the unintended and unexpected.* There are often good reasons for
  the unexpected to happen and for things not to go as planned, especially
  when adaptation involves tackling long-term challenges with a high degree
  of uncertainty. However, it is easy to ignore the possibility of unexpected
  outcomes in an attempt to stick to a budget or in order to keep things simple.
  Try to ensure that your evaluation design is sufficiently flexible to include the
  exploration of unexpected outcomes; do not limit M&E to asking, 'Did we do
  what I said we would do?'
- *An indicator is one measure of performance; it does not provide the full picture.* Use
  quantitative indicators in conjunction with other M&E tools (including
  quantitative research, comparative studies, stakeholder engagement and
  expert elicitation) to provide a richer picture of performance.
- *Quantify where possible, but ensure that indicators are meaningful* (what do they actually
  tell us?) *and achievable* (is the data available?). Indicators should track progress,
  while objectives drive actions – not the other way around.

14.1.6 Asking the right questions

Adaptation interventions are implemented at a range of spatial scales, crossing sectoral boundaries and involving a diverse range of activities and stakeholders. Adaptation is also rarely the sole focus of M&E activity. For example, the evaluation of an irrigation scheme might consider a range of social, economic and environmental objectives in addition to considering whether the scheme will be effective under changing climatic conditions. Consequently, there is no one-size-fits-all monitoring and evaluating framework that can be applied in all circumstances. A question-based approach, such as UKCIP's AdaptME toolkit, does not seek to provide practitioners with the answers, but instead presents questions that challenge users to better understand how adaptation relates to the subject of their evaluation. Figure 14.1.2 shows how these questions can be grouped under six key themes (i.e. purpose, subject, logic and assumptions,

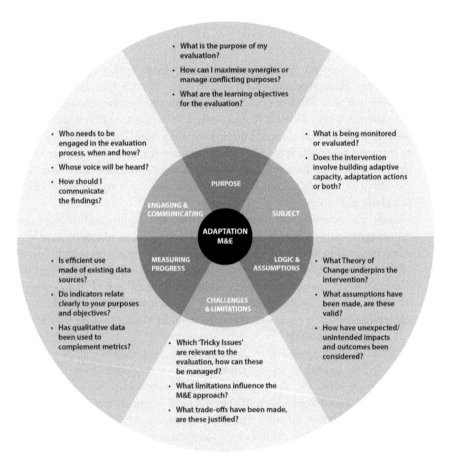

FIGURE 14.1.2 A question-based approach to adaptation M&E (source: Pringle & Street 2012)

challenges and limitations, measuring progress, engaging and communicating) and used by practitioners to inform the design of M&E approaches to adaptation.

This approach is sufficiently flexible to permit its application in existing M&E systems. It enables users to reflect on their own objectives and take into account local climate impacts, social and economic circumstances, values and beliefs.

14.1.7 Concluding thoughts

The monitoring and evaluation of adaptation is inherently challenging, and creative methodologies are required if we are to better understand our progress in adapting to climate change. Effective approaches must go beyond measuring the performance of a particular project or programme and place learning at the centre of an adaptation process characterised by continuous improvement and innovation. We are still learning about how best to monitor and evaluate adaptation; as the number of approaches, tools and frameworks grow, it will be increasingly important to refine the guiding principles for good adaptation M&E and to ensure that we are asking the right questions. This is particularly important given the fact that no one-size-fits-all solution can be applied to the wide variety of adaptation activities and the range of spatial scales at which they are delivered.

14.2 Adaptation indicators: A basis for monitoring implementation and effectiveness

Jelle van Minnen, Mike Harley, Kaj van de Sandt and Willem Ligtvoet

14.2.1 General background

Climate change adaptation seeks to address both the positive and negative implications of changing climatic conditions. Many European countries have developed or are in the process of developing national adaptation strategies, which include policies, measures and actions to increase the resilience of those systems that are vulnerable to climate change. Despite these developments, climate change adaptation as a policy area is still at an early stage of development. Nonetheless, progress has been made by a number of institutions towards establishing indicators to monitor the adaptation process and its outcomes. In this section, we examine some of these initiatives and draw out the lessons learned.

14.2.2 Adaptation indicator frameworks in Europe

Indicators are used to simplify, quantify, standardise and communicate complex and often disparate data and information. Good indicators will communicate a simplified reality in a meaningful way and should (EEA 2003):

- Be scientifically well founded
- Be based on sound statistics
- Be representative of the issue or area under consideration
- Be comparable with other indicators that describe similar areas, sectors or activities
- Show developments over time (i.e. a time interval during which changes can be seen)
- Include a reference value for comparing changes over time
- Include an explanation of causes behind trends
- Match the interest of the target audience
- Be easy to interpret
- Invite action (e.g. further investigation, practical activity)
- Be attractive to the eye and accessible.

The primary purpose of adaptation indicators is to monitor the implementation and outcomes of adaptation policies, measures and actions. Adaptation indicators can also be used to (Harley & van Minnen 2009):

- Justify and monitor funding for adaptation
- Mainstream adaptation within and between sectors
- Compare adaptation achievements across sectors, regions and countries.

A number of frameworks to facilitate the development of adaptation indicators have been produced, and some are presented as examples here. A conceptual framework produced for the European Environment Agency (EEA) (Harley et al. 2008, Harley & van Minnen 2009) considers planned adaptation to climate change impacts, capturing the 'processes' associated with the development of adaptation policies, the delivery of adaptation measures and the 'outcomes' of adaptation actions (cf. Figure 14.2.1). Process-based indicators seek to monitor the key stages that lead to choices regarding end-points or outcomes and should inform and justify decisions. Outcome-based indicators seek to explicitly monitor the end-points or outcomes of measures and actions and should focus on the long-term effectiveness of decisions. The framework has also been used to show how adaptation indicators might be defined for the biodiversity policy area (Harley & van Minnen 2010).

The Adaptation Sub-Committee (ASC) of the UK's Committee on Climate Change is charged with providing expert advice on adaptation policy for England. The ASC has developed an adaptation assessment toolkit that includes monitoring and evaluation frameworks. The monitoring framework (Figure 14.2.2) considers the biophysical impacts of climate change and weather-related hazards in addition to other contextual factors that determine exposure and sensitivity to risk. Indicators of residual risk and realised impact are used to assess the 'action' taken to reduce those impacts and risks. The framework uses existing and proxy indicators where appropriate, creating new ones when these

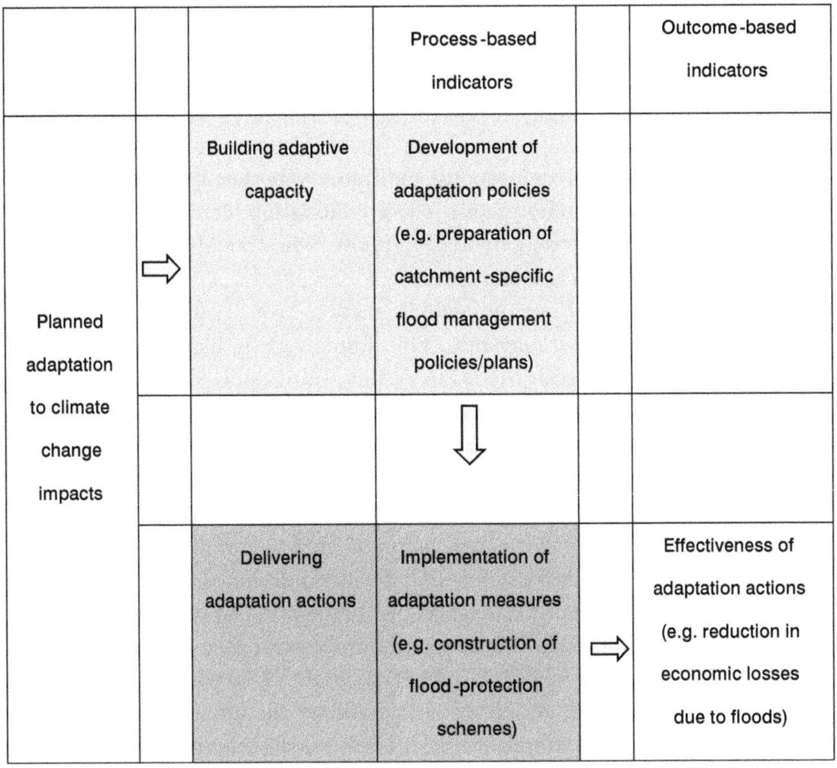

			Process-based indicators		Outcome-based indicators
Planned adaptation to climate change impacts	⇨	Building adaptive capacity	Development of adaptation policies (e.g. preparation of catchment-specific flood management policies/plans)		
			⇩		
		Delivering adaptation actions	Implementation of adaptation measures (e.g. construction of flood-protection schemes)	⇨	Effectiveness of adaptation actions (e.g. reduction in economic losses due to floods)

FIGURE 14.2.1 EEA conceptual framework for adaptation indicators (source: Harley et al. 2008, Harley & van Minnen 2009)

are not available. Given the multi-faceted nature of climate change, the ASC has concluded that it is not necessary to develop indicators for every possible impact or risk. The process of selecting a subset of impacts on which to focus is the most important and significant decision in the development of adaptation indicators. A similar indicator framework is currently being developed for the Scottish government by ClimateXChange (the government's centre of excellence on climate change).

In Germany, an indicator system to support adaptation has been developed by the Umweltbundesamt (UBA), the German Federal Environment Agency (UBA 2010, 2011). The system focuses on identifying the most important climate change impacts in 15 sectors and showing adaptation measures and actions that have already been undertaken. A number of core requirements were formulated in its development:

- The system should be based on the DPSIR (Driving Forces-Pressures-State-Impacts-Responses) approach
- It should effectively prioritise the themes to be covered (the National Adaptation Strategy provides a clear focus)

- It should be adaptive and kept up-to-date
- It should use existing data and indicators
- The indicators identified should be recognised and accepted by stakeholders and other experts.

An initial set of impact and response indicators were then selected on the basis of their relevance, data status, cause–effect relationship, comprehensibility and spatial explicitness. The development of response indicators was found to be more challenging due to significant data barriers.

A group of European Environmental Protection Agencies (EPAs) is developing an adaptation indicator framework. This framework is based on the ASC's adaptation assessment toolkit (ASC 2011), but also includes the involvement of stakeholders. When applying the framework in various sectors, it should be possible to identify the key stakeholders of relevance to the sector, the different issues to be addressed in terms of adaptation and the associated information needs for the monitoring of adaptation processes and outcomes. The inclusion of stakeholders will be important not only in the application of the framework (once completed), but also in its continuing development and the identification of specific indicators.

Finally, there have been initiatives in the Netherlands to develop monitoring systems for adaptation, including adaptation indicators. One of these is the so-called 'adaptive capacity wheel' (Gupta et al. 2010). This wheel is essentially a research tool for use with stakeholders to facilitate the assessment of adaptive capacity. It consists of six parameters (inner circle) and 22 criteria/indicators (outer circle), with scores aggregated to communicate the level of adaptive capacity (cf. Figure 14.2.3). The wheel can help assess the current level of adaptive capacity of an institution and whether and how it should be redesigned to enable more effective adaptation to climate change. Data collected from document reviews, observations and interviews are analysed, and the strengths and weaknesses of the institution

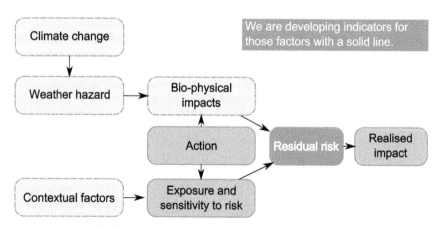

FIGURE 14.2.2 Monitoring framework in ASC adaptation assessment toolkit (source: Based on Miller et al. 2012)

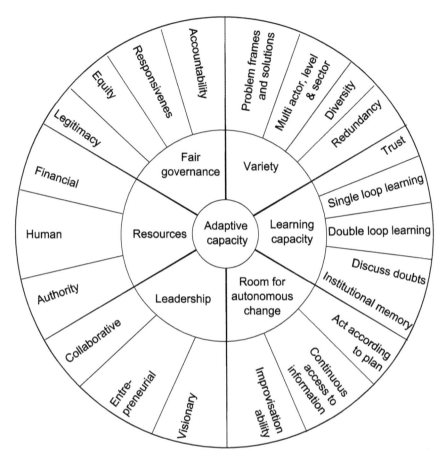

FIGURE 14.2.3 Dutch adaptive capacity wheel and scoring scheme (source: Gupta et al. 2010)

are then defined in terms of its adaptive capacity, with the outcomes being fed back to participants. The wheel has been used to assess the adaptive capacity of institutions involved in water management in the polder of de Wijde Wormer and the cities of Delft and Zaandam (Bergsma et al. 2009). The wheel can also be used to assess the strengths and weaknesses of policies in relation to adaptation and to identify obstacles to improving their adaptive capacity. Although not designed for this purpose, the systematic categorisation of the elements of adaptive capacity also provides a structure for organising a monitoring and evaluation system for those elements.

14.2.3 Concluding thoughts and lessons learned

Although some progress has been made in the identification of adaptation indicators and a number of indicator frameworks have been produced, most are at an early stage of development and have not been systematically tested to prove their feasibility and usefulness. However, some essential elements can be identified:

- Ideally, both process-based and outcome-based indicators will be needed to provide information for monitoring the implementation and effectiveness of adaptation policies, measures and actions. Process-based indicators seek to monitor the key stages that lead to choices regarding end-points or outcomes and should inform and justify decisions. Outcome-based indicators seek to monitor explicit end-points or outcomes and should focus on the long-term effectiveness of decisions.
- Process-based indicators can be used, for example, to monitor activities that relate to the availability of national, regional or sectoral adaptation strategies/action plans or activities that bring together local knowledge and experience to inform local action, including activities in situations for which no national guidance exists.
- Outcome-based indicators will be of importance in the longer term, as policy goals become more strategically targeted and the evidence that actions have reduced vulnerability and increased resilience becomes more apparent. However, it will be difficult at the outset to identify robust outcome-based indicators that will enable precise monitoring of the effectiveness of adaptation actions over long time horizons.
- The indicator(s) selected will depend on spatial and temporal factors. At larger spatial scales, process-based indicators could be more relevant than outcome-based indicators. At such scales, the key stages that lead to choices regarding end-points or outcomes and that inform and justify decisions are likely to be more significant than the end-points or outcomes themselves.
- Consequently, contextual information and indicators are needed to evaluate progress through the various stages of the adaptive management cycle and improve the monitoring of change over time. The indicator(s) selected will depend on the overall objective of an adaptation policy, measure or action (cf. Figure 14.2.1). Process- and outcome-based indicators should provide data with which actors can (i) monitor the effectiveness of adaptation actions and the impacts of these actions, (ii) review progress and the direction of travel in relation to the delivery of adaptation actions, and (iii) re-assess adaptation actions and the overall long-term vision in light of new evidence, scenarios and projections.
- The type of indicator(s) used will also depend on the nature of the policy, measure or action and on related aspirational goals or desired end-points. In the case of flood protection, for example, the overall goal of 'floods will cause no loss of life' is different from 'everyone should be protected equally from flooding', and so the types of indicators chosen would need to be correspondingly different. Similarly, the performance of a sea defence measure may be framed technically (in terms of its height) or socially (in terms of human benefit). Again, the indicators used to monitor outcomes in each

case would necessarily be quite different. Because adaptation must address a range of vulnerabilities across many sectors, an extensive set of indicators might ideally be utilised to provide a detailed assessment of effectiveness. However, it is not feasible, or indeed necessary, to capture the entire spectrum of measurable parameters. Instead, the challenge is to prioritise, combine or aggregate indicators to present an overall picture of preparedness. The indicator system developed for use in Germany is an example of a system in which such challenges have been addressed (UBA 2010).

In conclusion, Table 14.2.1 sets out a number of recommendations to support the future development and use of adaptation indicators.

TABLE 14.2.1 Recommendations to support the future development and use of adaptation indicators

Agree on focus and vision

- Working definition of successful adaptation for sector/area of responsibility in relation to the impacts to be addressed
- Policy goals for each sector/area/impact
- Nature of baseline for each sector/area/impact
- Desired outcome for each sector/area/impact

Secure and legitimise implementation

- Identify the party responsible for developing, implementing and maintaining indicators, and the resources required
- Identify collaborators and stakeholders and any resources needed to facilitate their input
- Identify a 'champion' for each indicator to advance its development
- Agree on time-scales for the development of the indicator set and time horizons for the impacts and adaptation actions under consideration

Ensure seamless integration

- Language and terminology used must be familiar and meaningful to stakeholders
- Indicators should complement existing monitoring and reporting and, where possible, use data already being collected
- Indicators should be consistent with indicator development across policy areas and sectors

Design the system to provide the information that users need

- Indicators should show the effects of and justify proposed adaptation actions
- Indicators should be linked with other evaluation systems to enable the justification of actions and investments

Guarantee credibility

- Identify and maintain relevant data sources in conjunction with sector experts
- Devise clear quality criteria for indicators and underlying data

Source: Based on Miller et al. 2012

14.3 The French approach to monitoring an adaptation plan

Bertrand Reysset

14.3.1 Background: Adaptation policy making in France

Since 2006, France has had a national adaptation strategy that provides general guidelines and raises awareness of the need to mainstream adaptation into our current decision-making processes and policies.

In order to extend this general framework approach and to follow up the national climate impact assessment undertaken in 2007–2009, it was decided to develop a detailed action plan. The French Programme Law 2009-967 made provision for the preparation of a cross-cutting National Adaptation Plan (NAP) by 2011 (French Republic 2009). The NAP has been developed in two phases. First, a multi-stakeholder environmental forum was held from December 2009 to October 2010 to determine policy and action recommendations. These recommendations were reported to the Ministry of Ecology, Sustainable Development and Energy. In the second phase, the minister asked the various other ministries to translate these recommendations into concrete actions to be implemented through a national adaptation plan as stipulated in the law.

The NAP will be gradually implemented over a five-year period (2011–2015) through 240 measures addressing 20 different sectoral or thematic areas (Ministry of Ecology 2011). The measures are of various natures, including research funding, infrastructure investments, reviews of norms and standards and institutional rearrangements.

To enable the monitoring of this complex cross-sectoral national adaptation plan, a comprehensive process was required to monitor the degree of implementation of the proposed actions. This was the first time that national adaptation planning had been carried out and thus it was considered important to learn from and improve the process as well as ensure that actions would be effective.

14.3.2 Establishing a monitoring system for the French NAP

The establishment of a monitoring system was identified as a crucial necessity at the beginning of the NAP's development and thus the design of indicators for monitoring accompanied the entire process. The primary objective was to avoid delaying the monitoring issue until after the end of the participatory planning exercise and thus to prevent the creation of an *ex-post* monitoring process to fill the gap.

During the action planning process, stakeholders were asked to frame their proposals using a common adaptation framework. One part of this framework asked for the specification of monitoring indicators and an implementation timetable for each action.

Furthermore, at the outset of the action planning process, it was explained that an implementation report would be made publicly available each year.

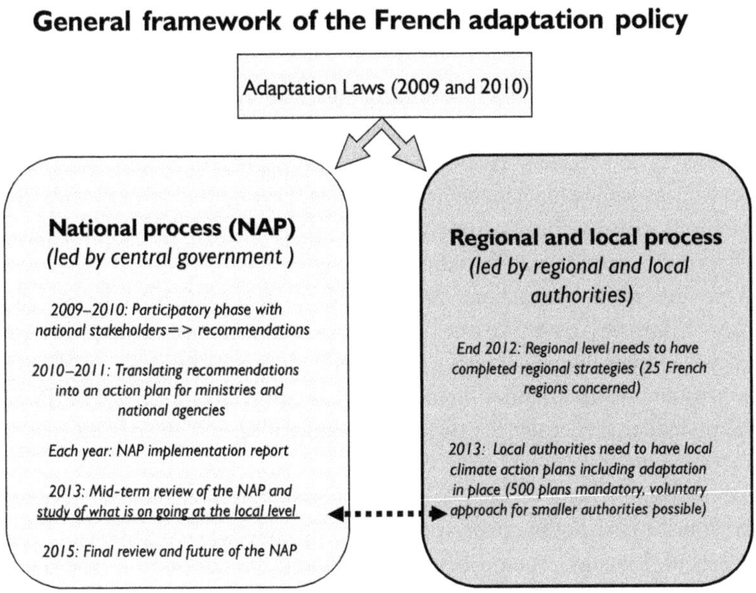

General framework of the French adaptation policy

Adaptation Laws (2009 and 2010)

National process (NAP)
(led by central government)

2009–2010: Participatory phase with national stakeholders=> recommendations

2010–2011: Translating recommendations into an action plan for ministries and national agencies

Each year: NAP implementation report

2013: Mid-term review of the NAP and study of what is on going at the local level

2015: Final review and future of the NAP

Regional and local process
(led by regional and local authorities)

End 2012: Regional level needs to have completed regional strategies (25 French regions concerned)

2013: Local authorities need to have local climate action plans including adaptation in place (500 plans mandatory, voluntary approach for smaller authorities possible)

FIGURE 14.3.1   Interlinkage of the national adaptation process with regional and local processes (source: author's research)

Thus, contributors and organisations in charge of the future implementation of adaptation actions (i.e. action leaders) were motivated to follow the common procedure. The monitoring indicators have thus been designed collectively by the stakeholders and the action leaders responsible for implementing the measures.

The French NAP is focused on national issues; as a result, the monitoring is undertaken at the national level and is based on indicators available nationally (e.g. data that are collected at the national level). For example, in areas with adaptation-friendly agricultural practices, these data might include the percentage of leakage in water networks or expenditures dedicated to specific actions.

However, the national adaptation policy framework makes provisions not only for national planning (i.e. NAP), but also for adaptation at the regional and local level (cf. Figure 14.3.1). This process is led by regional and local authorities, respectively, who have been asked to put adaptation plans into place. Although these processes are independent from the national level, a general progress report concerning actions planned at regional and local levels is requested in the NAP and will be carried out in 2013. This will allow comparison of the implementation progress of adaptation at the national level versus regional and local levels.

Despite the fact that national and local/regional planning are two independent processes, several actions implemented through the NAP provide tools for sub-national decision-makers to facilitate their activities (vulnerability assessment guidelines, downscaled climate data, etc.).

14.3.3 Practical experiences with planning and implementing a monitoring scheme

The aim of the French NAP is to implement adaptation actions on the ground and to define how and when such actions will be implemented. Thus, the NAP functions as a roadmap that strives for very concrete actions and takes into account the monitoring issue at an early stage. During the participatory process, a common framework has been developed to ensure that every action proposal will have its own implementation indicators.

Drawing from the actions and indicators recommended by stakeholders, various ministries have set out a roster of actions that fit into this common framework, identifying implementation indicators and result-oriented indicators as well as an implementation timetable. There are 20 sectors in the NAP; for each, a theme leader (generally the directorate-general of a ministry) has followed the same procedure proposed by the NAP.

Because each theme leader was responsible for identifying one set of monitoring indicators for each action (to be approved by the general NAP coordination, i.e. the Ministry of Ecology), the process has facilitated the adoption of the monitoring dimension of this first NAP by different ministries.

As touched upon earlier, the organiser first considered trying to identify a single indicator for each sector, i.e. an indicator that could quantify how much a sector has successfully adapted due to the implementation of an action. However, this approach proved to be infeasible: in practice, it is difficult to assign one indicator to represent the implementation of several actions carried out within a sector, as the indicator cannot be specific enough, and the actions may not be comparable.

Considering the timeframe of the NAP development process (18 months), it was decided to focus on indicators for the implementation of action and to plan a comprehensive review at the mid-term (2013) and at the end of the NAP in 2015 (Ministry of Ecology 2011).

However, some of the current NAP indicators can sometimes appear to be too simplistic. To enable a cross-sectoral monitoring process, a pragmatic approach was necessary for this first NAP: 20 thematic leaders had to be able to easily handle indicators without adding any overwhelming burdens. Thus, the NAP process has remained dynamic and will likely benefit from the experience of this first exercise and from that of other initiatives that are currently trying to develop such 'degree of adaptation' indicators.

To assist in the task of elaborating indicators, the monitoring timetable and all the indicators are spelled out within the NAP official document. Concrete examples of indicators include an action of the NAP in the health sector to increase awareness of skin cancer overseas. The indicator for this action is the availability of a UV index-based warning system for French overseas regions. The warning system has been functioning since 2012, and thus the action is considered to be implemented. The implementation indicator (the availability of the system) can be checked each year.

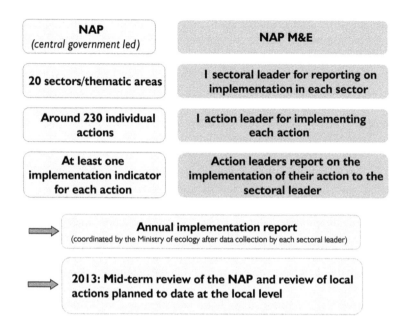

FIGURE 14.3.2   Monitoring and evaluation framework of the French NAP (source: author's research)

Another action in the NAP is the identification of building codes for transport infrastructure that are sensitive to climate parameters; for example, road building standards differ between northern and southern regions of France, and thus it is relevant to determine how climate change interacts with the current climate-based standards. The implementation indicator for the identification of climate-sensitive standards is the number of codes identified. To date, 80 codes have been identified. Another action will review whether these climate-sensitive codes require any update in order to better cope with climate change impacts.

In general, this first NAP will test the French capacity to implement adaptation actions and also test the efficiency of the flexible monitoring system. Each year, an annual implementation progress report is produced. These reports are sent to the national multi-stakeholder environmental forum to describe what has been accomplished during the year to implement their initial recommendations. In February 2012, the first draft implementation progress report for the first 6 months of implementation was published. It confirmed that the process was operational, showing that 80% of all planned actions were already on track. However, the report only focuses on monitoring progress in implementation, not on evaluating the level or effectiveness of adaptation.

The annual implementation report for the year 2012 was due by mid-2013; this was followed by a mid-term review that provided first evaluative elements (cf. Figure 14.3.2).

Between July 2011 and February 2012, only a few of the NAP actions were completed. For instance, downscaled climate projections (8 km range) are now freely available and easily accessible to help support adaptation initiatives. This is the main result of one of the first actions of the NAP, for which the monitoring indicators are:

1   the output of the action (i.e. effective freely accessible data) and
2   follow-up statistics for the downloading of the data.

Although it would be interesting to know how many actions/projects will be implemented, encouraged or informed by these climate projections, we are currently unable to track or evaluate this. Another example of the limited capacity of evaluating impacts can be found for another action implemented in 2012: a climate change adaptation module has now been included in the training curriculum for forestry technicians. Although we can monitor the number of students trained, this does not enable us to quantify the net impact of this measure in terms of adaptation, because we will have to wait for many years to check its effectiveness. This is another example of an indicator that is useful but not complex enough to capture the adaptive benefit resulting from the training action. It is a process-based indicator rather than an outcome-based indicator, which is easier to implement (see also Section 14.2).

14.3.4 Concluding thoughts

The monitoring component of the National Adaptation Plan (NAP) was anticipated at the beginning of the plan's development. Stakeholders and leaders in the adaptation actions were responsible for designing specific indicators to monitor each action. The indicator-oriented framework applied during the NAP's development process has led to the design of more concrete and result-oriented actions that strive for adaptation. This pragmatic approach enables a regular update on the implementation of the NAP and facilitates the mainstreaming of adaptation within the numerous sub-themes.

It is clear that although adaptation measures (such as knowledge development or shifts in afforestation) can be monitored in the short term (implementation, outputs or process), their final impacts and efficiency cannot be evaluated for 10 or 20 years to come. This indicates that it might not be possible to monitor policy processes and actual vulnerability reduction at the same time. Building indicators that are able to track or quantify the degree and effectiveness of adaptation is a challenge for both policy-makers and researchers.

As monitoring indicators are currently insufficient to objectively evaluate the effectiveness of actions in improving the level of climate adaptation, the French NAP will be evaluated through a mid-term review process, but without the help of any comprehensive set of adaptation-specific indicators. The evaluation will focus more on the relevance of actions and processes in enhancing adaptation;

it will not quantify how much the overall level of adaptation has increased. Two evaluations were scheduled: the mid-term review in 2013 and the final review in 2015 at the end of this first NAP.

Measuring the baseline degree of adaptation or the improvement in the level of adaptation using the same metric for different sectors is theoretically the best way to monitor an adaptation policy; however, more research is still needed. By prioritising low-regrets options and knowledge development actions, we can overcome this lack of a true adaptation metric and the problems associated with uncertainty, at least enough to enable a first set of actions to be implemented.

14.4   Lessons learned for the monitoring and evaluation of adaptation

Andrea Prutsch

All three of the cases presented in this chapter confirm that the monitoring and evaluation (M&E) of adaptation is still at an early stage of development, and thus learning what works in M&E and what does not is essential. As Pringle describes in Section 14.1, our M&E efforts should be underpinned by two main questions: 'Are we doing the right things?' and 'Are we doing things right?'

In order to be successful in M&E, a number of challenges specific to climate change adaptation need to be acknowledged. Pringle (14.1) summarises these 'tricky issues' as follows: uncertainty, long time-scales, the counterfactual, attribution, and defining and measuring success. Interestingly, most of these challenges also had to be addressed in France when an approach was developed to monitor the progress of the adaptation action plan (see 14.3).

As adaptation activities are very context-specific (e.g. dependent on the spatial scale, sectors, stakeholders involved, socio-economic situation), there is no one-size-fits-all monitoring and evaluation framework. Consequently, Pringle presents a question-based checklist for M&E of adaptation processes with the aim of helping users to better understand their adaptation situations and informing the design of M&E approaches. He also recommends a number of guiding principles for an M&E system: (i) think outside the project box, (ii) challenge assumptions, (iii) make use of existing systems, (iv) acknowledge trade-offs, (v) engage and involve, (vi) consider the unintended and unexpected, (vii) an indicator is only one measure of performance and (viii) quantify where possible, but ensure that indicators are meaningful and achievable.

In the study presented by van Minnen et al. (14.2), a number of existing indicator-based frameworks for M&E are highlighted in order to showcase the broad range of possible approaches that can be taken. In Section 14.3, Reysset summarises the key steps of the French M&E framework and the lessons learned.

The use of indicators is central in the French approach in order to monitor the implementation of the NAP and the outcomes of adaptation policies, measures and actions. When this approach is linked to the classification of process-based and outcome-based indicators provided by van Minnen et al., the French indicators may eventually seek to monitor the outcomes of policies (the level of adaptation), but in practice they initially focus on process-oriented indicators. The French M&E requirements have accompanied the entire development of the national adaption action plan. The French experiences show that the use of a single indicator for each sector does not provide sufficient information. As van Minnen et al. assert, an extensive set of indicators might ideally be utilised to provide a detailed assessment of effectiveness in adaptation. However, different sets of indicators can reduce comparability across sectors. The French M&E system has developed a set of indicators that reflect the progress made by specific actions within a specific sector. The involvement of stakeholders (who are also in charge of developing the adaptation actions) was crucial for the development of the indicators. Thus, the development of the French adaptation action plan and the development of the monitoring indicators were carried out by the same group of stakeholders. In the case of France, this pragmatic approach led to a result-oriented monitoring system. The question now is how to evaluate the effectiveness of the adaptation actions. This will be the next step to be addressed in the French M&E process.

In general, the practical cases included in Chapter 14 focus on monitoring, touching only briefly on the issue of evaluation. As a result, the lessons learned presented above also focus on monitoring. With respect to the evaluation of adaptation effectiveness, experiences from similar policy fields (such as evaluation attempts in the field of sustainable development) might help to foster progress on this issue in adaptation.

In summary, the following lessons emerge from the three cases:

- Monitoring and especially evaluation of adaptation are still at early stages of development but nevertheless form a critical part of the adaptation process.
- A flexible, question-based approach can be a valuable starting point, and learning must be a core function of M&E.
- Although practical experience in the use of indicators is still limited, some progress has been made with regard to the production of conceptual frameworks for their development. Lessons have been learned about the typology of adaptation indicators, their temporal and spatial scales and aspirational goals, their limitations and the need for additional conceptual information. Actors should be aware that one indicator does not provide a complete picture of adaptation.
- Quantitative indicators should be used in conjunction with other M&E tools (including quantitative research, comparative studies, stakeholder engagement and expert elicitation) to provide a richer representation of performance.
- The design of a monitoring concept needs to be anticipated at the beginning of adaptation action planning. As adaptation planning is often participatory, the combination of action planning and monitoring can save time and increase stakeholders' willingness to engage.

- Efforts should be made to learn from early ongoing attempts at the implementation of monitoring systems elsewhere, taking into consideration their temporal and spatial scales, their specific objectives (why do we monitor and for whom?) and their limitations.

Note

1 A list of key resources can be found in the AdaptME toolkit and at the end of this chapter.

References

Adaptation Sub-Committee (ASC) (2011) *Adapting to climate change in the UK – measuring progress*, Adaptation Sub-Committee Progress Report, Committee on Climate Change, London.

Adaptation Sub-Committee (ASC) (2012) *Climate change – is the UK preparing for flooding and water scarcity?* London.

Adaption Sub-Commitee (ASC) (2013) *Managing the land in a changing climate*, London.

Anderson, S. (2011) *Assessing the effectiveness of climate adaptation*, Lessons from adaptation in practice, IIED, London.

Bergsma, E., Gupta, J. and Jong, P. (2009) *Institutions for climate change – case study on individual responsibility in adaptive capacity*. Working Document 7, Report number W09/10, Institute for Environmental Studies, Amsterdam.

EEA (2003) *Environmental indicators: typology and use in reporting*. Copenhagen.

French Republic (2009) 'Programme Law n°2009-967 relating to the implementation of the Grenelle Environnement', Official Journal of the French Republic n°0179, 5 August 2009, p. 13031, Article 42.

Gardner, J., Parsons, R. and Paxton, G. (2010) *Adaptation benchmarking survey: initial report*, Climate Adaptation Flagship Working Paper No. 4, CSIRO. http://www.csiro.au/resources/CAF-working-papers.html.

Gupta, J., Termeer, K., Klostermann, J., Meijerin, S., van den Brink, M., Jong, P., Nooteboom, S. and Bergsma, E. (2010) 'The adaptive capacity wheel: a method to assess the inherent characteristics of institutions to enable the adaptive capacity of society', *Environmental Science & Policy* 13 (6), pp. 459–471.

Harley, M. and van Minnen, J.G. (2009) *Development of adaptation indicators, European Topic Centre on Air and Climate Change*, Technical Paper 2009/6, European Environment Agency. http://acm.eionet.europa.eu/docs//ETCACC_TP_2009_6_Adaptation_Indicators.pdf.

Harley, M. and van Minnen, J.G. (2010) *Adaptation indicators for biodiversity, European Topic Centre on Air and Climate Change*, Technical Paper 2010/15, European Environment Agency. http://acm.eionet.europa.eu/reports/docs//ETCACC_TP_2010_15_Adap_Ind_Biodiv.pdf.

Harley, M., Horrocks, L., Hodgson, N. and van Minnen, J.G. (2008) *Climate change vulnerability and adaptation indicators, European Topic Centre on Air and Climate Change*, Technical Paper 2008/9, European Environment Agency. http://acm.eionet.europa.eu/docs/ETCACC_TP_2008_9_CCvuln_adapt_indicators.pdf.

Lamhauge, N., Lanzi, E. and Agrawala, S. (2012) *Monitoring and evaluation for adaptation: lessons from development co-operation agencies*, OECD Environment Working Papers, No. 38, OECD Publishing.

Miller, K., Harley, M., Kent, N. and Beckmann, K. (2012) *Climate change adaptation-related indicators*, SNIFFER Report ER23. http://www.sniffer.org.uk/files/7213/4631/9842/Sniffer_ER23_Phase_1_final_report.pdf.

Ministry of Ecology, Sustainable Development, Transport and Housing (2011) *National climate change adaptation plan*, Paris.

Preston, B., Westaway, R. and Yuen, E.J. (2011) 'Climate adaptation planning in practice: an evaluation of adaptation plans from three developed nations', *Mitigation and Adaptation Strategies for Global Change* 16 (4), pp. 407–438.

Pringle, P. (2011) *AdaptME, adaptation monitoring and evaluation toolkit*, UKCIP, Oxford.

Pringle, P. and Street, R. (2012) *AdaptME toolkit, adaptation monitoring & evaluation*, a poster presentation for the Climate Adaptation Futures conference, Arizona.

Spearman, M. and McGray, H. (2011) *Making adaptation count; concepts and options for monitoring and evaluation of climate change adaptation*, World Resources Institute.

UBA (2010) *Establishment of an indicator concept for the German strategy on adaptation to climate change*, Report of the German Federal Environment Agency, Dessau-Roßlau, Germany.

UBA (2011) *Entwicklung eines Indikatorensystems für die Deutsche Anpassungsstrategie an den Klimawandel (DAS)*, Dessau-Roßlau, Germany.

Villanueva, P.S. (2011) *Learning to ADAPT: monitoring and evaluation approaches in climate change adaptation and disaster risk reduction – challenges, gaps and ways forward*, SCR Discussion Paper 9, Strengthening Climate Resilience.

PART III

State of art outside Europe

15

ADAPTATION EXPERIENCES IN OTHER INDUSTRIALISED COUNTRIES AND IN DEVELOPING COUNTRIES

Rob Swart

This manual has been developed around the ten principles that Prutsch et al. (2010) proposed to the European Environment Agency to support the development of adaptation policies and processes. In the preceding chapters, these principles were illustrated by providing examples of how they have been implemented in practice in a number of case studies contributed by experts from across Europe. The collection of these examples was facilitated by the emerging network of experts, policy makers and practitioners that has emerged since the turn of the millennium. Not only do these actors work on adaptation in their own countries, they also meet in various research programmes and policy fora at the European level. At the time of writing, a majority of European nations had adopted National Adaptation Strategies that provide the context and incentive for stakeholders at regional and local levels to assess their vulnerabilities and develop and implement adaptation policies and measures. At the European level, in April 2013 a European strategy was published that addresses important issues such as the integration of climate change adaptation in European policies, the development of a sound knowledge base to support decision-making and a set of institutional arrangements to further the development of adaptation policies at the transnational and European levels involving various stakeholders. Europe is a very diverse continent that features a variety of climatic regions, landscapes, cultural values, governance structures and economic circumstances. Therefore, we believe that the principles and lessons from the cases in this manual are applicable to a broad set of circumstances in industrialised countries and perhaps also in the developing world. However, it is interesting to explore the situation with respect to climate change adaptation in other regions of the world to identify similarities and (in particular) differences that may provide additional insights and ideas for the effective development and implementation of adaptation policies that will reduce vulnerability to climate change. Chapter 15 discusses the current state of affairs with regard to adaptation in the United States, Australia and Japan, and assesses the applicability of the principles laid out in this manual to developing regions.

In their summary of the state of adaptation policy development in the United States in Section 15.1, Richard Moss, Tom Wilbanks and Sherry Wright observe that due to limited stakeholder interest, vocal scepticism on the part of an active and politically potent minority, uncertainties regarding relatively near-term risks and perceptions that national responses may have negative consequences for some important economic sectors, climate change adaptation is not a high-priority issue in the US. It remains to be seen whether this will fundamentally change as a result of the increased attention devoted to climate change in the aftermath of the hurricane disaster that struck the country's east coast in 2012. At the same time, at the state and local levels, an increasing number of adaptation activities can be observed, and (as in Europe) a growing adaptation research community actively seeks to develop an adaptation science. Even taking into account the budget restrictions that have followed the global economic crisis, US research programmes remain sufficiently large to produce results that will be of great interest to other world regions. The section provides suggestions for further explorations of the rich research and policy developments in the US.

The contribution from Australia focuses on the processes of co-creation and co-ownership of knowledge that have been developed to address the serious problems related to Australia's vulnerability to climate change. Rohan Nelson and Mark Howden argue in Section 15.2 that such an approach can help to shift the policy focus from climate science and impact analysis to the implementation of economic policies and regulatory reforms to engenderment of action. Because also in Europe it is increasingly acknowledged that a classical top-down approach is unlikely to be the most effective way to develop adaptation policies, these Australian experiences that could help remove the institutional barriers to effective climate change responses could be particularly useful in furthering the adaptation debate.

The Japanese central government is very aware of the country's own vulnerability and that of the surrounding Asian countries with which the nation has key economic ties. Thus, as Makoto Tamura, Kazuya Yasuhara, Nobuo Shirai and Mitsuru Tanaka discuss in their contribution in Section 15.3, it is no surprise that Japan has embarked upon the systematic development of national adaptation policies and research programmes, following the concept of 'wise adaptation'. *Wise adaptation* on the one hand encompasses an effective, efficient and flexible approach that takes into account deep uncertainty, and on the other hand considers adaptation in the wider context of sustainability, contributing to other social goals. The Japanese combination of a top-down scientific approach, based on downscaled projections of climate change and impacts, with bottom-up incorporation of local stakeholders in developing strategies, appears to be similar to the situation in Europe and further exchanges of experiences can be fruitful.

Is the situation in developing countries really so different from industrialised countries, or might our principles be valid also in developing world circumstances? Kelly Levin explores this question in Section 15.4, and finds that this manual's principles basically represent common sense and are generically applicable. However, their specific application in a developing country context can be different. Even if Europe is currently in an economic and financial crisis which makes the development and implementation of adaptation policies difficult, still resources are often a lesser problem than in most developing countries. So, in a developing country context, strengthening of available resources is even more important than in industrialised countries. She particularly points at the issue of thresholds which may be exceeded earlier in developing countries due to lower coping capacities, but which may also be more relevant than currently recognised in industrialised countries.

In the last contribution to this chapter, Section 15.5, Britta Horstman discusses to what extent knowledge developed in industrialised countries is transferable to developing countries, taking into account the fact that many aspects are likely to be context-dependent. She observes an extensive overlap between adaptation to climate change and the development process itself. It is important to recognise the main characteristics of adaptation before judging transferability. This includes the core questions (Who or what adapts? Adaptation to what? How does adaptation occur?) but also the institutional dimension, time and spatial scales, the recognition of uncertainties and reflexivity and learning. These dimensions suggest that it is not problem definitions or solutions that can be transferred, but rather the knowledge that enables the problem definitions and the identification of solutions. As investments in adaptation research and measures in the industrialised world can be expected to increase, it is important that the knowledge developed is also used (to the extent applicable) to support developing countries in reducing their far greater vulnerability. The author offers suggestions on how this can be done.

15.1 The state of the art in adaptation science, policy and practice in the United States

Richard H. Moss, Thomas J. Wilbanks and Sherry B. Wright

15.1.1 Introduction

Adaptation science, policy and practice in the United States present a very mixed picture, combining a great deal of diverse activity at state and local scales and across various sectors of the economy with a limited and tentative response on the part of the national government. With fiscal crises looming on the horizon and continued political controversy over climate change, progress in developing a linked set of national and state/local policies and programmes is likely to be limited and episodic. This article provides a brief overview of processes in the United States that are building awareness of the need for adaptation, of the efforts at different levels of government and in the private sector to plan and implement adaptation measures and policies, of the development of policies that support adaptation and, finally, of the state of science supporting adaptation.

15.1.2 Toward an awareness of adaptation needs and options

Awareness of the need for climate change adaptation in the United States has emerged from a variety of driving forces. First, a series of national climate change assessments – combined with the periodic reports issued by the Intergovernmental Panel on Climate Change (IPCC) and assessments by non-governmental institutions such as the Union of Concerned Scientists – have raised awareness of climate change risks and the possible consequences. Second, in part in response to requests from the US Congress, the US National Academy of Sciences/National Research Council has issued recent reports that have been widely discussed in the media. Third, a wide variety of parties at the local and state scales and in some key economic sectors have become active in promoting awareness of climate change issues (see below).

Periodic national climate change assessments are a major avenue for building awareness of climate change risks and impacts in the United States. These are mandated under the Global Change Research Act, passed by Congress in 1990, and to date, two national assessments and a large number of scientific synthesis reports have been completed by the US global change research enterprise.

- The first US national assessment, conducted between 1997 and 2001,[1] was a large national effort that viewed assessment as a two-way communication process between researchers and potential users (Clark & Dickson 2003), with the creation of a network of scientists and decision makers considered to be as important as the actual report itself.
- During the George W. Bush presidencies (2001–2009), the national global change research effort produced 20 'state of science' assessments on a range of topics of interest to climate change impact and adaptation discussions in the US.[2]
- Toward the end of this period, in June 2009, a synthesis of these reports and other emerging research-based information was issued under the title *Global Climate Change Impacts in the United States.*[3]

The US has currently conducted its third national climate assessment,[4] with a report due in 2013, and is engaged in the creation of sustainable infrastructure that will facilitate an ongoing distributed national assessment process over the longer run. For the first time, the assessment includes a review of adaptation in the United States, drawing upon approximately 30 submissions from independent organisations as well as information from chapters covering regions and economic sectors (cf. Bierbaum et al. 2012 for a description of the process and preliminary conclusions of the national assessment adaptation chapter, due to be completed in 2013).

The US National Academy of Sciences/National Research Council (NAS/NRC) has also been active in reviewing, advising and contributing to raising awareness of climate change vulnerabilities and research needs. Most notably, the NAS/NRC produced four panel reports and an overview report on *America's Climate Choices*, including a panel report, *Adapting to the Impacts of Climate Change* (NRC

2010a), that reviews a large number of possible options for seven sectors but notes that the knowledge base on adaptation options is still very limited. These findings and the bottom-up interests and concerns of stakeholders were the main focus of the National Climate Adaptation Summit in Washington, DC. This Forum endorsed the NAS/NRC report's recommendation that adaptation be pursued not as a responsibility of the federal government alone but as a partnership across scales and sectors, with each partner contributing what it does best.

Other important sources of insight on climate change impacts in the US include several regional assessments (e.g. *Arctic Climate Impact Assessment* (ACIA, 2008), *Northeast Climate Impacts Assessment* (UCS 2007), *Climate Change Impacts in the US Southeast* (Stratus Consulting 2010)), the North American chapter in the IPCC Fourth Assessment Report from Working Group II (IPCC 2007a) and the recently completed IPCC report on *Managing the Risks of Extreme Events and Disasters to Advance Climate Change Adaptation* (IPCC 2012).

15.1.3 Adaptation planning and implementation in the United States

As a growing number and range of climate impacts can be attributed to climate change (IPCC 2007a, USGCRP 2009) and global GHG emissions continue to grow (IPCC 2007b), there is increasing recognition in the US that climate change impacts in many regions, sectors and systems are inevitable. Indeed, even if emissions were drastically curtailed and greenhouse gas concentrations stabilised, serious impacts will still occur (NRC 2011). Adaptation is therefore no longer a response option; it is a necessity. The question for some in the US, not yet fully answered, is: 'So what? The climate has always changed – why is this different?' For others who acknowledge the risks, the critical questions are more action-oriented and practical, often focusing on the identification of measures that will support adaptation but also serve additional objectives, and thus are justifiable for reasons other than climate change.

In general, the emerging practice in the US involves using a risk management approach as a strategy to prepare for an uncertain future, usually including considerable stakeholder participation: consideration of a range of possible future climate conditions in order to assess vulnerabilities (rather than basing strategies and actions on a particular set of projections), identification of adaptation options that will reduce vulnerabilities, implementation of adaptation measures that make sense now and the development of more adaptive planning strategies for the future. A common theme is an emphasis on co-benefits, i.e. actions that reduce vulnerabilities to climate change impacts in the longer term but that also provide benefits for economic and social development in the short term; this is usually a requirement for sustained public and policy-maker support.

Most of the climate change adaptation planning and practice in the United States to date, both public and private, has occurred at scales below that of the national government. A growing number of state and municipal governments are evaluating potential adaptation needs and developing adaptation plans. Adaptation planning is also in its early stages in the federal agencies, although what can be done is currently limited by political controversy, especially in Congress, which authorises and appropriates funds for government operations. Adaptation activities in private-sector firms have been considerable, especially in the sectors of insurance, tourism and agriculture and in especially vulnerable regions; however, in most cases these efforts are not publicly reported and are difficult to document and summarise. Firms in the energy and agricultural sectors are particularly sensitive to climate fluctuations and are aware of the need for adaptation. Examples include Entergy, a multi-state electric utility in the Gulf Coast region of the US, which has produced a major report on 'Building a Resilient Energy Gulf Coast' intended to inform sensible approaches to addressing climate change vulnerabilities in its region that could threaten the sustainability of their customer base,[5] and ConAgra Foods, which focuses on supply chain risk and minimising reliance on single-sourced ingredients and suppliers to avoid disruption in production (reported in Bierbaum et al. 2012).

State/municipal actions

Climate change adaptation in the US is still very much in the planning stage, developing bottom-up rather than top-down. Many state governments see a need for broad-scale adaptation planning. Thirteen states out of 50 are developing adaptation plans (Alaska, California, Connecticut, Florida, Maine, Maryland, Massachusetts, New Hampshire, New York, Oregon, Virginia, Washington and Wisconsin); eight other states (Arizona, Colorado, Iowa, Michigan, North Carolina, South Carolina, Utah and Vermont) cite the need for adaptation planning within their climate action plans. In some cases, acceptance of the need for adaptation planning transcends disagreements about the *causes* of change; for example, former Alaska Governor Sarah Palin ramped up that state's planning efforts (C2ES 2011). However, although planning has started, implementation is lagging; in addition, many measures are not explicitly or solely intended for adaptation, although they are certainly relevant to it.

There has been activity at the municipal level in cities of varying sizes. Of nearly 300 local US governments that responded to a global survey, 59% indicated they were engaged in some form of adaptation planning (Carmin et al. 2012). While this figure seems promising, by comparison, 95% of Latin American and 92% of Canadian cities were engaged (admittedly, sample sizes for these regions were much smaller and thus could be non-representative). In the US, large cities including New York, Chicago and Seattle (King County, WA) are leading the way in developing and beginning to implement comprehensive adaptation plans. New York City and King County (Seattle) have both conducted major impact assessments. PlaNYC is a sustainability and growth management initiative for New York City that includes a Climate Change Adaptation Task Force that has encouraged the participation of a variety of private and public-sector stakeholders.

The Task Force uses a common set of climate change projections associated with the concept of 'acceptable levels of risk' to stimulate discussions among the stakeholders. This risk management approach has led to the development of adaptation strategies embedded within a continuing process of reassessment, including requirements for periodic progress reports and revisions of climate adaptation plans (NRC 2010a). King County formed a climate change adaptation team to build capacities in county departments for the consideration of climate change vulnerabilities in plans, policies and investment decisions. Climate change is now incorporated into all country emergency response and resource management plans, consolidated in a King County Climate Plan that includes commitments to both mitigation and adaptation. Many state and local adaptation plans are referenced on the Georgetown University Law Center's climate change website.[6] ICLEI USA[7] and the US Conference of Mayors Climate Protection Agreement provide useful resources for state and local governments (cf. C2ES 2011, Moser 2009, Georgetown Climate Center[8]).

Executive orders and early results at the federal level

Because of the ongoing controversy in the US Congress over climate change, new national initiatives to foster mitigation or even adaptation seem unlikely. However, the Obama administration has initiated several modest efforts that seem possible within existing Congressional authorisations and funding. The orientation of these initial measures is practical and focuses on encouraging federal agencies to conduct initial vulnerability assessments and develop and incorporate adaptation plans in their operating plans.

- Executive Order 13514 is a broad mandate for agencies to incorporate sustainability objectives (e.g. energy and water use efficiency) into their routine planning and operations. It includes a directive to agencies to "evaluate agency climate change risks and vulnerabilities to manage the effects of climate change on the agency's operations and mission in both the short and long term". Implementing instructions to the agencies regarding adaptation planning were issued in early 2011, with deadlines for the submission of initial assessments and plans in 2012.
- An Adaptation Task Force has been established, with leadership from senior officials at the Office of Science and Technology Policy and the National Oceanographic and Atmospheric Administration (NOAA).[9] This task force is mandated to provide guidance to federal adaptation actions.
- The Securities and Exchange Commission (SEC), which provides oversight for stock and options exchanges and other investment markets, issued guidelines in early 2010 to corporations stating that companies must analyse their potential risks and opportunities related to climate change and disclose these risks to investors. A report reviewing compliance found generally weak climate risks disclosure to date by businesses (Ceres 2011).

15.1.4 Adaptation policy in the United States

As of mid-2011, climate change is not a high-priority issue for US policy. Climate change actions are characterised by limited stakeholder interest, vocal scepticism on the part of an active and politically potent minority, uncertainties regarding relatively near-term risks and perceptions that effective national responses might have negative consequences for some important economic sectors. Meanwhile, the nation is struggling with its national debt combined with the cost of government programs at national, state and local levels, engagement in wars in Iraq, Afghanistan and elsewhere, a high level of unemployment, controversies over social welfare support programmes, hot-button issues such as immigration policy, divisive issues related to social values and other problems that demand decision-maker attention and action. Even for leaders who care about climate change as a global and national challenge, other issues have tended to overshadow effective action in response to that challenge.

The role of the federal government in a truly *national* adaptation strategy (in contrast to a *federal* government strategy) would include facilitating cooperation and collaboration across different levels of government and between the government and other parties; providing technical and scientific resources to the range of parties carrying out vulnerability assessments and adaptation planning; re-examining policies that may inhibit adaptation (such as land management guidelines that assume climate stationarity, e.g. the legal definition of the 100-year floodplain); supporting scientific research on climate change adaptation to strengthen risk management by providing better options, better information about options and better tools for informing decisions; and practising adaptation in its own federal government agencies and programmes (NRC 2010a). However, especially in the current national policy atmosphere of climate change scepticism and budget-cutting, progress in most of these directions has been slow.

Efforts are underway to evaluate both the potential for adaptation to reduce climate change impact costs and stresses and adaptation benefits in relation to costs; however, these efforts are still at a very early stage. Highly relevant actions to develop or promote policy are taking place at state and local levels of government, in a number of economic sectors and within the non-governmental organisation (NGO) community, but many of the same impediments that block action at the federal level also present challenges in some states and municipalities – for instance, existing land use and development regulations and codes.[10] Some states, for example, have banned the use of sea-level rise projections in coastal planning out of concern for the implications of such projections for property values and coastal development.

Of course, the US continues to participate actively in processes associated with the UN Framework Convention on Climate Change (UNFCCC), including meetings of the Conference of Parties (COP), and the current leadership of the executive branch supports the pursuit of global agreements on moderating greenhouse gas (GHG) emissions and assisting climate change adaptation in developing countries.

15.1.5 Adaptation science in the United States

Adaptation *research* is increasingly active in the United States, although sources of financial support are often limited; however, adaptation *science* is at best still in a formative stage.

The research community in the United States focused on adaptation is vibrant and growing. Based at universities, other non-profit institutions and private-sector organisations, this community has identified an evolving research agenda focused on boundary organisations that facilitate interactions between researchers and users. This agenda includes the identification and testing of adaptation options; the development of methods and tools to support risk assessment, communication and management; the development and use of techniques such as scenario planning; the incorporation of uncertainty; the creation of information systems for integrating diverse and diffuse experiences with adaptation at local to regional scales; transformative adaptation; evaluation of the effectiveness of adaptation options; and monitoring, learning and iterative decision-making. The importance of this research agenda is reflected in recent developments at the US National Academy of Science, which include a lively discussion of adaptation science needs in the science panel report on *America's Climate Choices* (NRC 2010b) and the establishment of a new Board on Environmental Change and Society that will build on past efforts focused on the 'human dimensions' of global environmental change.

However, adaptation science – as a robust body of science with validated bodies of theory, agreement on standards for the analysis and treatment of issues such as uncertainty, and time series of fundamental data as the basis for evaluating outcomes – is an aim rather than a reality. In the United States, the federal government entity responsible for climate change science is the US Global Change Research Program. Its mandate is to produce research to "understand, assess, predict, and respond" to global change. Historically, this programme has focused on the first portion of this mandate; more attention to 'response' is needed. The recently released 10-year strategic research plan for the USGCRP[11] articulates an agenda that seeks to accomplish this by increasing the attention devoted to developing and conveying science in a way that is useful to users. This plan articulates a new vision and mission with four goals: "advance science, inform decisions, conduct sustained assessments, and communicate and educate". To support implementation, several interagency working groups have been established, including a working group on 'adaptation science' that is tasked with identifying the needs, existing resources and research gaps for adaptation decision-making. There are many remaining challenges to the realisation of this vision, including a lack of clarity regarding how it will be funded and how strongly federal agencies will be encouraged to take it seriously.

15.1.6 Implications and next steps

The principles for adaptation enumerated in this volume would generally be applicable to practice in the United States, although practitioners and researchers alike are hesitant to draw broad conclusions about good practice. An emphasis on iterative risk management, including sustained engagement with decision makers, joint framing and fact finding, the consideration of co-benefits and social learning are recognised as important components of effective adaptation practice in the United States, but the nascent state of adaptation science precludes agreement on principles at this stage.

Predicting the potential evolution of climate policy and progress in adaptation is a risky business anywhere, and especially in the United States, where public opinion has fluctuated and policy has oscillated. A risk-averse adaptation strategy would seem to involve the assumption that very little federal government policy on reducing emissions will be enacted, placing a much larger burden on adaptation policy as the viable approach to managing the risks posed by anthropogenic climate change. Forces are at play – including a growing body of observations that supports arguments that climate change is real – that suggest that attention to adaptation will continue to increase. Whether this will translate into commitments to reduce US emissions remains to be seen.

15.2 Doing it right: Getting science into Australian adaptation policy

Rohan Nelson and Mark Howden

15.2.1 Short general background

Participatory approaches to science-policy engagement are widely acknowledged to be best practice by policy-oriented researchers in many fields. Underlying the well-intentioned rhetoric of use-inspired co-development is the common-sense idea that if science is going to be relevant to policy, scientists need to work closely with policy advisers. This is especially true of climate change adaptation, which is an activity conducted throughout society often beyond the observation and influence of either policy advisers or scientists (Jasanoff & Wynne 1998). The ubiquitous nature of adaptation means that there are far more adaptation activities occurring throughout society than policy advisers can or need to monitor and evaluate in terms of potential policy intervention. This, in turn, means that there are far more policy processes than adaptation researchers can support, regardless of how interdisciplinary or holistic we attempt to be.

So how do well-intentioned interdisciplinary groups of researchers faced with this complexity prioritise which adaptation policy processes to engage with? The argument that this decision itself can only be made through participatory engagement is circular logic that does not resolve the problem. In initiating science-policy research, researchers decide which issues to tackle, which policy makers to work with and what methods to employ. In this section, we contend that determining priorities by means of a transparent and well-thought-out set of criteria is likely to honour the use-inspired co-development objectives of participatory action research much more than decisions made using a set of (often implicit) researcher-held values would.

The purpose of this manual is to look beyond the principles of participatory action research (which most adaptation researchers readily accept), exploring how these principles can be implemented. The idea that a gaping disconnect is emerging between rhetoric and practice in adaptation research was central to the 2012 Climate Adaptation Futures conference. Over a period of two decades, we have sought to implement variations of the ten guiding principles around which this manual is based. We advocate (and hope to demonstrate) a degree of *humility* in reporting on our efforts to support the reform of adaptation policy in Australia. By humility, we mean acknowledging the important but limited role that science can hope to play in building adaptive capacity amongst policy makers and society more generally (Meinke et al. 2009).

15.2.2 The basic idea

Climate adaptation research has always needed to involve stakeholders, and the practice of doing so has led to a core set of operating principles guiding decision-centred, outcome-oriented interventions based on participatory action research (e.g. Howden et al. 2007, Meinke et al. 2009). These principles include the need to start with the values, aspirations, opportunities and decision-contexts of stakeholders (including the institutional environment). The next step is to assess the decisions that are within the existing decision horizon, the sensitivity of these decisions and their systems of interest regarding possible climate-related impacts; the stakeholders' existing capacities to deal with these impacts and how this capacity can be increased; the array of adaptation options and their consequences, risks and costs; the implementation path for selected adaptations (including facilitating factors, barriers and limits); and monitoring to determine what works, what does not work and why for active adaptive management (e.g. Howden et al. 2007, Meinke et al. 2009).

Participatory action research for climate adaptation focuses on co-design, co-production of knowledge and co-ownership of the results, strongly prioritising the knowledge held by stakeholders (e.g. Cash and Buizer 2005:11204). It uses risk management but is not limited by narrow interpretations of this approach (Nelson et al. 2007), embracing robust approaches to managing uncertainty (Lempert and Collins 2007) and opening up options for more systemic and transformational change (Howden et al. 2010:101). This approach attempts to align the supply of and demand for adaptation science (Sarewitz and Pielke 2007:5) and differs strongly from the 'standard' linear model that places climate change projections and their methodological uncertainty as the primary science focus (e.g. NRC 2009).

15.2.3 Australian adaptation policy

Most engagement with stakeholders in adaptation research in Australia has focused on engagement with industry and community groups to support adaptation action. There has been much less engagement with policy stakeholders.

There has been a slow, incomplete, but growing awareness in Australian public-policy circles of the need to refocus adaptation policy away from incremental reaction to climate impacts towards more holistic anticipation of uncertain future climate conditions. A feature of climate policy in Australia has been that the government agencies that formulate policy also fund the science of climate prediction. Consequently, it has been difficult for policy advisers with a predisposition towards impact science to shift from a linear model of targeted action based on prediction to a broader set of public-policy measures to manage climate uncertainty throughout society.

15.2.4 Principles in action

A small, informal, open community of practice has been working to refocus adaptation policy. This applied policy-research community has relied on an eclectic mix of activities funded by diverse and often oblique sources to try to bridge the inevitable gap forming between research dominated by predictive climate science and policy agencies struggling to manage the more holistic dimensions of uncertainty. Some of the evolving goals of this community have been to 1) build a community of practice in outcome-oriented adaptation science, 2) promote cultural change and build capacity within research agencies in order to shift the perception of adaptation as primarily a science problem toward an understanding of adaptation as a social and economic public-policy issue and 3) rewrite the policy narrative on adaptation from a science-gap model (knowledge will lead to action) to an economic policy agenda (government interventions targeted at overcoming market and other barriers to adaptation).

A key ongoing issue for this policy-research community has been the prioritisation of policy processes. The criteria used by researchers to choose which of these policy processes to engage with are usually implicit, part of a difficult-to-define set of higher-order professional skills that includes personal attributes, preferences, institutional mandates and networks. Institutional pressures such as requirements for promotion and funding often play a pervasive role. These institutional pressures are important because of the high and often under-acknowledged overheads associated with participatory action research. The time and effort required for this type of close engagement must be traded off against incentives and opportunities to focus on more conventional research activities.

An early activity within this community of practice was the explicit definition of a set of criteria for prioritising and guiding interactions with policy agencies (see Table 15.2.1). These criteria were used to implement the first principle on which this manual is based – *Initiate adaptation, ensure commitment and management* –

TABLE 15.2.1   Criteria used for selecting policy processes in Australian adaptation research

Criteria	Indicators
Relevance	*National* – The policy process concerns an issue of existing or emerging community and policy debate and is consistent with national research priorities. *Corporate* – The policy process involves issues consistent with organisational priorities, mandates and funding opportunities. *Research team* – The policy process is consistent with the interests and capabilities of the researchers involved and their evolving research programmes.
Tractability	*Outcomes* – Successful engagement by the research team with the policy process has the potential to deliver improved outcomes for Australian society by increasing the quality of the information used to make policy decisions. *Institutional support* – Engagement between the research team and the policy advisers concerned is supported by both the policy and research institutions involved. *Effective relationships* – There is sufficient trust, goodwill and compatibility between the values, professional cultures and ideologies of the researchers and policy advisers to enable an effective working relationship to be established.
Capability	*Adoption* – The operating culture of policy advisers supports the kinds of strategic thinking that enable the effective use of contributions from scientists. *Impact* – Scientists have skills, knowledge and approaches with the potential to significantly enhance policy processes.
Benefits	*Incentives for policy advisers* – Policy advisers have clear incentives to engage with the research team in terms of their own policy success, interests and job satisfaction, and successful engagement is recognised and rewarded by their peers and institutions. *Incentives for scientists* – Scientists have clear incentives to engage with policy advisers in terms of their own research success, interests and job satisfaction, and successful engagement is recognised and rewarded by their peers and institutions. *Research demand* – Engaging with the policy process is likely to significantly increase the quality of research demanded and applied by policy advisers, as judged against outcomes.

and we suggest that success with all of the subsequent principles is dependent on getting this one right.

15.2.5 Results: Policy outcomes

The results of this approach and the application of the principles learned from stakeholder engagement were immediate and substantial in terms of increased demand for our climate adaptation science. While coincident contextual factors always confound the direct attribution of policy outcomes, engagement based on these principles resulted in scientific publications that foreshadowed significant changes to, for example, drought policy (e.g. Nelson et al. 2008). As a result of this policy-relevant research, one of us (RN) was seconded by a national policy agency as an economic adviser to build capacity and reformulate adaptation policy from a science-gap to a more holistic economic policy agenda.

This secondment as a policy leader for adaptation enabled a partnership to form between policy advisers and adaptation researchers in our community of practice. A mechanism for facilitating these interactions – the Adaptation Policy Forum – was designed to meet the agency's capacity-building and policy-development objectives. The goals of this forum were 1) to explore the concept and nature of adaptation as a public-policy problem and what this means for the role of government(s), 2) to promote a policy design culture in which ideas are encouraged and knowledge is shared to support innovation and 3) to build the skills and capacity of staff to pursue the broader public-policy dimensions of adaptation policy.

The Adaptation Policy Forum was a series of weekly lunchtime discussions in the lead-up to the development of new policy proposals. Topics for discussion emerged from a dynamic evaluation of the capacity required for policy development relative to the existing capacity of staff. A snow-balling effect took place, with additional topics added to meet the emerging demand for discussion on previously unanticipated topics. Topics included the limits of predictive modelling to support policy development, practical approaches to managing uncertainty, the economics of adaptation, and policy options for overcoming the market failures that contribute to maladaptation. The forum was supported by a collaborative intranet workspace that served as a structured and easily accessible knowledge management system on diverse topics relevant to adaptation policy.

The forum contributed directly to the agency's response to a major enquiry on 'Barriers to Effective Climate Change Adaptation' conducted by the Australian Productivity Commission.

The need for humility on the part of the research community was brought into focus during the selection of topics and speakers to act as discussion leaders in the forum. The community needed to acknowledge that not all researchers have the personal attributes and capabilities necessary to engage with and support policy advisers. The criteria used to select the science presenters to the Adaptation Policy Forum centred on their 1) scientific credibility, 2) capacity to understand the policy context and to frame their science accordingly, 3) preference to engage in outcome-focused discussion, 4) ability to operate beyond the bounds of their immediate organisational affiliations and personal interests and 5) willingness and ability to communicate with and build the capacity of the policy community with whom they were engaging. Scientists can have egos, and these can get in the way of effective engagement.

Science-policy roles such as this one present a risk to the employment security of the individuals concerned, since staff that span the boundaries between research and policy agencies may not always be perceived as serving the core interests of either. Career objectives need to be carefully weighed against impact objectives. For example, the Adaptation Policy Forum was one of the first activities discontinued when funding for government adaptation programmes was not renewed. The risks to the individuals involved can be managed to some extent through existing mechanisms, such as secondments that provide ongoing employment security. A much rarer alternative is for this type of science-policy dialogue to be funded as a core activity by either science or policy agencies. Employment security can also be provided informally through employment opportunities created across communities of practice that value boundary-spanning activities.

15.2.6 Concluding thoughts

Our experiences with the Adaptation Policy Forum highlight the advantages of setting clear goals for science-policy interactions and including explicit criteria for the selection of adaptation issues and policy partners (cf. Figure 15.2.1).

We hope our experience demonstrates that 1) applying the principles of participatory action research laid out in this manual to adaptation policy can lead to successful outcomes, but also that 2) reaching beyond the rhetoric to actually implement these principles is challenging. Because adaptation is an activity that people throughout society engage in, adaptation research is meaningless unless it guides and supports these activities. Identifying which activities we can most usefully support requires explicit judgements to be made – with humility – regarding our own capacity, the utility of our science to decision makers and our willingness to accept risk in order to take advantage of opportunities to have an impact. A seemingly never-ending challenge is the development of alternative policy narratives that bypass the barrier to adaptation posed by an institutionalised focus on predictive certainty in order to support climate adaptation.

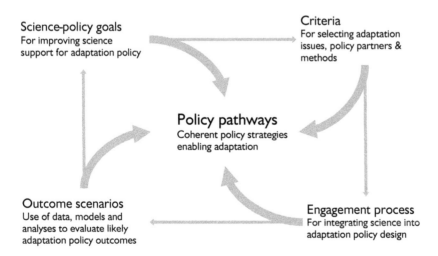

FIGURE 15.2.1   Participatory development of adaptation policy

Partnerships between researchers and policy advisers can lead to innovative mechanisms for linking science with policy development. These science-policy interactions are not solely or even primarily about providing tools or information: the relationship element is the critical aspect. Creative mechanisms are required to build relationships and facilitate the transfer of knowledge and ideas across the science-policy divide and to support the people who do this well. The success of informal efforts suggests significant benefits could flow from institutional support for boundary-spanning activities.

15.3 Wise adaptation to climate change: Japan

Makoto Tamura, Kazuya Yasuhara, Nobuo Shirai and Mitsuru Tanaka

15.3.1 Short general background

The impacts of climate change are already evident in every sector in Japan. Undoubtedly, threats due to climate change will increase in the 21st century (e.g. Project Team for Comprehensive Projection of Climate Change Impacts 2008). In addition, the tremendous damage caused by the 2011 Great East Japan earthquake and tsunami has spurred a renewed call for a safe and secure society. Japan needs to re-examine its disaster prevention standards and risk management policies as well as its approach to climate adaptation.

To address the challenges emerging from climate change, there has been a succession of efforts, ranging from comprehensive research on climate change to the development of policies for adaptation as well as mitigation. Japan has recognised the significance of impacts of climate change and is institutionalising adaptation strategies, such as the 4th Basic Environmental Plan revised in 2012.

The Japanese government and research communities have studied the projected impacts of climate change and investigated adaptation (MOE 2008, 2010, MEXT et al. 2009, CSTP 2010). The Ministry of Environment (MOE 2008) has proposed the notion of 'wise adaptation' based on the following concepts:

1   As a policy development operating under uncertainty, adaptation should be based on an effective, efficient and flexible approach. In spite of significant progress in research and policies, uncertainties still exist in the projections of future climate change, the associated impacts and social trends. Because rapid advances are occurring in global observations and climate projections, adaptation plans that can be revised every few years should be adopted rather than unchangeable measures.

2   Wise adaptation considers climate change adaptation in the wider context of sustainability and the well-being of society. Adaptation to climate change should contribute to other social goals, such as mitigation of climate change,

the creation of an environmentally friendly, safe and secure society and responses to the aging of society.

This section explains the current state of Japanese research and policies on climate adaptation and discusses the roles of scientific and regional communities.

15.3.2 Practical experiences with planning and implementing the addressed aspects of adaptation

Dual approach for climate change adaptation

Two main approaches for adaptation have been developed to address adaptation in Japan: a top-down scientific approach and a bottom-up regional approach (cf. Figure 15.3.1). A scientific or top-down approach involves long-term adaptation measures for both national and local governments, including climate projections, their downscaling, vulnerability assessments and planning of adaptation (cf. e.g. Klein et al. 1999). In contrast, community-based adaptation addresses challenges at the local level and seeks to promote the participation of stakeholders (especially in the local community) in the process of formulating adaptive measures. This regional or bottom-up approach encourages people in the community to recognise their future risks and to participate in the planning and implementation of adaptation measures (cf., e.g. Adger et al. 2005).

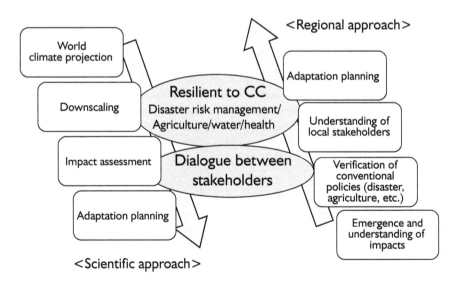

FIGURE 15.3.1 Dual approach for climate change adaptation

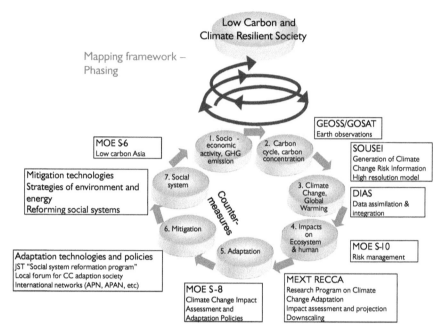

FIGURE 15.3.2 Research map of climate change issues in Japan (source: Adapted from Hiramatsu 2011)

15.3.3 Scientific approach

Climate change issues can be divided into seven categories, ranging from the science of climate change to society's responses (Hiramatsu 2011). The Japanese scientific community has studied climate change for many years, as is shown in Figure 15.3.2 (see, e.g. GOSAT, DIAS, SOUSEI and RECCA). Japan's research on climate impacts and adaptation is summarised in that figure.

Methods for climate projections and impact assessments have been developed. The work of the Project Team for the Comprehensive Projection of Climate Change Impacts (2008, 2009), the so-called Japan MOE S-4 project, includes assessments in which various impacts have been integrated using both physical and economic values. This reveals the relationships between integrated impacts on Japan and stabilisation concentrations, and suggests the level at which climate stabilisation must be achieved.

Nevertheless, there are remaining issues in these assessments; for example, the uncertainty in the projections and the vulnerability assessment, which includes socio-economic conditions and adaptive capacities in the region. Under the successor project to the MOE S-4, the MOE S-8 project (2010–14), an advanced impact/adaptation assessment model, is being developed with the goal of obtaining a more detailed understanding of the physical and economic impacts by field as well as assessments of vulnerability and the effects of implementing adaptation and planning.

In order to implement adaptation at the local level, more detailed projections (such as the 1 km or 100 m mesh downscaling) and support for adaptation planning are required, as the impact of climate change varies depending on location. Consequently, the scientific community should obtain this knowledge through monitoring, projections and assessment methods, whereas local authorities should understand the impacts, make plans and implement adaptation.

15.3.4 Regional approach

As a follow-up to the recommendation of wise adaptation, MOE (2010) has developed a guideline for local authorities. An important proposal in this guideline is a two-stage approach for short- and long-term adaptation, which serves as a framework to make the policy more flexible. The short and long terms are defined as less than 10 years and 10 to 100 years, respectively. Because impacts due to climate change are already occurring, the initiation of responses to prevent or mitigate their damage is urgently required. In this approach, major actions include the expansion of monitoring for early warning and the strengthening of existing policies in each sector and region. In long-term adaptation, the major goal is to enhance society's adaptive capacity through the periodic assessment of future risks, reducing vulnerability and strengthening resilience.

This guideline is designed primarily for local authorities to enable them to take the initiative in formulating responses to climate impacts. However, they may find it difficult to start adaptation planning due to the lack of detailed projections of the future climate and its impacts in their own regions, in addition to the lack of policy guidance. Therefore, the guideline is also formulated so that local authorities can overcome these barriers to adaptation planning and implementation. The guideline proposes two tracks: Track A for full planning and implementation, and Track B for simplified initial steps to follow in the early stages, as shown in Figure 15.3.3.

In this context, the Japan Local Forum for Climate Change Adaptation Society (Forum CCAS) was established in 2011 by the MOE S-8 project. The forum seeks to disseminate the results of its research on adaptation and climate change impacts (e.g. around 10 km mesh downscaling) and information on policies related to climate change adaptation, in addition to facilitating information exchanges. In 2011, the Forum CCAS started trial studies in cooperation with Nagano prefecture and created the 'Guideline for local adaptation', which discusses suitable plans and the implementation process for their local situation, referring to MOE (2010). Other prefectures with characteristics different from Nagano are also planning local adaptation strategies. It is noteworthy that some local governments (rather than the national government) are already undertaking initiatives in collaboration with scientific communities.

These trials have also revealed the challenges that local governments may encounter in their adaptation process. For instance, (1) the decision-making

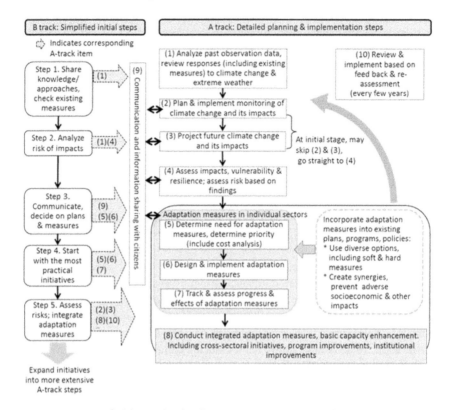

FIGURE 15.3.3   Guidance for local governments in the initiation of adaptation planning and implementation (source: MOE 2010)

procedure under uncertain projections for climate impacts is not fully developed, (2) the introduction of adaptation may be regarded as a barrier for mitigation because some governments have already implemented mitigation and (3) adaptation cannot be well defined because some 'hidden adaptations' have already been implemented in conventional sectors (such as high temperature-tolerant plants in agriculture and disaster risk management for heavy rainfall).

However, most of the obstacles for local communities originate from a lack of understanding of adaptation. It is therefore necessary to (1) institute laws or national strategies for local adaptation planning, (2) identify an adaptation framework based on model studies and (3) provide scientific knowledge that meets local needs.

15.3.5 Further issues

The Asia-Pacific region is among the most vulnerable to climate change due to its high level of poverty, extreme dependence on primary sectors such as agriculture, its rapidly growing unplanned urbanisation and poor developmental governance. In addition, climate change represents a serious risk due to the extensive interdependency of Japan's economy with those of Asia-Pacific countries.

In order to validate and promote adaptation strategies in conjunction with mitigation, the MOE S-8 project studies assessment methods for vulnerability and adaptation effectiveness, applying them to case studies in the Mekong and Ganges Deltas. In addition, APN and APAN are attempting to construct a platform by collecting good practices and creating a database for adaptation.

15.3.6 Concluding thoughts

This section has discussed the current state of Japanese scientific and regional communities with regard to climate change, focusing on the concept of 'wise adaptation'. Wise adaptation to climate change, when implemented in the local context, will result in changes in regional and social modalities and possibly in the simultaneous resolution of other problems. Therefore, the high-level objective is the adoption of a flexible response to climate change in order to sustain a dynamic and vital society. To formulate effective adaptation, it is important to integrate both scientific and regional approaches.

Acknowledgements

This section was supported by the Environment Research and Technology Development Fund (S-8: 2011–2014) of the Ministry of the Environment (Project leader: Nobuo Mimura, Ibaraki University) and a Grant-in-Aid for Scientific Research to Makoto Tamura (Ibaraki University).

Appendix

APAN (Asia-Pacific Adaptation Network): http://www.apan-gan.net

APN (Asia-Pacific Network for Global Change Research): http://www.apn-gcr. org

DIAS (Data Integration and Analysis System): http://www.editoria.u-tokyo. ac.jp/dias/

GOSAT (Greenhouse gases Observing SATellite): http://www.gosat.nies.go.jp/

MOE S4: http://www.nies.go.jp/s4_impact/

MOE S8: http://www.nies.go.jp/s8_project/

RECCA (Research Program on Climate Change Adaptation): http://www.mext-isacc.jp/

SOUSEI (Program for risk information on climate change): http://www.jamstec. go.jp/sousei/index.html

15.4 Principles for adaptation decision-making in developing countries

Kelly Levin

15.4.1 Introduction

This manual posits that ten principles can be used to guide the effective efforts of developed countries to adapt to a changing climate. What are the implications for developing countries? To be sure, there are many similarities between the risks that a changing climate poses for developed and developing countries. However, the way in which these risks manifest themselves on the ground are quite different, given that developing countries are more vulnerable to climate change and have fewer resources to adapt. This section[12] examines the ten principles put forward in the manual as they relate to developing countries' adaptation. It is organised as follows. First, each principle is examined for its implications for developing countries; second, additional principles that are critical to developing country adaptation are proposed with accompanying rationales.

15.4.2 The implications of the ten principles for developing country adaptation

Explore potential climate change impacts and vulnerabilities and identify priority concerns

Taking vulnerability into account when designing interventions can help governments target the poor and vulnerable members of society and link climate adaptation to mainstream development efforts. An important first step for adaptation planning is the integration of climate risks and vulnerability into existing decision-making tools. For example, many countries require environmental assessments before implementing programmes or approving projects with potentially significant environmental impacts in order to evaluate the possible consequences for ecosystems and communities. These assessments could be strengthened by requiring additional consideration of potential climate risks and the associated vulnerability of populations and ecosystems. For example, South Africa has successfully pioneered such an approach. Government mapping of the climate risks posed to ecosystems has been integrated into environmental impact assessments of proposed land-use planning activities (World Resources Institute 2011:117). As a result, local planners have been steered away from activities that would increase the vulnerability of ecosystems to climate change.

Initiate adaptation, ensure commitment and management

Communities are more likely to help implement adaptation efforts if they understand the value of participating and if the initiatives address their needs. For example, in Mali, government-led activities to integrate climate risks into the agriculture sector included working with farmers to develop climatological profiles for their individual fields, not just the surrounding agricultural land (World Resources Institute 2011:64). Farmers received 10-day bulletins on hydrological, meteorological, agricultural and pest conditions, with more specific, downscaled data delivered every one to three days. This 'farmer-centred approach' earned the support of rural communities, as they could apply this information to crop production activities and there was a clear pay-off.

Build knowledge and awareness

Useful information is the lifeblood of effective decision-making, and this is particularly true for adaptation, given the uncertainty that surrounds future impacts. Many developing countries lack the basic infrastructure and capacity to gather and distribute the information necessary for decision-making in a changing climate (Bueti & Faulkner 2010). Investments in weather-monitoring stations and other data-collection systems are of great value for collecting information about changes on the ground and for providing the raw data for forecasts.

In addition, at times information dissemination must be rapid, especially in cases of extreme events, and capable of reaching remote communities and informing government strategies. Innovative information and communication technologies, such as the use of radio and weather bulletins (as mentioned above), exist in developing countries and demonstrate potential for supplying the information necessary to protect vulnerable lives, livelihoods and ecosystems.

One instrument that can increase knowledge and awareness that holds promise, especially in situations characterised by high uncertainty, are scenario exercises (World Resources Institute 2011:98). The scenarios provide alternative views of the future against which plans and policies can be tested; the exercises bring stakeholders together to talk through acceptable levels of risk and policy objectives. These exercises can clarify the resources and conditions required and the trade-offs involved in a nation's effective preparation for future change. Role-playing can be used to facilitate dialogue by encouraging participants to adopt new positions, explore a range of solutions and seek consensus.

Identify and cooperate with relevant stakeholders

Engaging the public can help define adaptation needs, leading to better outcomes and – given that resources are limited in developing countries – informing government thinking on how to choose among various priorities and define acceptable levels of risk. For example, the importance of public engagement was a key lesson that emerged from efforts in Nepal to prevent deadly glacial lake outburst floods, which can be triggered when the water dammed by a glacier or a moraine is released (World Resources Institute 2011:46). When initial interactions between government officials and the community lapsed, there was little follow-up. Thus, public engagement should be viewed as an asset for governments, not a burden to be avoided.

Explore a wide spectrum of adaptation options

After identifying target areas for action and acceptable levels of risks, the next step for decision makers is to choose the options that will be implemented in order to address the risk at hand. This will require additional information on the effectiveness, costs and benefits and consequences of each course of action, as well as on public attitudes toward and the constraints involved in their implementation. An assessment of possible interventions, including those that address both prevention and preparedness and hard and soft options (among others), will need to be conducted. There are several databases of interventions[13] that can assist in this regard. In addition, in order not to waste scarce financial and other resources, decision makers should determine whether the measures are flexible enough to withstand changing climatic conditions and whether they should be implemented in advance or be adopted quickly as needed.

One important tool that can provide a useful aid as planners look to take incremental short-term action that does not preclude the option of taking more aggressive action later is a decision route map. This tool is particularly useful for contending with long-term uncertainty and advancing robust adaptation strategies over years or decades. The map identifies different options for policymakers depending on how climate impacts unfold.

Prioritise adaptation options

For resource-constrained developing countries that are already struggling to meet their basic needs, taking measures to address climate risks will inevitably require setting priorities. The integration of climate risks into commonly used tools, such as environmental impact assessments, cost-benefit analyses and multi-criteria analyses, and innovative tools, such as the scenario planning exercises mentioned above, can play a vital role in helping officials navigate the complexities of decision-making in a changing climate.

It is important to note, however, that costs and benefits alone will not provide government officials with the complete picture they need to make equitable and effective decisions. In Namibia and Tanzania, for example, economic analyses concluded that climate change could impact GDP by less than 1%, but equity and distributional analyses revealed that the burdens of these impacts would fall heavily on smallholder farmers and the urban poor (Berger & Chambwera 2010).

Additionally, in negotiating tricky terrain, well-established participatory approaches can help governments pursue fair and effective processes. The process of engagement can also facilitate the creation of support for tough decisions and may serve as a conflict management mechanism in certain situations.

Work with uncertainties

Much uncertainty surrounds how certain climate impacts will play out. For example, projected rainfall change in Ghana by 2050 ranges from a 49% increase to a 66% decrease from 2010 levels, making planning for sectors such as agriculture and hydroelectric power highly challenging (The World Bank Group 2010:48). However, uncertainty cannot become an excuse for inaction. For many decisions, short-term courses of action can be taken that will keep future options open as circumstances change. Measures can include 'low-regrets' measures that further development and poverty reduction or enhance ecosystem resilience, thus reducing vulnerability over the long term. For example, Yemen's Nationally Appropriate Mitigation Actions (NAPAs) include measures that fulfil development objectives while strengthening the country's capacity to adapt to a changing climate; one of Yemen's NAPAs involves water conservation through reuse and improved irrigation, thereby increasing the efficiency of water-intensive practices as well as preparing for a range of climatic scenarios (Dessai & Wilby 2010).

Avoid maladaptation

Maladaptation must be avoided to prevent resources from being wasted and communities and ecosystems being exposed to greater climate risks. This is especially true in developing countries, where vulnerabilities are higher and resources are scarcer. For example, if poorly designed, insurance can also lead to maladaptation by providing incentives for the adoption of risky behaviour (Ranger 2010). Crop insurance can lead to the perverse effect of farmers allowing crops to die in order to collect insurance payouts that are higher than a poor growing season's earnings (UNDP 2007/2008).

Modify existing and develop new policies, structures and processes

Addressing climate risks can provide a unique opportunity to confront other poverty-related risks and reduce overall vulnerability. Governments can take advantage of this confluence of interests by integrating climate risk management into ministries for economic development, finance and other relevant sectors and establishing a central agency to coordinate their adaptation efforts. Furthermore, many countries have long-term plans that inform future decisions over several decades, but have yet to take climate change into account. To promote their effectiveness, such plans should integrate climate risks and also allow for periodic assessment and revision.

In addition, coordination among national-level government agencies and with other stakeholders and institutions at local, sub-national, regional and international levels is a prerequisite for successful adaptation efforts. At present, the planning for risks posed by climate change is often divided between different ministries and lacks a coordinating authority. A 2010 survey of 45 countries by UNDP (Lim and Baumwoll 2010) found that only 46% had inter-ministerial committees or councils to manage climate issues. Of these countries, 52% of these committees fell under the jurisdiction of the Ministry of the Environment, 43% under the President, Premier or Prime Minister's office and 5% under the Ministry of Planning and Development. Overall, many of these committees lacked high-level political support.

However, some countries are taking coordination quite seriously. In Burkina Faso, a permanent secretariat has been created in the country's National Council for Environment and Sustainable Development. This secretariat is responsible for coordinating the implementation of climate change adaptation projects. Cross-sectoral and multi-sectoral projects, however, are still supervised by the Ministry of the Environment.[14]

Institutional mandates must also be reformed to better contend with different types of climate risks, e.g. through the creation of long-term goals that are better suited to the time frame of climate impacts. For example, South Africa's National Transportation Master Plan spans from 2005 to 2050 (Department of Transport, Republic of South Africa 2009).

Monitor and evaluate systematically

Public engagement is crucial in the monitoring and evaluation of adaptation initiatives, especially in the development of an understanding of how effective a given activity is in responding to climate impacts. For example, adaptation efforts may fail if those affected are not engaged in monitoring, operations and oversight. This was a key lesson that emerged from efforts in Nepal to prevent deadly glacial lake outburst floods (GLOFs), which can be triggered when the water dammed by a glacier or a moraine is released (World Resources Institute 2011:46). GLOFs have the potential to cause significant destruction in downstream valleys. The Tsho Rolpa glacial lake is the largest in the Nepali Himalayas; a decade ago, the threat of it flooding downstream valleys led the government of Nepal to take proactive measures. These included lowering the lake's level by 3 m and setting up an emergency warning system. Although these measures were considered to be necessary to avert a catastrophic flood, public buy-in was limited. When initial interaction between government officials and the community lapsed, there was little monitoring and evaluation of the interventions. Although mountain village residents had helped construct the early warning system – keeping its operation and maintenance expenses low – the early warning devices eventually were pillaged and rendered non-functional. The Nepalese case demonstrates that public engagement in the initial implementation is not enough. Ongoing public engagement and community self-interest are required in the monitoring and evaluation of adaptation activities if they are to be effective.

15.4.3 Additional principles to be followed

In addition to the principles mentioned above, two other principles can help guide adaptation in developing countries.

Evaluate resource needs and strengthen resources accordingly

It will take considerable financial, human and social capital for developing country governments to pursue a comprehensive approach to making their economies and communities climate-resilient. Governments and donors will need to make long-term investments that mirror the decades-long lifespan of predicted climate impacts. Adaptation decision-making should also place a premium on protecting and sustainably managing ecological resources, which can protect people from climate-related hazards as well as provide livelihoods. Finally, national governments should be proactive in developing social resources, such as those that enable communities to act collectively, cope with adverse conditions and show reciprocity and mutual support in times of crisis. These resources can play a crucial role in building the adaptive capacity of vulnerable groups and populations.

Keep thresholds in mind

It is critical to ensure that decision-making takes into account thresholds that, if passed, could lead to significant adverse impacts. This is particularly relevant in developing countries, where a greater share of income and livelihoods, not to mention the basic survival needs of hundreds of millions of people, depends on climate-sensitive ecosystems and their (at times) uncertain thresholds. Research on thresholds should become a critical component of national adaptation efforts and international research priorities.

If the overshooting of such thresholds will lead to significant, irreversible harm, more proactive measures will be necessary to prevent them from being exceeded. The consequences of some decisions made today will be hard to reverse, such as the location of new infrastructure (power plants, landfills, drinking water reservoirs) or the location of housing developments for a growing coastal city. If climate risks are not taken into account in such decisions, investments may be lost and vulnerability may be increased.

In conclusion, the manual's ten principles are relevant for developing countries as well, but with different weighting, and at least two additional criteria (with regard to resources and thresholds) should be added. While many of these principles could be applied to other public-policy challenges, what calls attention to these principles is precisely the context in which they will be employed: the field of climate change and its disruptive impacts. It is this context that makes these principles so vital to uphold if we are to have any chance of effectively adapting to a changing climate.

15.5 Context-specific and yet transferable? A reflection on knowledge-sharing in the field of adaptation to climate change

Britta Horstmann

15.5.1 Introduction

People in developing countries are particularly vulnerable to the adverse effects of climate change and must adapt to its impacts. Europe can support the adaptation processes of these populations by sharing its knowledge in the field of adaptation policy and research. However, one of the core research findings is that adaptation to climate change is a context-specific and location-based process. If knowledge on adaptation processes is also context-specific, to what extent is climate change adaptation knowledge from one country at all relevant for other countries and contexts? From a theoretical point of view, are there characteristics and success factors of adaptation processes? And if so, what does the theoretical knowledge about adaptation processes imply for the design of useful knowledge-sharing and support?

15.5.2 What adaptation knowledge is relevant?

Knowledge-sharing in the field of adaptation to climate change requires that there be relevant knowledge to share. The relevance of knowledge can be determined by several criteria and perspectives. Two criteria are particularly important for knowledge-sharing, as they also offer perspectives on the 'transferability' of adaptation knowledge: context-independence and success.

The perspective of context-independence sheds light on the kind of knowledge that will be the subject of knowledge-sharing by asking to what extent knowledge is context-independent and of common interest. This means searching for generally valid information regarding an adaptation process in the form of characteristics, common questions and answers, patterns or typical processes, for example. It also entails analysis of the extent to which it is possible to standardise this knowledge and the extent to which it is socially constructed and embedded in cultural, social or institutional contexts and norms. The perspective relies on the assumption that knowledge that is context-independent is:

a   of interest to many actors in many contexts and therefore highly relevant for knowledge-sharing and support

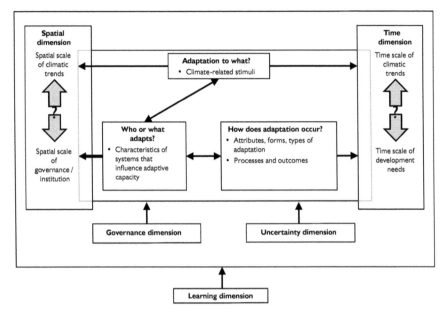

FIGURE 15.5.1 Characteristics of adaptation processes relevant for transferability (source: author's own, modified from Smit et al. 2001)

b   more easily 'transferable', as it has already been formalised, defined or described, i.e. it has been made explicit in some form and can therefore be more easily communicated (it is 'explicit knowledge'), and

c   more easily 'transferable' to other contexts, as it is less dependent on normative value judgements that are highly influenced by context variables.

The second perspective asks whether the knowledge at stake leads to successful adaptation processes. This perspective expands the analysis of the normative dimensions of the adaptation process with an ethical perspective. What do we know about the success factors of adaptation processes? Is the knowledge we have robust enough, and is it knowledge we can recommend to other actors? Or do the motivations of the transfer relate more to the pursuit of the developed country's research or business objectives?

Based on research results, the following section gives a brief overview of context-dependent and independent characteristics and the success factors of planned adaptation processes.

15.5.3 Characteristics and success factors of adaptation processes: What do we know?

The concept and definition of adaptation to climate change has changed over time and still remains rather broad. In the 1990s, adaptation was mainly conceptualised from an environment perspective focusing on the analysis of climate change impacts; however, over the past decade, adaptation has been increasingly viewed from a development perspective focusing on the analysis of livelihood vulnerability and vulnerability towards short-term climate variability (vulnerability-based adaptation). Each of these two perspectives emphasises a different type of policy response (see, e.g., O'Brien 2008). At the end of the 1990s, Smit et al. (1999, 2000) synthesised generally agreed-upon concepts and terms of adaptation, which at that time found broad acceptance in research and were discussed in the third and fourth IPCC assessment reports (Smit & Pilfosova 2001, Adger et al. 2007). Three questions are key to the adaptation concept:

1   *Who or what adapts?* Adaptation processes can take place in natural and human systems at various scales, e.g. in a community, at the firm level or in a sector.
2   *Adaptation to what?* What is the point of reference for this process of change? This can be a climate-related stimulus such as a change in moisture or temperature. In the case of vulnerability-based adaptation, it can also be a non-climatic factor that renders the actor (or the natural system) vulnerable to climate variability or change.
3   *How does adaptation occur?* What are the desired qualities of an adaptation process, e.g. the type of adaptation considered with regard to form (such as legal, institutional or technological adaptation processes) or with regard to function and purpose (e.g. retreat, accommodate, protect)?

These three basic questions illustrate the breadth of the concept of adaptation. The challenge for practitioners, therefore, lies in the definition of the problem itself, in narrowing it down to what adaptation means in a specific context. Thus, in addition to the three questions listed above, there are further aspects that can be considered as context-independent and characteristic to an adaptation process (Horstmann 2008). These questions have been widely addressed by scholars from various disciplines:

4   *Who defines and governs adaptation, how and why?* (*Governance dimension*) As the core challenge of adaptation lies in the definition of its context-specific meaning, the governance dimension is particularly important. Institutions, actors, decision criteria and procedures are critical to decisions regarding the type of adaptation activities that are ultimately undertaken. For example, who are the actors who should be involved in such definitions and governance, and what are their ideas of adaptation to climate change?
5   *What time period is an adaptation activity aligned with?* (*Time dimension*) Adaptation activities require definition of the time-scales of the process and determination of whether it is aligned with long- or short-term climatic trends. This decision must consider development needs and climatic trends, the availability of climate data and information, and political planning and financing structures.
6   *What is the spatial scale of an adaptation activity?* (*Spatial dimension*) Defining the spatial scale means defining the scope of an adaptation activity. For

example, does the adaptation activity coincide with the governance scale, or does the activity have implications for other actors at other scales and in other regions?

7 *How should uncertainty be addressed in adaptation processes? (Uncertainty dimension)* Uncertainty is a dominant feature of adaptation in a twofold sense. First, the impacts of climate change are uncertain, and no definite information on the associated adaptation needs will be forthcoming. In addition, there is a degree of uncertainty with regard to the determinants of vulnerability, and there is no guarantee that certain measures will lead to enhanced adaptive capacity or resilience. Both of these aspects of uncertainty imply the risk of maladaptation. These risks also point to the question of who is responsible if adaptation processes fail and to questions of legitimacy in decision-making processes (again underlining the importance of the governance dimension for adaptation).

8 *How should reflection and learning be integrated into the adaptation process? (Learning dimension)* Adaptation to climate change is a change process over time. Without reflection and learning, there will be no change or innovation. Monitoring and evaluation can be an important vehicle in this type of learning process.

Overall, these questions illustrate that there is a sizable overlap between adaptation and general development processes. Some of this is implicitly or explicitly addressed by the principles in this manual. It also illustrates that adaptation greatly depends on contextual factors that influence the kind of adaptation actions finally undertaken. Adaptation to climate change is thus not an objective, but rather a highly subjective and normative process embedded within and determined by social contexts. This implies that it is difficult if not impossible to standardise knowledge in many adaptation dimensions. As a result, it is not a problem definition or solution as such that can be transferred, but rather knowledge that will help developing countries to make such a definition, context-specific analysis or decision. One could refer to this as 'enabling knowledge' in the form of experience and knowledge about, e.g., instruments, methodologies or governance processes.

Whether this knowledge can lead to successful adaptation in practice is not clear. Some scholars have compiled principles of adaptation seeking to provide guidance for good adaptation decision-making and practice (e.g. Prutsch et al. 2010, Eriksen et al. 2011, this volume). However, the evaluation of the effectiveness of adaptation activities is still in its early stages, and the required metrics and criteria for such an evaluation are contested (because the concept of adaptation is vague, normative and complex). The attribution of an activity to an outcome can be challenging or, with regard to long-term climate changes, impossible. The metrics of 'successful adaptation' and success criteria depend on human value judgements, which vary within and across countries. This also implies that comparisons can be difficult, as an activity that was deemed successful in one country might not be in another.

15.5.4 What are the implications for knowledge-sharing and support?

Unless actors, content and context are defined more closely in practice, meaningful knowledge-sharing in the field of adaptation to climate change will be impossible and any recommendations will inevitably remain at the general level. Nevertheless, what conclusions can be drawn from a theoretical perspective?

This brief analysis has illustrated that adaptation is a complex process that significantly overlaps with development. Therefore, knowledge-sharing in the field of adaptation can build on extensive experience in development cooperation. This experience has been documented in various guidebooks that compile lessons learned and recommendations (e.g. Leonard et al. 2011, Fukuda-Parr et al. 2012, Task Team on South-South Cooperation 2011). However, adaptation is not identical to development. What further recommendations must thus be added or emphasised from an adaptation perspective?

- *First, adaptation knowledge-sharing should be designed based on the needs of developing countries and not on the offer of available knowledge (e.g. from European countries).* The core questions of the adaptation process listed above only provide broad theoretical guidance on relevant adaptation knowledge areas. Due to the overlap with development processes, knowledge-sharing in the field of adaptation can involve manifold actors and institutions. It is impossible to identify priority actors or target groups from a theoretical perspective of adaptation alone. The scope of adaptation knowledge needs can range from legal or policy-related aspects to meteorology and climate services to practical agriculture. It can take different forms, including collaborative research, technical cooperation, capacity-building (see, e.g., Korean Development Institute/World Bank 2011) or South-South knowledge-sharing (see, e.g., World Bank Institute 2012). The relevant questions and areas for knowledge exchange in developing countries must be defined more closely in practice.

- *Second, adaptation knowledge-sharing thus requires a partner in developing countries.* What seems obvious at first sight is not always common practice. In adaptation-related research, knowledge-sharing frequently plays a marginal role, as research is often conducted by 'external experts' and not in cooperation with, or led by, national or country-based researchers.

- Building partnerships can also be a means of dealing with the normative dimensions of adaptation. In the field of transboundary research, the Swiss Academy of Science regards partnerships as a necessity for complex problems and whenever "a research question necessitates a strong science-society interface" (Stöckli et al. 2012). Adaptation to climate change is this type of complex challenge.

- *Third, knowledge-sharing should focus on enabling developing countries to take adaptation decisions on the basis of context-specific analyses.* As the adaptation concept does not entail practical guidance and because it is difficult to standardise adaptation knowledge, actors and decision makers need to detail the meaning of adaptation in a specific context. For sound adaptation decisions, decision makers need context-specific data, information and analysis as well as context-specific decision support processes and tools. Knowledge-sharing support should therefore strengthen analytical capacities and focus on building the necessary prerequisites for good decision-making, including research infrastructure.

• *Fourth, adaptation knowledge-sharing requires the risks and normative dimensions of adaptation knowledge to be disclosed.* Adaptation knowledge entails many context-specific risks, uncertainties and normative dimensions that might need to be translated, adapted or appropriated by actors in a specific context. As a necessary first step, practitioners in adaptation knowledge exchange should therefore identify and discuss these normative dimensions. These can also be hidden in apparently scientifically neutral concepts and methodologies, as two examples illustrate. If some people in a country are affected by drought and others by flood, what is more important? Who is more vulnerable? As there is no universally accepted definition of vulnerability among scientists, and accordingly no agreed-upon metrics or sets of indicators (e.g. Birkmann 2006, Gallopín 2006) on which a decision could be based, it is primarily a political responsibility to determine these metrics (Klein 2009).

Another example is cost–benefit analysis (see Chapter 10). If such an analysis excludes the informal sector because data is difficult to obtain, it might exclude the most vulnerable people from the analysis. If these people are not part of the analysis, how can they be part of the solution?

An additional step in dealing with context-dependent knowledge could be to incentivise more research on adaptation-related standards or patterns (see Sietz et al. 2011 as an example) in these areas where possible. This is important; however, there are limits to such an approach, as many dimensions of the adaptation process remain socially constructed.

• *Fifth, general platforms of knowledge-sharing in the field of adaptation should be complemented by formats that will allow experience-based learning.* The background to this recommendation is the question of whether knowledge we can communicate and standardise is sufficient to lead to successful adaptation processes. Many of the available formats for adaptation knowledge-sharing (such as this manual, the Nairobi Work Programme under the UN climate convention and internet platforms) deal with explicit knowledge. However, experience in knowledge-sharing has shown that it is particularly important to consider tacit knowledge, knowledge that is "rooted in experience, insight and intuition" and that cannot be easily communicated (World Bank Institute 2012:7). As adaptation is highly socially embedded, this knowledge dimension plays an important role in successful adaptation processes. For the acquisition of tacit knowledge, experience (Lam 2000) and practice-based learning (which would also allow the "co-creation of knowledge" between actors; Regeer & Bunders 2009) are key.

The fifth recommendation results in a certain dilemma for those who would like to support adaptation knowledge-sharing for many actors in many developing countries. Put simply, the more standardised adaptation knowledge is, the more relevant it is for many actors. However, the more standardised adaptation knowledge is, the more irrelevant it is for practical adaptation decisions. Thus, designing adaptation knowledge-sharing that is simultaneously cost-effective and meaningful is very challenging. It requires political decisions to be taken on the questions of whom to share knowledge with, why and how.

Finally, the analysis of adaptation characteristics shows that capacities to manage change and transition processes are a key element of adaptation to climate change. This requires not only sound data and information but also sound analytical capacities to define adaptation processes, supportive infrastructure and, not least, possibilities for learning and innovation. Knowledge-sharing plays an important role in this setting. European countries should therefore give knowledge-sharing a central part in the support of adaptation to climate change in developing countries.

15.6 Lessons learned and differences between Europe and other regions

Rob Swart

With some notable exceptions, such as the small group of IPCC contributors, a global network of adaptation researchers, practitioners and policy makers is still lacking. 'Adaptation science' is at most in a nascent stage. Global research programmes that comprehensively cover adaptation do not exist. In the context of international climate negotiations, the overwhelming emphasis has been on adaptation in developing regions; much less attention has been devoted to adaptation in industrialised countries and the transfer of knowledge. However, beginning in 2009, a series of biannual global adaptation conferences have been organised, currently under the umbrella of the new and still small UNEP PROVIA (Programme of Research on Climate Change Vulnerability, Impacts and Adaptation) programme; in addition, in 2013, a series of European adaptation conferences was initiated.

The comparisons between the status of adaptation science and policy in different world regions in this chapter suggest that an intensified exchange of experiences can be very beneficial in accelerating the currently slow pace of effective adaptation knowledge development. In all regions, a key challenge is reconciling the need for context-specific adaptation at the local level or in certain economic sectors with the development of an overarching strategy that defines the problem and the framework and sets the broad objectives. Specific solutions for policy and research development originate from the different economic, institutional and cultural contexts in various regions, offering opportunities for cross-fertilisation and mutual learning between different industrialised regions and between industrialised and developing countries.

Although the cases in this chapter merely scratch the surface of the large body of knowledge already available from countries on all continents, we have made an attempt to derive a number of lessons learned:

- To improve the assessment of vulnerabilities, broaden the portfolio of options and identify opportunities for new methods and tools to evaluate these options, it can be helpful to not only look for experiences in similar situations at the national or regional level, but also beyond (i.e. in other regions of the world).
- However, the context dependency of many features of adaptation implies that it is crucial to consider to what extent it will be possible to standardise and share adaptation knowledge in a meaningful way. In many ways, "the more standardised adaptation knowledge is, the more relevant it is for many actors. However, the more standardised adaptation knowledge is, the more irrelevant it is for practical adaptation decisions" (Section 15.5).
- Knowledge that is context-independent is (a) of interest to many actors in many contexts and therefore highly relevant for knowledge-sharing and support; (b) more easily 'transferable', as it has been formalised, defined or described already, i.e. it has been made explicit in some form and can therefore be more easily communicated (it is 'explicit knowledge'); and (c) more easily 'transferable' to other contexts, as it is less dependent on normative value judgements that are highly influenced by context variables.
- All regions appear to have started adaptation analysis from a traditional top-down perspective, developing options to address the projected impacts associated with downscaled climate modelling results. Different regions are exploring approaches that attempt to move beyond this by replacing climate as the central focus with development objectives. Section 10.3 presents an example of new methodological techniques from Europe in the area of water safety, while Section 15.2 describes a stakeholder process to facilitate a development-oriented refocusing of adaptation in Australia. Japan is experimenting with the combination of top-down and bottom-up approaches and the associated guidelines. Such development-oriented approaches may be better suited to exploring the thresholds beyond which systems may collapse or change may reach unacceptable levels and current development strategies might not be adequate anymore.
- The ten generic principles in this manual appear to be equally valid in other regions, even if the actual applications may differ. Differences between regions in terms of resource availability represent a major issue in this respect. However, because adaptation is context-specific, it is not adaptation knowledge content as such that should be transferred, but rather the knowledge of how to produce that content.
- With respect to meaningful knowledge-sharing between industrialised and developing countries, five conditions are relevant: (a) adaptation knowledge-sharing should be designed for the needs of developing countries and not for the offers of available knowledge (e.g. from European countries), (b) adaptation

knowledge-sharing therefore requires a partner in developing countries; (c) knowledge-sharing should focus on enabling developing countries to take adaptation decisions on the basis of context-specific analyses; (d) adaptation knowledge-sharing requires the risks and normative dimensions of adaptation knowledge to be disclosed and (e) general platforms of knowledge-sharing in the field of adaptation should be complemented by formats that allow experience-based learning (Section 15.5).

- Mutual learning and knowledge-sharing are terms used to describe the two-way interactions that are required for advancing adaptation knowledge rather than knowledge transfer. The complex and normative challenges of climate change adaptation require extensive science-practice interactions and necessitate a transdisciplinary approach. This two-way interaction takes time and effort, and should be demand-driven, featuring the transparent and legitimate definition of user needs and the equitable involvement of all relevant stakeholders.

Notes

1 http://library.globalchange.gov/climate-change-impacts-on-the-united-states-the-potential-consequences-of-climate-variability-and-change-foundation-report
2 http://library.globalchange.gov/products/assessments/2004-2009-synthesis-and-assessment-products
3 http://library.globalchange.gov/products/assessments/2009-national-climate-assessment/global-climate-change-impacts-in-the-u-s-highlights-booklet
4 http://library.globalchange.gov/products/assessments?cat=10
5 www.entergy.com/.../GulfCoastAdaptation/Building_a_Resilient_Gulf_Coast
6 http://www.georgetownclimate.org/adaptation/state-and-local-plans
7 http://www.icleiusa.org/climate_and_energy/Climate_Adaptation_Guidance/free-climate-adaptation-resources
8 http://www.georgetownclimate.org/adaptation/state-and-local-plans
9 http://www.whitehouse.gov/administration/eop/ceq/initiatives/adaptation
10 http://www.georgetownclimate.org/adaptation/law-and-policy-work
11 http://library.globalchange.gov/u-s-global-change-research-program-strategic-plan-2012-2021
12 This chapter's conclusions are largely drawn from a research exercise undertaken as part of the 2010–2011 *World Resources Report: Decision Making in a Changing Climate*, a joint publication of the World Resources Institute, the United Nations Development Programme (UNDP), the United Nations Environment Programme (UNEP) and the World Bank, involving over 100 adaptation experts, public officials, practitioners and civil society representatives from more than 30 countries.
13 For example, see: http://unfccc.int/adaptation/nairobi_work_programme/knowledge_resources_and_publications/items/6227.php; http://www.undp-alm.org/.
14 Personal communication with the UNDP-AAP Programme Manager and the UNDP-AAP Senior Adviser, 23 May 2011, and 21 and 23 June 2011.

References

ACIA (2008) *Arctic Climate Impact Assessment*, Cambridge University Press. http://www.acia.uaf.edu/pages/scientific.html.

Adger, W.N., Hughes, T.P., Folke, C., Carpenter, S.R. and Rockstrom, J. (2005) 'Social-ecological resilience to costal disasters', *Science* 309 (5373), pp. 1036–1039.

Adger, W.N., Agrawala, S., Mirza, M.M.Q., Conde, C., O'Brien, K., Pulhin, J., Pulwarty, R., Smit, B. and Takahashi, K. (2007) 'Assessment of adaptation practices, options, constraints and capacity. Climate change 2007: impacts, adaptation and vulnerability', in Parry, M.L., Canziani, O.F., Palutikof, J.P., van der Linden, P.J. and Hanson, C.E. (eds.) *Contribution of Working Group II to the Fourth Assessment Report of the Intergovernmental Panel on Climate Change*, Cambridge, UK: Cambridge University Press, pp. 717–743.

Berger, R. and Chambwera, M. (2010) *Beyond Cost-Benefit: Developing a Complete Toolkit for Adaptation Decisions*, Briefing Document, International Institute for Environment and Development, London. http://pubs.iied.org/pdfs/17081IIED.pdf.

Bierbaum, R., Smith, J., Lee, A., Blair, M., Carter, L., Stuart Chapin III, F., Fleming, P., Ruffo, S., Stults, M., McNeeley, S., Wasley, E. and Verduzco, L. (2012) 'A comprehensive review of climate adaptation in the United States: more than before, but less than needed', *Mitigation and Adaptation Strategies to Global Change*, 18 (3), pp. 361–406. DOI 0.1007/s11027-012-9423-1.

Birkmann, J. (2006) *Measuring Vulnerability to Natural Hazards: Towards Disaster Resilient Societies*, Tokyo, New York, Paris: United Nations University Press.

Bueti, C. and Faulkner, D. (2010) *ICTs as a Key Technology to Help Countries Adapt to the Effects of Climate Change*, Expert Perspectives Series Written for the World Resources Report 2010–2011, Washington, DC. http://www.wri.org/publication/world-resources-report-2010-2011.

Carmin, J., Nadkarni, N. and Rhie, C. (2012) *Progress and Challenges in Urban Climate Adaptation Planning: Results of a Global Survey*, Cambridge, MA: MIT.

Cash, D. and Buizer, J. (2005) *Knowledge-Action Systems for Seasonal to Interannual Climate Forecasting: Summary of a Workshop*, Report to the Roundtable on Science and Technology for Sustainability, Washington, DC: National Academies Press.

CERES (2011) *Disclosing Climate Risks & Opportunities in SEC Filings: A Guide for Corporate Executives, Attorneys & Directors*, Boston, MA: CERES.

Clark, W.C. and Dickson, N.M. (2003) 'Sustainability science: the emerging research program', *Proceedings of the National Academy of Sciences* 100 (14), pp. 8059–8061.

CSTP (Council for Science and Technology Promotion) (2010) *Planning Technological Development towards Realizing a Society Adapting to Climate Change* (Final Report), Task Force for Planning Technological Development towards Realizing a Society Adapting to Climate Change, The Cabinet Office (in Japanese).

C2ES (2011) *Climate Change 101: Understanding and Responding to Global Climate Change*, Center for Climate and Energy Solutions, Vienna. http://www.c2es.org/docUploads/climate101-fullbook.pdf.

Department of Transport, Republic of South Africa (2009) *The National Transport Master Plan 2050: A Sustainable Transport for Africa*, Presentation to the UNHABITAT, UITP and UATP Seminar, Nairobi, Kenya. http://www.unhabitat.org/downloads/docs/7997_16763_Lanfranc_Wakhishi.ppt%20Final.pdf.

Dessai, S. and Wilby, R. (2010) *How Can Developing Country Decision Makers Incorporate Uncertainty about Climate Risks into Existing Planning and Policymaking Processes?*, Expert Perspectives Series Written for the World Resources Report 2010-2011, Washington, DC. http://www.worldresourcesreport.org/files/wrr/papers/wrr_dessai_and_wilby_uncertainty.pdf.

Eriksen, S., Aldunce, P., Bahinipati, C.S., D'Almeida, M., Molefe, J.I., Nhemachena, C., O'Brien, K., Olorunfemi, F., Park, J., Sygna, L. and Ulsrud, K. (2011) 'When not every response to climate change is a good one: identifying principles for sustainable adaptation', *Climate and Development* 3 (1), pp. 7–20.

Fukuda-Parr, S., Lopes, C., Malik, K. (ed.) (2012) *Capacity for Development: New Solutions to Old Problems*, London: Earthscan.

Gallopín, G.C. (2006) 'Linkages between vulnerability, resilience and adaptive capacity', *Global Environmental Change* 16 (3), pp. 293–303.

Hiramatsu, A. (2011) 'Structuring knowledge of climate change', in Sumi, A., Mimura, N. and Masui, T. (eds.) *Climate Change and Global Sustainability: A Holistic Approach*, Tokyo: United Nations University Press, pp. 10–29.

Horstmann, B. (2008) 'Framing adaptation to climate change: a challenge for building institutions', Bonn: Deutsches Institut für Entwicklungspolitik/German Development Institute. Discussion Paper 23/2008.

Howden, S.M., Crimp, S.J. and Nelson, R.N. (2010) 'Australian agriculture in a climate of change', in Jubb, I., Holper, P. and Cai, W. (ed.) *Managing Climate Change*, Melbourne: CSIRO Publishing, pp. 101–111.

Howden, S.M., Soussana, J.F., Tubiello, F.N., Chhetri, N., Dunlop, M. and Meinke, H. (2007) 'Adapting agriculture to climate change', *Proceedings of the National Academy of Sciences* 104 (50), pp. 273–313.

IPCC (2007a) *Climate Change 2007: Impacts, Adaptation and Vulnerability*, Contribution of Working Group II to the Fourth Assessment Report of the Intergovernmental Panel on Climate Change, Parry, M.L., Canziani, O.F., Palutikof, J.P., van der Linden, P.J. and Hanson, C.E. (eds.), Cambridge, UK and New York, USA: Cambridge University Press.

IPCC (2007b) *Climate Change 2007: Mitigation*, Contribution of Working Group III to the Fourth Assessment Report of the Intergovernmental Panel on Climate Change, Metz. B., Davidson, O.R., Bosch, P.R., Dave, R. and Meyer, L.A. (ed.), Cambridge, UK and New York, USA: Cambridge University Press.

IPCC (2012) *Managing the Risks of Extreme Events and Disasters to Advance Climate Change Adaptation*. A Special Report of Working Groups I and II of the Intergovernmental Panel on Climate Change, Field, C.B., Barros, V., Stocker, T.F., Qin, D., Dokken, D.J., Ebi, K.L., Mastrandrea, M.D., Mach, K.J., Plattner, G.-K., Allen, S.K., Tignor, M. and Midgley, P.M. (eds.), Cambridge, UK and New York, USA: Cambridge University Press.

Jasanoff, S. and Wynne, B. (1998) 'Science and decision making', in Rayner, S. and Malone, E. (eds.) *Human Choice & Climate Change Volume 1 The Societal Framework*, Columbus, OH: Battell Press, pp. 1–87.

Klein, R.J.T. (2009) 'Identifying countries that are particularly vulnerable to the adverse effects of climate change: an academic or political challenge?' *Carbon and Climate Review* 3 (3), pp. 284–291.

Klein, R.J.T., Nicholls, R.J., and Mimura, N. (1999) 'Coastal adaptation to climate change: can the IPCC technical guidelines be applied?', *Mitigation and Adaptation Strategies for Global Change*, 4 (3–4), pp. 239–252.

Korea Development Institute/World Bank Institute (2011) *Using Knowledge Exchange for Capacity Development: What Works in Global Practice? Three Case Studies in Assessment of Knowledge Exchange Programs Using a Results-focused Methodology*, a Joint Study.

Lam, A. (2000) *Tacit Knowledge, Organizational Learning and Societal Institutions: An Integrated Framework*, London, UK.

Lempert, R. and Collins, M. (2007) 'Managing the risk of uncertain threshold responses: comparison of robust, optimum, and precautionary approaches', *Risk Analysis* 27 (4), pp. 1009–1026.

Leonard, A., Kumar, S. and Kuplinski, D. (2011) *The Art of Knowledge Exchange. A Results-focused Planning Guide for Development Practitioners*, Washington, DC: Worldbank.

Lim, B. and Baumwoll, J. (2010) *Changing the Course of Development: UNDP's Role in Supporting National-Level Decision-Making Processes in a Changing Climate*, Expert Perspectives Series

Written for the World Resources Report 2010–2011, Washington, DC. http://www.wri.org/publication/world-resources-report-2010-2011.

Meinke, H., Howden, S.M., Struik, P., Nelson, R., Rodriguez, D. and Chapman, S. (2009) 'Adaptation science for agricultural and natural resource management – Urgency and theoretical basis', *Current Opinion in Environmental Sustainability* 1 (1), pp. 69–76.

MEXT (Ministry of Education, Culture, Sports, Science and Technology), JMA (Japan Meteorological Agency) and MOE (Ministry of Environment) (2009) *Synthesis Report on Observations, Projections, and Impact Assessments of Climate Change, Climate Change and Its Impacts in Japan* (version 2012), 2013, p. 84 (in Japanese).

MOE (Ministry of Environment) (2008) *Wise Adaptation to Climate Change*, The Committee on Climate Change Impacts and Adaptation Research, Ministry of Environment, Japan.

MOE (Ministry of Environment) (2010) *Approaches to Climate Change Adaptation*, The Committee on Approaches to Climate Change Adaptation, Ministry of Environment, Japan.

Moser, S. (2009) *Good Morning America! The Explosive Awakening of the US to Adaptation*, Charleston, SC, NOAA and Sacramento, Santa Cruz, CA, California Energy Commission.

Nelson, R., Howden, S.M. and Stafford Smith, M. (2008) 'Using adaptive governance to rethink the way science supports Australian drought policy', *Environmental Science & Policy* 11 (7), pp. 588–601.

Nelson, R., Kokic, P. and Meinke, H. (2007) 'From rainfall to farm incomes – transforming advice for Australian drought policy: Part II – Forecasting farm incomes', *Australian Journal of Agricultural Research* 58 (10), pp. 1004–1012.

NRC (2009) 'Restructuring federal climate research to meet the challenges of climate change (National Research Council)', in *Committee on Strategic Advice on the U.S. Climate Change Science Program*, Washington, DC: The National Academies Press.

NRC (2010a) *Adapting to Impacts of Climate Change*, Report of the Panel on Adapting to Impacts of Climate Change, NAS/NRC Committee on America's Climate Choices, Washington, DC: The National Academies Press.

NRC (2010b) *Advancing the Science of Climate Change, America's Climate Choices: Panel on Advancing the Science of Climate Change*, Washington, DC: The National Academies Press.

NRC (2011) *Climate Stabilization Targets: Emissions, Concentrations, and Impacts over Decades to Millennia*, Washington, DC: The National Academies Press.

O'Brien, K. (2008) 'Climate adaptation from a poverty perspective', *Climate Policy* 8 (2), pp. 194–201.

Project Team for Comprehensive Projection of Climate Change Impacts (2008) *Global Warming Impacts on Japan – Latest Scientific Findings*, Ministry of Environment.

Project Team for Comprehensive Projection of Climate Change Impacts (2009) *Global Warming Impacts on Japan – Long-Term Climate Stabilization Levels and Impact Risk Assessment*, Ministry of Environment.

Prutsch, A., Grothmann, T. Schauser, I., Otto, S. and McCallum, S. (2010) *Guiding Principles for Adaptation to Climate Change in Europe*, ETC/ACC, Technical Paper.

Ranger, N. (2010) *Mainstreaming Adaptation into Development: Serving Short- and Long-Term Needs in the Least Developed Countries*, Expert Perspectives Series Written for the World Resources Report 2010–2011, Washington, DC. http://www.wri.org/publication/world-resources-report-2010-2011.

Regeer, B.J. and Bunders, J.F.G. (2009) *Knowledge Co-creation: Interaction Between Science and Society. A Transdisciplinary Approach to Complex Societal Issues*, VU University Amsterdam, Athena Institute.

Sarewitz, D. and Pielke, R. (2007) 'The neglected heart of science policy: reconciling supply of and demand for science', *Environmental Science & Policy* 10 (1), pp. 5–16.

Sietz, D., Lüdeke, M.K.B. and Walther, C. (2011) 'Categorization of typical vulnerability patterns in global drylands', *Global Environmental Change* 21, pp. 431–440.

Smit, B. and Pilfosova, O. (2001) *Adaptation to Climate Change in the Context of Sustainable Development and Equity*, in McCarthy et al. (eds.) Climate Change 2001: Working Group II: Impacts, Adaptation, and Vulnerability, IPCC Third Assessment Report: 877–912.

Smit, B., Burton, I., Klein, R.J.T. and Street, R. (1999) 'The science of adaptation: a framework for assessment', *Mitigation and Adaptation Strategies for Global Change* 4, pp. 199–213.

Smit, B., Burton, I., Klein, R.J.T. and Wandel, J. (2000) 'An anatomy of adaptation to climate change and variability', *Climatic Change* 45 (1), pp. 223–251.

Stöckli, B., Wiesmann, U. and Lys, J.A. (2012) *A Guide for Transboundary Research Partnerships: 7 Questions*, Bern, Switzerland: Swiss Commission for Research Partnerships with Developing Countries (KFPE).

Stratus Consulting (2010) *Impacts of Climate Change in the Southeastern United States*, prepared for the Environmental Protection Agency.

Task Team on South-South Cooperation (2011) *Scaling Up Knowledge Sharing for Development – A Working Paper for the G20 Development Working Group, Pillar 9*. http://wbi.worldbank.org/wbi/Data/wbi/wbicms/files/drupal-acquia/wbi/4%20G-20%20DWG%20 Report%20on%20Knowledge%20Sharing.pdf.

UCS (2007) *Northeast Climate Impacts Assessment*, US. http://www.northeastclimateimpacts.org/.

UNDP (United Nations Development Programme) (2007/2008) *Fighting Climate Change: Human Solidarity in a Divided World*, Human Development Report, New York. http://hdr.undp.org/en/reports/global/hdr2007-2008/.

USGCRP – US Global Change Research Program (2009) *Global Climate Change Impacts in the US*.

World Bank Group (2010) *Economics of Adaptation to Climate Change: Synthesis Report*, Washington, DC. http://www-wds.worldbank.org/external/default/WDSContentServer/WDSP/IB/2012/06/27/000425970_20120627163039/Rendered/PDF/702670ESW0P1080 0EACCSynthesisReport.pdf.

World Bank Institute (2012) *The Art of Knowledge Exchange. A Primer for Government Officials and Development Practitioner.* http://wbi.worldbank.org/wbi/Data/wbi/wbicms/files/drupal-acquia/wbi/ArtofKEPrimerFinal.pdf.

World Resources Institute (WRI) in collaboration with United Nations Development Programme, United Nations Environment Programme and World Bank (2011) *World Resources 2010–2011: Decision Making in a Changing Climate – Adaptation Challenges and Choices*, Washington, DC: WRI.

PART IV

Lessons learned

16

LESSONS LEARNED FROM PRACTICAL CASES OF ADAPTATION TO CLIMATE CHANGE IN INDUSTRIALISED COUNTRIES

Rob Swart, Torsten Grothmann, Sabine McCallum, Andrea Prutsch and Inke Schauser

We know that adaptation is necessary to avoid or reduce the impacts resulting from climate change, also in industrialised countries. Climate change is already occurring due to past emissions and is expected to increase due to the lack of success in climate mitigation policy. What we want to know now is: how and when should we act in order to adapt effectively? Because of the great diversity of adaptation situations in terms of climate change impacts and vulnerabilities, adaptation options, actors and socio-economic and geophysical contexts, this question cannot be answered with a simple one-size-fits-all solution.

In this manual, some common recommendations are provided that are considered to be valid across different countries, in different economic sectors and in response to different climatic threats. The manual organises these recommendations along 10 guiding principles for good adaptation and provides examples of how these principles can work in practice. The principles were originally developed for the European Environment Agency, hence the focus is on the European context, but – as discussed in Chapter 15 – they are also relevant in industrialised countries outside Europe and to a large extent in developing countries as well.

The guiding principles are organised according to three phases in the adaptation process: *prepare the ground, plan the response, implement adaptation and review success* (cf. Table 16.1). In real life, adaptation will be an iterative, learning-by-doing process in which these phases will not be completed consecutively or only once. How the guiding principles should be applied depends on the specific adaptation situation. For example, the practical cases presented in this manual have shown that stakeholder engagement is useful for all three phases. However, because it is a very time- and resource-intensive process, it might only be possible to use it for specific questions arising during various phases of the adaptation process.

TABLE 16.1.1 Phases of adaptation processes and guiding principles for good adaptation

---

Prepare the ground for adaptation:
Explore potential climate change impacts and vulnerabilities
and identify priority concerns
Initiate adaptation, ensure commitment and management
Build knowledge and awareness
Identify and cooperate with relevant stakeholders

Plan for adaptation:
Explore a wide spectrum of adaptation options
Prioritise adaptation options
Work with uncertainties

Implement adaptation and review results:
Avoid maladaptation
Modify existing and develop new policies, structures and processes
Monitor and evaluate systematically

16.1 How to prepare the ground for adaptation

When preparing the ground for adaptation, it is important to carefully *explore the potential climate change impacts and vulnerabilities and identify priority concerns*, as well as to *build knowledge and awareness* on these issues in a particular region, sector, organisation or situation. While this may seem like a simple exercise, the cases in this manual suggest a number of issues that should be taken into account:

- Before assessing climate change impacts and vulnerabilities, *clarify the purposes of the assessments* in as much detail as possible to increase their usability in decision-making and practice (cf. Chapter 5).
- Pay attention to past and current vulnerabilities, as they can reveal future vulnerabilities to a large extent (cf. Chapter 5).
- To the extent practicable, *acquire a sound understanding of the evolving, state-of-the-art knowledge* on climate change impacts, vulnerabilities and adaptation options, based on scientific research and practical experiences, e.g. through inter- and transdisciplinary assessments (involving natural and social scientists as well as experts from practice and stakeholders) and generic portals such as Climate-ADAPT in Europe (cf. Chapters 5 and 7).
- When assessing future vulnerabilities, *consider future socio-economic changes* in addition to climate change, and consider a broad range of possible developments (in terms of greenhouse gas emissions and socio-economic changes) to avoid misguided adaptation strategies that could lead to maladaptation (cf. Chapters 5 and 13).
- To assess risks and to build knowledge and awareness for adaptation to climate change, *apply a target- or user-group specific approach* that also investigates user needs and learns from their practical experiences, using common and positive language as much as possible (cf. Chapters 5 and 7).

- *Present scientifically accurate knowledge in an appropriate form for the users.* To build trust, connect the knowledge to the experiences of the stakeholders with current vulnerabilities (cf. Chapters 5 and 7).
- In addition, to stimulate adaptation action, *address emotions and provide role models and good-practice examples* for adaptation (cf. Chapter 7).
- To prevent uncertainties in projections of climate change and its impacts from becoming a barrier to adaptation, *point out vulnerabilities to current climate variability,* discuss common trends in climate projections, foster an understanding of 'science as a debate' (instead of 'science as search for absolute truth'), frame uncertainties as messages about the possibility of damages and losses not materialising, communicate methods of dealing with uncertainties (e.g. through adaptive management) and apply dialogue and training formats that allow direct, preferably face-to-face interactions with stakeholders and decision-makers (cf. Chapters 5 and 11).
- Taking into account the context dependency of adaptation, carefully *consider the extent to which knowledge or experiences from elsewhere can be meaningful* for the specific adaptation situation at hand (cf. Chapters 9 and 15).

Preparing the ground involves not only the assessment of potential climate change impacts and vulnerabilities and good communication of the assessment results, but also *ensuring commitment and management* of the adaptation process and – possibly even more crucially – *identifying and cooperating with relevant stakeholders,* developing a good understanding of their capacities to adapt, engaging them early in the process and fostering their continued involvement during the entire adaptation process. This is time- and resource-intensive and may initially slow down the process, but it will pay off in the longer term. The main lessons learned in this respect are:

- *Establish a strong (preferably formalised) link between climate change research and decision-making* when starting the adaptation process. To kick-start adaptation processes in businesses, request information on risks and adaptation plans from businesses, particularly those that deliver public benefits (cf. Chapter 6).
- *Establish clear long-term commitments* to sustaining adaptation and building leadership, particularly legally binding commitments from governments and responsible organisations. Furthermore, foster continuous learning among adaptation actors (cf. Chapter 6).
- *Devote adequate attention to the effective and equitable design of the interaction process* between experts, policy-makers and stakeholders in order to build trust and foster mutual learning (cf. Chapters 7 and 8).
- *Clarify the objectives of stakeholder engagement* and *identify the appropriate stakeholders before designing and starting the process.* Differentiate between methods for stakeholder identification (e.g. by means of stakeholder mapping) and stakeholder engagement (e.g. using various workshop formats). Tailor the stakeholder engagement methods to the group(s) of stakeholders that will be involved (cf. Chapter 8).

- *Consider stakeholders from a wide range of contexts,* from different sectors, administrative levels, businesses and civil society groups. Try to avoid the 'asymmetric' stakeholder engagement that often occurs because particular stakeholder groups lack the financial or time resources to participate. Thus, consider reimbursement for those stakeholders lacking resources (cf. Chapter 8).
- *Motivate stakeholders by communicating how they could be affected by climate change impacts and/or adaptation actions.* Ensure the availability of appropriate and easily understandable information about impacts, vulnerabilities and adaptation options, sharing good-practice examples from elsewhere. Provide positive messages about adaptation options that relate to current development plans. Develop shared strategic goals to unite the diverse stakeholder population (cf. Chapters 7 and 8).
- *Consider potential failures in stakeholder engagement processes* that may be connected to inequality between stakeholders, entrenched viewpoints, fear of losing authority, missing knowledge or negative experiences with previous engagement processes (cf. Chapter 8).

16.2 How to plan for adaptation

When the ground has been prepared and stakeholders have agreed upon the need for adaptation action, it is time to explore ways of responding to the challenges. However, many of the activities undertaken to prepare the ground must be maintained. In particular, continued *stakeholder engagement* remains an important task in this phase.

A key challenge in this phase is *to identify and prioritise the available options.* This involves *an inventory of the wide spectrum of options,* including different adaptation approaches (such as grey, green or soft measures) and measures serving various objectives (such as building adaptive capacity or delivering actions). The cases in this manual emphasise in particular the following recommendations (cf. Chapter 9):

- *Don't forget that adaptation is much broader in focus than one might think.* It encompasses measures that help interrelated natural and human systems (in other words, social-ecological systems) at different geographic levels to cope with extreme weather events and climate variability and change.
- *Use readily available databases first, then look for solutions and approaches for generating and evaluating options elsewhere.* Create opportunities to develop innovative solutions by taking a longer-term perspective beyond the local level and by working in partnership with people inside and outside of the system at risk.
- *Search for positive adaptation options* that are relatively insensitive to uncertainty, such as increasing adaptive capacity or utilising options that will also be useful for reasons other than climate change, or that can flexibly be adapted over time as more knowledge becomes available; no- or low-regrets options can combine large benefits with relatively low costs.

- *Analyse and sort a menu of adaptation options that is as wide as possible* in order to create new solutions, combine existing options and find gaps in the current or planned adaptation activities. Tailoring adaptation options to the specific context adds meaning to the assessment and creates a sense of ownership of the results of the process.

After a menu of suitable options has been identified, these *adaptation options must be evaluated* in order for actors to be able to make informed adaptation decisions. Suggestions from the cases include (cf. Chapter 10):

- *Apply more than one method* when evaluating options to stimulate learning. Different methods of evaluating adaptation options represent different ways of problem framing, may be applicable in different situations, have different strengths and weaknesses and can lead to different solutions.
- *Select the preferred option by considering the strengths, weaknesses and data dependency of all options.* In addition to 'conventional' socio-economic evaluation methods such as cost-benefit, cost effectiveness and multi-criteria analysis, other methods may also be relevant for addressing uncertainties, such as real options analysis, robust decision-making, or portfolio analysis.
- *Ensure agreement regarding definitions and criteria among the stakeholders* when applying multi-criteria analysis, a popular and attractive method of prioritising options in many cases.
- *Avoid maladaptation,* defined as actions that increase vulnerability or decrease coping capacity elsewhere or in the future, actions that increase greenhouse gas emissions, actions that use resources unsustainably, actions with high opportunity costs, actions that lead to increased inequality and actions that limit choices for future generations (cf. Chapter 13).

For the phase of adaptation planning, the uncertainties associated with future climate change, impacts and vulnerabilities can pose a significant barrier to adaptation decisions. Different approaches are available that *take uncertainties into account,* including the precautionary principle, adaptive management, robustness and resilience. In addition, many flexible options, no- or low-regrets options that serve other purposes, and options with safety margins (or redundancy) provide attractive opportunities, regardless of how climate change eventually materialises. The main lessons learned from the cases of adaptive management and robust decision-making included in this manual are (cf. Chapter 11):

- *Establish and maintain long-term monitoring for adaptive management.* This is vital for the prevention of poor adaptation decision-making based on incorrect assumptions about the impacts of climate change.
- *Consider robust decision-making* (RDM) as a particularly appropriate method of aiding long-term decision-making in an uncertain climate. RDM implies the selection of a project or plan that meets its intended goals across a variety of

plausible futures. However, although RDM principles are simple in theory, its practical application can be computationally intensive and time-consuming.

16.3 How to implement adaptation and review success

After adaptation options have been selected (a process that includes the consideration of existing uncertainties), they must be implemented. When the options have been implemented, monitoring and evaluation to review their success are critical. Making adaptation options and plans a reality poses a great challenge. In this manual, we have suggested two important guiding principles for the implementation of adaptation: *avoid maladaptation* and *modify existing and develop new policies, structures and processes*. Building on the practical cases presented for these two principles, we have identified the following main lessons learned (cf. Chapters 12 and 13):

- *Consider adaptive management as a flexible approach* to adopting adaptation measures at various time-scales that addresses the problem of deep uncertainty regarding future climatic and socio-economic changes.
- *Carefully evaluate how the measures may lead to different types of maladaptation* when planning and implementing adaptation measures; this requires attention to complex linkages (notably, increased vulnerability elsewhere or later, unequal impacts on vulnerable actors, unsustainable resource use, high opportunity costs and limits on future choices).
- *Apply a systematic, integrated approach for climate-proofing policies, structures and processes* to ensure coherence in mainstreaming efforts, avoid overlaps and utilise synergies.
- *Consider how the special characteristics of adaptation might require new institutional arrangements.* Do not hesitate to establish new institutions or collaborations where required.
- *Collaborate across sectors and consider spatial planning* as a tool to integrate the different land uses associated with adaptation measures. New arrangements may be derived from the integrated nature of adaptation.
- *Ensure continuity beyond short-term political cycles;* foster long-term governance commitments in order to support the long-term efficacy of adaptation plans.

Finally, when adaptation measures are implemented, the process of *monitoring and evaluation* (M&E) is important to allow us to learn from successes and failures and (when necessary and possible) adjust the measures. Although the cases presented in this manual confirm that M&E for adaptation is still at an early stage of development, we can derive the following lessons learned from initial explorative initiatives (cf. Chapter 14):

- *Anticipate the design of a monitoring concept at the beginning of adaptation action planning.* As adaptation planning is often participatory, combining action planning and monitoring can save time and increase stakeholders' willingness to engage.

- *Adopt a flexible, question-based approach with learning as a core characteristic* of the M&E system.
- *Learn from early ongoing attempts at implementing monitoring systems elsewhere,* considering their temporal and spatial scales, their specific objectives (why do we monitor and for whom?) and their limitations.
- *Build on existing monitoring systems* to limit additional work and costs and to facilitate progress reporting.
- *Use quantitative indicators in conjunction with other M&E tools* (including quantitative research, comparative studies, stakeholder engagement and expert elicitation) to provide a richer representation of the performance of adaptation measures.
- *Be aware that one indicator does not provide the complete adaptation picture.* Although the use of indicators is still at an early stage of development, some progress has been made with the production of conceptual frameworks for their elaboration. Lessons have been learned regarding the typology of adaptation indicators, their temporal and spatial scales and aspirational goals, their limitations and the need for additional conceptual information.
- *Develop simple systems with understandable indicators* that capture important dimensions such as implementation, mainstreaming, communication of adaptation and data availability.

The lessons learned identified in this manual – not only in this final summary but also in the other chapters – show that despite similarities to other planning processes, adaptation to climate change poses mutually reinforcing, multi-level, complex and therefore 'new' challenges for planners and policy-makers. In the introductory chapters of this manual, a number of specific challenges were noted that distinguish adaptation from many other decision-making fields: the deep *uncertainties* involved in climate change, the *mismatch* between the long-term and global nature of climate projections versus short-term local information needs, the fact that climate change is *just one challenge* amongst many (and often not the dominant one), the *complexity* of adaptation, the requirements with regard to achieving *social and ecological justice* and the various *barriers* to adaptation. These challenges must be properly addressed to ensure the climate-resilient development of cities, watersheds, rural areas and countries. Adaptation research, policy and practice have considerably evolved in recent years, both in industrialised and developing countries. Much experimentation and learning are still required to root adaptation more stably and effectively in everyday decision-making. The guiding principles, the presented cases of good adaptation practice and the recommendations drawn from them can help us to meet these challenges. We hope this will lead to new experiences and an effective and shared learning process as we move towards a more climate-robust future.

INDEX

Milton Keynes UK
Ingram Content Group UK Ltd.
UKHW020317111024
449327UK00040B/1350

9 780415 660341